The 1999 Elections to the European Parliament

Also by Juliet Lodge

DIRECT ELECTIONS TO THE EUROPEAN PARLIAMENT 1984 (*editor*)

DIRECT ELECTIONS TO THE EUROPEAN PARLIAMENT: A Community Perspective (*with Valentine Herman*)

EUROPEAN UNION: The European Community in Search of a Future (*editor*)

INSTITUTIONS AND POLICIES OF THE EUROPEAN COMMUNITY (*editor*)

TERRORISM: A Challenge to the State (*editor*)

THE EUROPEAN COMMUNITY: Bibliographical Excursions (*editor*)

THE EUROPEAN COMMUNITY AND THE CHALLENGE TO THE FUTURE (*editor*)

THE EUROPEAN PARLIAMENT AND THE EUROPEAN COMMUNITY (*with Valentine Herman*)

THE EUROPEAN POLICY OF THE SPD

THE NEW ZEALAND GENERAL ELECTION OF 1975 (*with Stephen Levine*)

THE THREAT OF TERRORISM (*editor*)

THE 1989 ELECTION OF THE EUROPEAN PARLIAMENT (*editor*)

THE 1994 ELECTIONS TO THE EUROPEAN PARLIAMENT (*editor*)

The 1999 Elections to the European Parliament

Edited by

Juliet Lodge
Director
The European Parliament International Study Unit
Faculty of Law
University of Leeds

Editorial matter, selection and Chapters 1, 20 and 21 © Juliet Lodge 2001
Chapters 2–19 © Palgrave Publishers Ltd 2001

All rights reserved. No reproduction, copy or transmission of this publication may be made without written permission.

No paragraph of this publication may be reproduced, copied or transmitted save with written permission or in accordance with the provisions of the Copyright, Designs and Patents Act 1988, or under the terms of any licence permitting limited copying issued by the Copyright Licensing Agency, 90 Tottenham Court Road, London W1P 0LP.

Any person who does any unauthorised act in relation to this publication may be liable to criminal prosecution and civil claims for damages.

The authors have asserted their rights to be identified as the authors of this work in accordance with the Copyright, Designs and Patents Act 1988.

First published 2001 by
PALGRAVE
Houndmills, Basingstoke, Hampshire RG21 6XS and
175 Fifth Avenue, New York, N. Y. 10010
Companies and representatives throughout the world

PALGRAVE is the new global academic imprint of
St. Martin's Press LLC Scholarly and Reference Division and
Palgrave Publishers Ltd (formerly Macmillan Press Ltd).

ISBN 0–333–80042–7

This book is printed on paper suitable for recycling and made from fully managed and sustained forest sources.

A catalogue record for this book is available from the British Library.

Library of Congress Cataloging-in-Publication Data
The 1999 elections to the European Parliament / edited by Juliet Lodge.
    p. cm.
Includes bibliographical references and index.
ISBN 0–333–80042–7
  1. European Parliament—Elections, 1999. I. Lodge, Juliet.

JN36 .A195 2000
324.94'0559—dc21

                                                      00–055695

10 9 8 7 6 5 4 3 2 1
10 09 08 07 06 05 04 03 02 01

Printed and bound in Great Britain by
Antony Rowe Ltd, Chippenham, Wiltshire

For Europeans

especially

Keri-Michèle, David and Chris; Tom, Hannah and Laura; Claude and Jenny; Maximilian and Iris; Alex H; Ellen and Inge; Emmanuel and Nathalie; William, Colin, John, Catherine and Alison; Insa; Christopher and Emily; Alkiviadis and Achilleas; Beth; Christine Huyn Mi and Maria Hee; Nuno; Joseph and Florence; Moa and Alva; Alex N; Dawn, Jack and Alex McC.

and

in memory of
Arthur Mayer
a true democrat and iconoclast

Contents

List of Figures and Tables	ix
Preface	xi
Notes on the Contributors	xiii
List of Abbreviations	xvii

Part I Introduction

1	Invisible, Irrelevant but Insistent? Euro-elections and the European Parliament *Juliet Lodge*	3
2	The European Parliament's Progress 1994–99 *Richard Corbett*	13
3	The EP since 1994: making its mark on the world stage *Christopher Piening*	21

Part II Country Case Studies

4	Austria *Josef Melchior*	33
5	Luxembourg and Belgium *Derek Hearl*	45
6	Denmark *Hans Jorgen Nielsen*	60
7	Germany *William E. Paterson and Simon Green*	72
8	Greece *Kevin Featherstone and George Kazamias*	89
9	Finland *Tapio Raunio*	100
10	France *David Howarth*	117

11	Ireland *Edward Moxon-Browne*	139
12	Italy *Philip Daniels*	149
13	The Netherlands *Henk van der Kolk*	160
14	Portugal *José Magone*	171
15	Spain *John Gibbons*	185
16	Sweden *Karl Magnus Johansson*	199
17	The United Kingdom *Janet Mather*	214

Part III Overview

18	The Party System of the European Parliament after the 1999 Elections *Tapio Raunio*	231
19	European Electoral Systems and Disproportionality *Mark Baimbridge and Darren Darcy*	252
20	Making the EP Elections Distinctive: A Proposal for a Uniform *e*-electoral Procedure *Juliet Lodge*	263
21	A Union Fit to Govern? The Role of the European Parliament and the Commission *Juliet Lodge*	284

Index   311

List of Figures and Tables

Figures

17.1 UK European election turnout, 1994 and 1997 — 223
17.2 English regions: EP election turnouts, 1994 and 1999 — 223

Tables

4.1 The EP elections in Austria (votes and seats) — 42
5.1 Luxembourg Chamber of Deputies election, 13 June 1999 — 50
5.2 EP election in Luxembourg, 13 June 1999 — 50
5.3 Principal Belgian political parties by linguistic community — 52
5.4 Belgian Chamber of Representative election, 13 June 1999 — 57
5.5 Results of EP election in Belgium, 13 June 1999 — 58
6.1 The Danes and the EU — 62
6.2 The result of the European election: vote (%) — 68
6.3 Attitudes towards the EU and the vote at the European election — 70
6.4 Distribution of seats — 70
7.1 Results of the German EP election, 13 June 1999 — 83
7.2 Results of the German EP election, 13 June 1999, by West and East and compared to the 1998 Federal election — 85
8.1 The Greek elections to the EP, 13 June 1999 — 96
9.1 Distribution of votes in Finnish national parliamentary elections in the 1990s and in the 1996 EP elections — 101
9.2 Results of the 1999 EP elections in Finland — 108
9.3 The 16 Finnish MEPs elected to the EP — 111
10.1 Results of the French EP election, 13 June 1999 — 129
11.1 Euro-election results in Ireland — 146
12.1 EP election results, 1994 and 1999, and the Chamber of Deputies, 1996 — 153
12.2 Membership of EP party groups — 157
13.1 Outcome of national and EP elections in the Netherlands, 1994–99 — 168
14.1 The main political parties — 173
14.2 What is your opinion on the performance of the EP? — 175
14.3 Electoral turnout (and abstention) — 179
14.4 The results — 180

14.5	Voters' mobility between 1994 and 1999	181
14.6	Continuity of Portuguese MEPs	181
15.1	Results of the 1999 EP election in Spain	196
16.1	Results of the Swedish EP election, 13 June 1999	208
17.1	European election results in the UK, 10 June 1999	224
17.2	European election results in Scotland	224
17.3	European election results in Wales	225
17.4	European election results in Northern Ireland	225
18.1	Party groups in the 1994–99 EP	233
18.2	Committees in the 1999–2004 EP	236
18.3	Party groups in the EP, September 1999	247
19.1	Summary of electoral systems	256
19.2	Disproportionality measurements	258
19.3	Ranking of disproportionality	258

Preface

The 1999 Euro-elections marked a watershed in the history of the European Parliament (EP). Not only had the EP consolidated its hard-won powers, but on the eve of these elections the European Union's polity showed itself to be mature and resilient. However, even with further enlargement on the horizon, the Commission's resignation, and the prospect of more institutional change, a sense of excitement was missing. The 1999 Euro-elections seemed simultaneously mundane and an accepted feature of the political life of a system which only twenty years earlier had seen its politicians fiercely disputing the wisdom and desirability of the people directly electing the European Parliament at all.

The seemingly run-of-the mill campaign belies, however, a deep transformation in the place of the EU in the national psyche of the member states. Its existence is accepted. Those who contest EU membership, no matter how vociferously, are seen as not altogether credible and as relics from bygone days. This is not to say that their views do not find an echo among the populace. It is to suggest that national political life has become Europeanised almost imperceptibly. From the first Euroelections in 1979, when creeping Europeanisation was something to be checked, to each IGC which confirmed and deepened European integration, the EU has seeped into and permeated much of our daily lives. This almost revolutionary change has occurred with the minimum of turbulence. Yet, while it is apparently accepted, even at a passive level, it is quite properly subject to challenge. The targets for political action and political change have altered: the European level is no longer immune from the voice of the people, however articulated.

This book examines the Euro-elections against this background. Part I surveys the European Parliament between 1994 and 1999. Part II focuses on the elections in each member state and sketches in the specific national dimension to those elections. Part III assesses the Euroelections' impact on the EP party system, speculates on the future of Euro-elections and offers some ideas as to how the electoral system might be reformed via electric-voting to encourage greater participation in general.

We are grateful to those people who debated the arguments with us; to Christopher Piening and Christian Andersson; to those who kindled Euro-wide levity; to the many who kindly provided data and interviews.

We accept responsibility for their interpretation. Thanks too to Tim Farmiloe, who, against the anti-European tide in Britain in the 1980s, launched the first in this series of Euro-election studies; Natalie Hey; Dimitris Kavakas; Hana Kadhom, and Martin and Sue Ainley for the wise-cracks and much more.

JULIET LODGE

Notes on the Contributors

Mark Baimbridge is Director of the Economics Programme at the University of Bradford. His main research interests are economic integration within the European Union and British party politics. He recently co-edited *The Impact of the Euro: Debating Britain's Future* (Macmillan, 2000) *and EMU in Europe: Theory, Evidence and Practice* (Edward Elgar, 2001), and is co-author of *Britain and Europe: The Great Divide* (Pinter, 2001).

Richard Corbett is Labour MEP for Yorkshire and Humberside. He is spokesman on constitutional affairs both for the Labour Party and for the Socialist Group in the European Parliament (EP). Having been elected MEP for the first time in 1996, he was Deputy Secretary General of the EP Socialist Group and a former civil servant. He has written extensively on European affairs. His books include the standard reference book on the EP: *The European Parliament* (4th edn, 2000) with Francis Jacobs and Michael Shackleton; *The Role of the European Parliament in Closer EU Integration* (1997); and *The Treaty of Maastricht: From Conception to Ratification* (Longman, 1994).

Philip Daniels is a Lecturer in European Politics at the University of Newcastle upon Tyne. He is a specialist in Italian politics and has published widely in this field. He has published widely on Italy and European integration, Italian political parties and Britain and the European Union. He is currently completing a co-authored book on Italian parties and elections.

Darren Darcy is a Researcher in the Department of European Studies at the University of Bradford. His research interests include the socio-economic backgrounds of legislators and British party politics. He is the co-author of the chapters 'New Labour's Parliamentarians' in *New Labour in Power* (Macmillan, 2000) and 'Mr Blair's loyal opposition? – the Liberal Democrats in parliament' in *British Elections and Parties Review* (Frank Cass, 2000).

Kevin Featherstone is Professor of European Politics and Jean Monnet Professor of European Integration Studies at the University of Bradford. He has written extensively on the politics of the EU – most recently in books on EMU and on relations with the USA – and on politics in modern Greece.

John Gibbons is Senior Lecturer in Politics at Manchester Metropolitan University. He has recently authored *Spanish Politics Today* (Manchester University Press, 1999) and 'Spain: a Semi-Federal State', in D. McIver, *The Politics of Multinational States* (Palgrave, 1999). He has also written on the Common Agricultural Policy (CAP) in F. McDonald and S. Dearden, *The Economics of European Integration* (Longman, 1999).

Simon Green is Lecturer in European Studies at the University of Portsmouth. He has published a number of articles on German electoral politics and is currently writing a monograph on immigration and citizenship policy in Germany.

Derek Hearl is Lecturer in Politics at the University of Exeter specialising in Europe, EU institutions, parties and party systems and German politics. Author of various articles and book chapters on election manifestos, party competition and Belgian and Luxembourg politics, his current research is on the theory and practice of party competition and regional voting patterns throughout Europe and in the politics of small countries and of divided societies.

David Howarth has been a Lecturer in European Politics at Aston University in Birmingham since 1997. Author of *The French Road to EMU* (Palgrave, 2000) he has co-edited (with Alex Warleigh) *The State of Art: Theoretical Approaches to the EU in the Post-Amsterdam Era*, 2000. He is currently working on two books on French economic policy and the European Central Bank (Palgrave).

Karl Magnus Johansson is Lecturer in Political Science and Director of the European Programme at the University College of Southern Stockholm, and affiliated with its Institute of Contemporary History and the Swedish Institute of International Affairs, Stockholm. His publications include *Transnational Party Alliances: Analysing the Hard-Won Alliance Between Conservatives and Christian Democrats in the European Parliament* (1997) and *Sverige i EU* (1999). His research focuses on the role of political parties, notably the transnational coalition of Christian Democrats, in and for the process of European integration.

Notes on the Contributors xv

Henk van der Kolk studied Public Administration and Public Policy at the University of Twente. His thesis was on the effects of electoral control upon the behaviour of local council members. He was co-director of the Dutch Parliamentary Election Studies of 1998. He is currently involved in the European Candidate Study of candidates to the European Parliament and a study about European issues and Dutch MPs.

George Kazamias is Lecturer in Politics at the University of Bradford; his research interests lie in twentieth-century Greek politics and history and the international relations of the wider South-East Europe.

Juliet Lodge is Professor of European Studies and Jean Monnet Professor of EU Policy Studies at the University of Leeds, where she is director of the European Parliament International Study Unit. European Woman of the Year in 1992, she has published extensively on the European Parliament, Euro-elections, democratic legitimacy and on the EU. Her current research focuses on EU justice and home affairs, and on EU crisis management under the common foreign and security policy (CFSP). She is part of the *Britain in Europe* team and prepared a report in 1998 on the impact of Economic and Monetary Union (EMU) on national parliaments for the European Parliament.

José Magone is Lecturer in European Politics at the University of Hull. He was Karl W. Deutsch Guest Professor 1999 at the Wissenschaftszentrum Berlin fuer Sozialforschung; and Robert Schuman Scholarship Holder at the European Parliament in 1990. Among his publications are *The Changing Architecture in Iberian Politics* (Edwin Mellen Press, 1996) and *European Portugal. The Difficult Road to Sustainable Democracy* (Macmillan, 1997).

Janet Mather is Lecturer in Politics at Manchester Metropolitan University. She is currently writing a book, *The European Union and British Democracy: Towards Convergence* (Palgrave, 2000). An article, 'The European Parliament: a model of representative democracy?', was published in *West European Politics* in 2000.

Josef Melchior is Assistant Professor at the Institute for Political Science at the University of Vienna. He is General Secretary of the Austrian Political Science Association and co-editor of the *Austrian Journal of Political Science*. He has written on constitutional issues and institutional reform of the EU. His current research focuses on European citizenship, identity and democracy.

Edward Moxon-Browne is Jean Monnet Professor of European Integration, and Director of the Centre for European Studies, at the University of Limerick. He was Reader in Politics at the Queens University of Belfast until 1992. His main research interests are EU citizenship; Irish membership of the EU; and the ethnic roots of political violence. Among his books are *Nation Class and Creed in Northern Ireland* (Gower, 1983), *Political Change in Spain* (Routledge, 1989), *European Terrorism* (Dartmouth, 1994) and *A Future for Peacekeeping?* (ed., Pergamon, 1998)

Hans Jorgen Nielsen is Lecturer at the Institute of Political Science, University of Copenhagen. His main field of work is in electoral behaviour. He has published on the Danish Maastricht referendums in 1992 and 1993 in *EF på valg* (1993) and commented on the Danish Amsterdam referendum in 1998 for the information service of the Danish Parliament.

William E. Paterson, OBE, is Professor of German Politics and Director of the Institute for German Studies, University of Birmingham. The author of numerous books and articles both on Germany and its relations with the European Union, he was recently awarded the Federal Republic of Germany's Order of Merit for his services to British-German relations.

Christopher Piening has spent much of his career as an official of the European Parliament and currently heads its UK office in London. Prior to that he was a specialist in the EU's external relations, holding the post of head of the secretariat for relations with non-EU countries in Brussels for almost ten years. Author of *Global Europe: The European Union in World Affairs* (Lynne Rienner, 1997), he was in 1995–6 EU Fellow and visiting professor at the Jackson School of International Studies at the University of Washington, Seattle.

Tapio Raunio is a post-doctoral researcher in the department of political science at the University of Helsinki. He is the author of *The European Perspective: Transnational Party Groups in the 1989–94 European Parliament* (Ashgate, 1997), and has published articles in journals such as *Government and Opposition, Journal of Legislative Studies, Party Politics* and *West European Politics* on parliamentary democracy in the European Union.

List of Abbreviations

AFO	Anti-Fraud Office
ALE	Free European Alliance
CAP	Common Agricultural Policy
CFSP	Common Foreign and Security Policy
EC	European Community
ED	European Democrats
EDA	European Democratic Alliance
EDD	Group for a Europe of Democracies and Diversities
EEC	European Economic Community
EFA	European Free Alliance
EFGP	European Federation of Green Parties
ELDR	European Liberal, Democrat and Reform Party
EMU	Economic and Monetary Union
EP	European Parliament
EPP	European People's Party
ERA	European Radical Alliance
ERM	Exchange Rate Mechanism
EU	European Union
EUL–NGL	Confederal Group of the European United Left
IGC	Intergovernmental Conference
JPC	Joint Parliamentary Committee
MEP	Member of the European Parliament
NATO	North Atlantic Treaty Organisation
PDEV	Personal Direct Electronic Voting
PES	Party of European Socialists
PfP	Partnership for Peace
PPC	Parliamentary Partnership Committees
QMV	Qualified Majority Voting
SEA	Single European Act
STV	Single Transferable Vote
TEA	Treaty of Amsterdam
TEU	Treaty on European Union
UEN	Union for the Europe of Nations Group
UEP	Uniform Electoral Procedure
UFE	Union for Europe
WEU	Western European Union

A list of the individual member states' national party abbreviations can be found at the end of the chapter concerning that state (or states).

Part I
Introduction

1
Invisible, Irrelevant but Insistent? The European Parliament and Euro-elections

Juliet Lodge

Euro-elections have always been contentious: in their conception, their advocacy before 1979 and their impact thereafter. They remain symptomatic both of malaise within the system as to the adequacy and sufficiency of its democratic credentials, practices and aspirations, and of a deep-rooted sense of 'unfinished business' that affects the European Parliament and the European Union (EU) more generally. The constantly changing face, roles, powers, authority, influence and public esteem (such as it is) of the European Parliament mirror the attendant changes within the EU without addressing them directly enough to enable the public fully to believe in their legitimacy. So long as legitimacy is seen to be nascent and fragile (even if less so now than in the past) there is room in public debate for the seeds of doubt to be sown and grown in a way that prejudices informed debate and serves ultimately anti-and undemocratic and unthinking platitudes masquerading as political discourse. Criticism of the EU and Members of the European Parliament (MEPs) should have been the currency of the election debate. Constructive argument over policy options and possibilities should have assumed centre stage. Once again, they did not. The cause of elections in a supranational setting was accordingly ill-served. Their purpose remained, if not obscure, then of apparently little direct consequence to those entitled to vote. The fact that the European Parliament (EP) had evolved steadily since the first elections from a 'talking shop' to a 'fraud-buster' had demonstrated its ability and intention to require the political accountability of those responsible for initiating and over-seeing policy but apparently failed to persuade the voters to vote in large enough numbers to satisfy those for whom a higher turnout than in 1994 was important. Holding the elections itself remains important, and not merely for symbolic reasons.

On the twentieth anniversary of the first direct elections by universal suffrage to the European Parliament, the same questions were posed that had been asked since their inception: were the EP elections 'European' or not? Were they supranational or second-order national elections? Or were they a mix of both European and national? In many respects, the impression was created of nothing much having changed since 1979. There was still a high degree of public ignorance as to the nature, purpose and effect of voting. Voter apathy remained marked. The campaigns lacked coherence both within the member states and across their boundaries. As in the past, polls suggested that there were European-wide concerns over specific policy areas but that these were often those where the EU lacked any real competence. They were often a mix of national and European preoccupations and these were reflected in the party manifestos of the transnational parties (and less so in many of those of their component national parties). In 1999, top priority for the public in eleven member states was employment, with action to combat drugs and crime coming second in eight member states (but first in the UK, Sweden and the Netherlands), and the environment and consumer policy third in five. National electoral cycles continued to roll forward with member states continuing to be divided over the desirability of allowing domestic elections to coincide with the fixed Euro-election date. National political leaders' engagement in the campaigns varied as much as ever from member state to member state, and the EU continued to be lambasted as 'too distant' from the very citizens to whom it was supposed to be getting closer.

At the same time, however, a good deal had changed since 1979. The EP's authority had grown steadily. MEPs increased their power by adopting a dual strategy, simultaneously combining Spinelli's maximalist strategy of a qualitative leap forward in the early 1980s with the minimalist small steps approach which gradually nipped away at the Council of Ministers' shell of autonomy. The former strategy let people see the future by providing an indicator, in the shape of a treaty, of a future polity based on the realisation of the principles of democracy, efficiency, effectiveness, accountability, openness and transparency. The latter employed the tactic of fully exploiting existing treaty provisions in order to expand EP participation in the legislative process before securing its *de jure* consolidation and entrenchment in subsequent treaty reforms. The EP's supervisory and advisory competence expanded and it moved from being a legislature in waiting to becoming a key part of the legislative process – through the co-operation procedure – and then a partner with the Council in the legislative process via the co-decision procedure. Its budgetary, assent, scrutiny and legislative powers had all grown. They should not be exaggerated. MEPs now can amend policy proposals. They can affect expenditure and the budget in ways not open to national parliaments either at EU or national levels. They can insist on a right to fill the democratic gap left by national parliaments' absence and exclusion from the exercise of effective control over member governments in the Council and European Council, as well as in respect of the organs of the European Central Bank. The EP has still some way to go before it will be satisfied with the role it is able to play in holding the executive fully accountable and in setting and influencing the policy agenda and outcomes. Nevertheless, it is a different body from what it was when the public was first called upon to elect its members in 1979. Moreover, from being a shadow institution on the sidelines of constitutional discourse, it has edged towards the middle: by the time of the 1996 Intergovernmental Conference (IGC), two MEPs were at the table (albeit with a very limited brief) but the symbolic importance of being there should not be under-estimated. It showed that governments no longer disputed the legitimacy of the EP.

It is perhaps all the more surprising, then, that the elections should have engendered so little enthusiasm and interest among national political parties and the public alike. Habitual voters, those who vote in elections whatever their level (national, regional, municipal, local, European) exercised the franchise. Education, socio-economic background, political interest and political attachment tend to correlate with propensity to vote. Whereas, in the past, an individual's attitude towards the EU was seen as an important determinant of voting behaviour, attitudes towards the EU are now seen to be a less accurate predictor of inclination to vote in most member states. Anti-Europeanism or Euro-scepticism does not necessarily imply a greater likelihood of abstention: the opposite may be the case depending on the broader political environment and electoral agenda in given member states. Low turnout – while a matter of concern – is no longer specifically an issue partly because turnout at elections is no longer used to justify an increase in the EP's legislative powers. Turnout is not the only measure of the EP's relevance to EU decision-making. Instead, rendering the legislative process effective, efficient and open and enhancing the quality of legislative outputs has become more important. An increase in the EP's powers may, then, be a by-product of steps to achieve this.

However, it would be foolhardy to suggest that the level of turnout is no longer important, or a matter of concern, not least when compared to US Presidential turnout or turnout for Congressional elections which dip below 35 per cent. It is probably the case that there is recognition of a more general malaise over apparent public disenchantment with traditional political processes, and that the locus of concern is more diffuse. It has shifted from within the walls of the EP and its parties to include the seat of national governments confronting the prospect of having to persuade the electorate of the desirability of the EU in general, and of the desirability of the reform (whether through enlargement and/or an expansion of competence) in particular. Disenchantment with the Maastricht Treaty scarred some member governments and sensitised them all, and the Commission, to the imperative of bringing Europe closer to the citizen, and making it more accessible, open and understandable.

Along with this comes recognition of their past negligence of the public and of the failure of political parties at all levels, and of national governments as well as the Commission (whom national governments were anxious to exclude from Euro-election campaigns in the past) to use the Euro-elections to present the electorate both with information about the EU and with meaningful choices over its future direction and policy priorities. There is a sense in which, unless the EP party groups can credibly lay claim to 'ownership' of the political agenda, of policies made, adopted and implemented, the locus of responsibility for EU legislative initiatives and outcomes will remain obscure in the public eye. Inevitably, this must compromise the ability of MEPs to campaign subsequently on a record of EU legislative achievement, as opposed to purely local interests of concern to their domestic constituencies. This must further confuse the public as to the purpose of Euro-elections. Mobilising turnout will therefore remain problematic.

The significance of this failure becomes more acute when speculating on the future: enlargement will bring in new members with different traditions, expectations, priorities, ambitions, mixed objectives and sometimes relatively crude institutional procedures compared to western practice. European integration is about choice. Elections provide an opportunity for rehearsing alternative options and engaging in wider debates about its direction and purpose. MEPs have a particular responsibility in this respect. What this means is that whereas in the 1980s they could focus more on constitutional change and ensuring that treaty reform took place, now they need to concentrate on the procedural steps that are essential to turning the EU into a genuine, functioning participatory democracy. It remains to be seen whether they can rise to the challenge.

The next IGC offers an important opportunity for addressing institutional reform and constitutional issues in a way that is relevant, visible and intelligible to the public. It is clear that the EU is a very different organisation in 2000 from the one it was when the first Euro-elections were held in 1979. Equally clear is member governments' continuing reluctance to acknowledge the logical consequences of this change. If democratic governance is to be sustained, then the relationship between the Commission and the EP (and the Council) needs to be examined robustly. Separating responsibility for policy initiation from public accountability for it has its own logic, but it is not one understood by the public and neither is it one that is readily translated into electoral processes derived from different premises. It may be that Commission autonomy remains essential to drive European integration forward in a wider Europe: it must not be 'captured' by political parties. At the same time, it is important to ensure that the public is able to discern and vote out – if necessary – those whom they believe have not adequately advanced an agenda corresponding to the public's needs. The indirect method (of EP censure of the Commission) is cumbersome. That is not to say that it is irrelevant or cannot be adapted to be more relevant to the people.

How the EP and the Commission behave remains important in the public mind. This is because they are probably the two most visible (if insufficiently visible and poorly understood) of the EU's institutions. If the public is concerned about the political efficacy of its institutions, then while people are unlikely to be liable to equate tests of efficacy with those institutions, they do equate political failures of various sorts with them. It is in this sense that openness becomes important. If access to information is to be curtailed, as has been suggested and as is certainly necessary to ensure operational effectiveness in some policy areas (notably under pillars II and III), then the reasons for it must be explained sufficiently well to ensure public confidence in the system. Any curbs on states which are relatively more open towards their citizens than others will not have the desired effect in terms of maintaining shaky public confidence or boosting the EU's democratic legitimacy. It is not surprising, therefore, that the democratic audit in Sweden has stressed constitutional-political engineering and advocated a constitutional court as final guarantor of citizens' freedom and rights. Democratic participation lies at the heart of the EU's conception of citizens' political rights in the European Union. This conception of citizenship, initially at least, necessarily rests on an exclusive idea of citizenship derived from holding the nationality of a member state and bestows socio-economic and political privileges accordingly. MEPs need to politicise and problematise the consequences to render the idea of EU citizenship itself relevant to the public. The charter of Fundamental Rights is but a first step in the right direction.

At the same time, however, it behoves the EP genuinely to present itself as the credible guardian of liberty and democracy at EU level. It needs, at the very minimum, to go on the offensive in respect of a uniform electoral procedure (Anastassopoulos Report, 1998). This would be the most salient and directly relevant way of showing EU citizens that they are equal, politically at least, in terms of participation and the weight that each individual vote carries. It may also help to address the obstacles to acting in a European way that arise out of being either a relatively new member state and/or a relatively small or weak one. Familiarity with the EU bred of the duration of a member state's membership inclines MEPs to adopt a European focus of representation rather than one simply based on their member state. Since this also correlates with state size (Wessels, 1998), it may be predicted that MEPs from states which feel that their voice is not sufficiently heard or is not sufficiently weighty (for example, within the Council of Ministers and/or Commission) will feel obliged to advance national interests within the EP. Again, the importance of this will increase as the EU rapidly enlarges over the next ten years. It is incumbent on the EP to develop and organise itself in ways that aggregate common interests and avert excessive fragmentation.

At the next IGC, those examining institutional reform need to go beyond a rehearsal of how the key institutions might be altered to accommodate more member states. The issue of the ratio of seats or posts to population is important and needs inventive responses, possibly considering the feasibility (let alone political acceptability over the longer term) of a system based on a geographic/regional pooling: such as that suggested by the idea of a Benelux Commissioner, for example, as opposed to three Commissioners for those states. It needs something which goes beyond size to considerations that embody principles of good and effective government. Should, for example, political heads of the Commission be elected from within the membership of the European Parliament? Or should the heads of the major party groups accept and assert political responsibility and public accountability for the work of a particular Commission department? Or should they be the point of contact for other groups (e.g. of citizens, voluntary sector bodies or nongovernmental organisations, as well as lobbyists and interest groups) seeking to influence the political agenda and legislative outputs? However this conundrum is explored, the Commission and the EP need to be seen to be working more closely together.

In addition, the relationship between the EP and the Council needs to be portrayed as more equal. The public has been encouraged – not least in Euro-election campaigns – to view MEPs and the EP as their national protagonists. That does not reflect reality in that the Council is the repository of national interest and the institution is ultimately where – voting procedures permitting – national interests can be negotiated, compromised or made to prevail and influence legislative content. While MEPs might be expected to perform a role as Euro-legislators, and while their self-conception might in part at least embody such a role conception, so far they lack a genuine Euro-constituency. Depicting the Council as the Council of the Member States might be sensible. Portraying the EP as the guardian of a European-wide interest and as the guardian of democratic practice might also find an echo with the public and dispel some confusion over who – or which institution – does what, or is responsible, accountable and open for election in the EU. Whether this will boost turnout will depend in part on how each institution performs its roles, and how the EP is seen by the public.

Obviously, there are many elements to the issue of ensuring greater public participation in the European electoral process. At successive EP elections, the national offices of the European Parliament have engaged in information activities designed to ensure that turnout is not deterred owing to a lack of relevant, non-partisan, information. In 1999, similar information campaigns were run. Often relatively low-key to avoid charges of partisan interference in domestic politics, these campaigns had negligible impact on turnout no matter how imaginative they were. As the chapters' overviews of each member state reveal, there were common arguments advanced to account for this: lack of public interest and acceptable information, lack of motivation, vague, ambiguous or weak campaigning by the parties, a weak Euro campaign either lacking in focus or addressing the wrong issues, poor presentation of the EP, and a lack of conviction that voting mattered much in terms of who would assume 'power' in the EU. In short, a public forum – a European public space – appeared to be missing in all the member states, both within the member states and at EU level where EU issues were discussed. But this appearance masked a deeper reality.

In practice, there were innumerable seminars and briefings for the media and the interested public. For example, in the UK, a hot-air balloon toured the country from late April 1999 and had made 22 appearances and visited every electoral region, including Northern Ireland, by polling day (10 June). Emblazoned with 'Use your Vote' and 'European Parliament Elections', it attracted media attention and served as a focal point for the mobile information unit that accompanied it and went round the country in bad weather. The 'Use your Vote' theme also featured on advertisements on the back of city buses in Scotland, Northern Ireland, parts of Wales and selected parts of England. Scotland-specific information was also produced in the light of the May elections to the Scottish Parliament, but equivalent resources and efforts in Wales (where the Welsh Assembly was elected on 6 May) were not possible. Sweden had a similar campaign which sought to use networks and multipliers (as others did) to open up seminars and information to the media and public. These seminars had a positive effect on the networks and on media coverage of the election. In addition, a Job Train (run between 17 May and 3 June jointly with the Commission, the Swedish Structural Funds Office and the Labour Market Board) aimed to communicate with citizens in regions. It stopped at, and exhibited in, 28 cities across Sweden. An agreement between the National Tax Authority (responsible for organising the election) and the Postal Service (charged with organising voting at post offices) resulted in a mobile post office in a tent where citizens could experience voting. A major radio campaign was conducted aiming at 18–35 year old voters, those traditionally less inclined to vote and more inclined to listen to commercial radio. Websites, leaflets, brochures, posters, bookmarks, postcards and stickers were also available throughout the member states. In short, there was hard information to be had. What was lacking was the effective presentation of policy options and viable, interesting *alternatives*.

It is perhaps the absence of a forum or arena in which alternatives were publicised effectively and discussed with the degree of passion common to major debates about political options in the domestic political arena that deprived citizens of the sense that, in voting in the EP elections, they were voting and recording choices over future policy directions. This may highlight both the inadequacy of the political parties' engagement and their failure to perform adequately educative and communication functions *vis-à-vis* the electorate. It also underlines the possibility for system change and a refinement of the agenda-setting roles of the EP and the Commission. The Commission's autonomy has been vital and remains so. But its significance may now lie less in autonomy to expand the scope of integration to new areas where supranational action and advancing the common good by upgrading the common interest is only attainable by virtue of Commission initiatives to shape and define policy. It may be that the Commission needs to be free to ensure that political choices fit with a broadly defined future vision and that appropriate steps, appropriately funded, are taken to realise it. The inevitable competition over policy priorities and directions is entirely appropriate and should be fought out in a public arena. The EP needs to develop its agenda-setting role to facilitate this, even if gradually. The EU itself, in the immediate past, appeared to lack direction, vision and an ability to discern and present alternatives in a credible, relevant and meaningful way. This is not to excuse MEPs or low turnout, but it does suggest that MEPs do need to be bold in their advocacy of political choices, bold in their questioning and follow-up of legislative proposals, and audacious in their defence of the EP on the occasion of the next Euro-elections. They need to define and present their role and their accomplishments in a critical way that enables the public to see the future in terms of political choices to be made, discussed and voted. Disappointing as turnout was in 1999, it could not justify abolishing the elections or returning to the previous system of nominating MEPs from among MPs. The Euro-elections may not yet be sufficiently relevant or salient to individuals to inspire them. As yet, there is but a rudimentary sense of mutual responsibility among EU citizens, and it is often ill-defined, poorly or not convincingly presented to the public. Small wonder that a sense of political community, let alone civil society, is fragile and contested: enabling the people to take possession of the political project of European integration remains a challenge for both MEPs and national politicians. In asserting the idea that the EU should embrace, develop and sustain an identity built on the values of democracy, social justice, equality and solidarity, political participation in a Euro-wide democratic process is central. Assuming that public participation and an active citizenship remain important to our sense of democratic practice and good government, it is essential that MEPs ensure that the IGC takes a bold look at the system of governance the EU should develop, sustain, implement and promote over the next generations. The member governments must be equally participatory: partial or complete opt-outs are inexcusable. The political processes must be sufficiently reasonable to command consent. The public must see and believe the EU's political processes to be democratic and reasonable. Euro-elections present an opportunity for testing publicly and affirming or withholding consent. They – and MEPs – are therefore central to the future well-being of the EU.

References

Anastassopoulos Report, European Parliament Resolution on a draft electoral procedure incorporating common principles for the election of Members of the European Parliament A4-0212/98, June 1998.

Linklater, A. (1998) *The Transformation of Political Community* (Oxford: Polity).

Marquand, D. (1997) *The New Reckoning: Capitalism, States and Citizens* (Oxford: Polity).

Wessels, B. (1998) 'Institutions Matter: National Setting, Electoral Mechanism and Representational Roles of the Members of the European Parliament Compared', paper for the NIG seminar, University of Twente, Enschede, July 1998 (forthcoming Boston: Rowan & Little, 2000)

2
The European Parliament's Progress 1994–99

Richard Corbett

One of the ironies of the declining turnout in European elections is the fact it has been simultaneous with an *increase* in the powers of the European Parliament. In the first European elections in 1979, voters were invited to elect a body with next to no influence on the policies of the European Community. By 1999, the Parliament had gained co-decisional powers on the bulk of European legislation, as well as a significant role in the appointment and scrutiny of the European Commission.

Of course, some would argue that the decline in electoral turnout should not be over-dramatised. A European wide average of 50 per cent is higher than the 38 per cent turnout in the last US Congressional elections. The downward trend (loss of 13 percentage points from 1979 to 1999) mirrors a trend in recent national and local elections in many member states: electoral turnout in national elections fell, for instance, by 14.8 points in the Netherlands between the 1977 and the 1998 elections, by 14.4 points in France between the 1978 and 1997 elections, by 13.1 points in Germany between 1976 and 1990 elections, and 10.2 points in Ireland between 1979 and 1996. As was the case in four out of the five European elections that have taken place so far, the lowest turnout of all in 1999 was in the UK. But in a by-election held the same day for a vacancy in the House of Commons, and under the traditional first-past-the-post system, a turnout of merely 19 per cent was achieved, again illustrating that the decline in turnout was by no means restricted to the European elections. None the less, the relatively low turnout, especially in some countries, was certainly a disappointment for MEPs, who felt, perhaps, that they had been able to demonstrate their relevance more than ever before in the preceding years and were poised to be able to do so still more with the newly gained Amsterdam Treaty powers.

Indeed, the 1994–99 European Parliament was remarkable in that it coincided with the first exercise of the powers obtained by Parliament under the Maastricht Treaty, the negotiation of new powers under the Amsterdam Treaty (coming into effect just before the 1999 elections) and the first effective exercise of Parliament's power to dismiss the Commission. More than ever before, MEPs felt that they had been able to show that they mattered.

Legislative powers

One of the first acts of the newly elected EP in 1994 had been to make use for the first time of its newly acquired power to reject legislation outright under the co-decision procedure, causing it to fall. Parliament did this on the voice telephony directive, when disagreements persisted between EP and Council on the issue of scrutiny over the Commission's implementing powers ('comitology'). The co-decision procedure had been introduced by the Maastricht Treaty for a limited number of areas (some 15 treaty articles). It provided for Commission proposals for Community legislation to be subjected to two readings each in Parliament and Council. If, by then, the two institutions had not agreed on identical texts, a conciliation committee would be convened to negotiate compromises. The committee would be composed of representatives of every member of the Council on one side and an equivalent number of MEPs on the other (then 12 + 12 now 15 + 15). The compromise, to be negotiated within six weeks (extendable to eight), would then have to be approved by both sides within a further six (or eight) weeks. The procedure as it stood under the terms of the Maastricht Treaty included two safeguards from Council's point of view. First, if the EP wished to reject a text outright before the conciliation phase, it would have to do so with two votes: the first announcing its intention to do so and the second, following an optional conciliation meeting with Council, as a definitive rejection, both having to be approved by an absolute majority of MEPs. Second, if the conciliation committee failed to agree on a common text, Council would have the option of re-confirming its own position, which would become law unless the EP rejected it by an absolute majority within six weeks.

Parliament disliked these provisions, and particularly the second one which it felt tilted the balance of power in co-decision towards Council. The EP hinted, and put procedures in its Rules of Procedure to encourage such a conception, that it would automatically reject any text adopted by Council in this way. In fact, the voice telephony directive was the first time the Council had tried this. Parliament duly rejected the text and the Council never again attempted to adopt a text unilaterally. The above provisions were eventually deleted by the Amsterdam Treaty, which thereby transformed the co-decision procedure into one in which the Council and the EP act like equal chambers in a bi-cameral legislature. During the EP's 1994–99 five-year term of office, and until the entry into force of the Amsterdam Treaty, 165 co-decision procedures were completed. In 99 cases (60 per cent) agreement was reached without convening the conciliation committee (63 cases where the common position was accepted by the EP without amendment plus a further 36 where it did amend Council's common position but Council accepted all Parliament's amendments). This leaves 66 cases where the conciliation committee was convened, of which 63 were completed successfully.

Inside the conciliation committee, a process of exchange has developed with both sides ready to make concessions but at a price that only becomes clear in the course of the negotiations. This exchange provides an opportunity for the EP to press its point in a much more intensive way than it ever could under the consultation or co-operation procedures. Evidence shows that it can get through conciliation changes to the common position that do make a substantial difference to the content of legislation. Two examples from 1998 – the Auto-Oil package and the 5th Framework Programme for Research and Technology – show this clearly.

The Auto-Oil package concerned measures to be taken against air pollution by emissions from cars and light commercial vehicles as well as the quality of petrol and diesel fuels sold in the EU. The EP adopted more than 100 amendments at second reading which contributed to the complexity of the negotiations but also provided the material for trade-offs between the two parties. The EP found itself facing a British Council Presidency which had a strong domestic interest in finding a solution. The Deputy Prime Minister, John Prescott, had returned from the Kyoto Summit at the end of 1997 committed to reducing the level of CO_2 emissions in Europe. He was thus eager to ensure that a deal was reached on the Auto-Oil package as a way of showing to the outside world that the EU respected the Kyoto deal. The British Presidency therefore made it clear that it wanted to find a compromise with the EP. The central element of that compromise was an acceptance by the EP of the limit values laid down in the Council position but on the basis that those standards should be *compulsory* by the year 2005, rather than optional as Council wanted. Thus the bargaining process led to an outcome which resulted in an EU-wide obligation for improved standards (which would have a marked effect for the general public as well as for the car and oil industries).

A second example of the changes that can be secured in conciliation is the 5th Framework Programme for Research and Technology which was adopted in December 1998. This was a particularly unpromising case for EP influence as it was an area where agreement in the Council had to be unanimous. The Austrian Presidency did not have the option of working towards a qualified majority. This was particularly important as the Spanish government had insisted on including in the common position what became known as the 'guillotine clause' under which any budgetary allocation agreed for the new programme would be subject to review pending the outcome of the negotiations on the Agenda 2000 revision of EU funding. These negotiations also made other delegations, notably the Netherlands and the UK, very reluctant to contemplate any significant increase beyond the common position which had set aside 14 billion euro for the four-year period of the programme; they were also able to veto it. The EP insisted that the current level of Community research could not be maintained under the terms of the common position. At second reading it voted for a figure of 16.3 billion euro and so the two institutions entered the negotiations more than 2 billion euro apart. Remarkably, they finally agreed a figure just short of 15 billion thereby effectively providing for a sharing of the spoils.

Again there was a degree of EP dissatisfaction with the outcome but, without co-decision, it would have been impossible to reach such an agreement. The co-operation procedure, for example, would have allowed the Commission to propose the EP's starting figure after second reading but there would have been unanimity in Council to re-establish the common position at a far lower level than that finally agreed. Furthermore, the process of negotiation allowed the two parties to revise the 'guillotine clause': it was accepted that the EP would be involved through co-decision in any revision of the figures arising from the new agreement on future financing under Agenda 2000. So, despite the undoubted difficulties posed by unanimity in Council, it does not make it impossible for the EP to influence the outcome.

In 1995, the issue of biotechnological inventions provided the first and so far only case where an agreement made by the EP's delegation in conciliation was rejected by the plenary. When a revised version of the same directive came back for consideration in 1997, there was a strong desire on both sides to avoid a similar outcome. Contacts between the two institutions took place even before Council's first reading to see how the concerns of the EP could be accommodated by the Council. This resulted in the willingness of the Council in its common position to accept, in one form or another, almost all the EP's 40 or so amendments. As a result, when the EP came to vote its second reading, a majority felt there was no need for any further amendments.

Scrutiny of the executive

The Maastricht Treaty had changed the nature of the relationship between the EP and the Commission, in particular as regards the latter's appointment. Prior to 1994, the Commission had normally been appointed for a four-year term of office by the national governments without EP involvement. Maastricht changed the term of office to five years to coincide with the EP's electoral cycle, and changed the appointment procedures to involve EP. It laid down that, after every European election, member governments would first agree on a candidate for President of the Commission who would be submitted to Parliament for a vote. Subsequently, the rest of the Commission would be put together in consultation with the President-designate and subjected to a vote of confidence by the EP in which it required majority approval to take office.

The first vote, on the President of the Commission, was merely consultative under the terms of the Maastricht Treaty. In political terms, however, many felt it to be binding: it would be inconceivable for a President to proceed to take office if he or she had been rejected by Parliament. Such a view was, indeed, confirmed both by the candidate in 1994, Jacques Santer, and by the then President of the European Council, Helmut Kohl. Both stated that if the EP were to reject the nomination, a new candidate would have to be found. In the event, the vote on Jacques Santer was very close, and he obtained his majority after considerable political effort to win support from the EP's main political groups. The episode highlighted the fact that the vote was not a formality. The Amsterdam Treaty then changed the vote on the President into a legally binding requirement for majority EP support.

For the second vote, on the Commission as a whole, the EP insisted that candidates all appear before the parliamentary committee that corresponded to their prospective portfolio for confirmation hearings. This required the Commission, for the first time ever, to agree on a distribution of portfolios before it took office. The Commission did not like this procedure. It was not provided for in the treaty, but the Commission had little choice when EP made it clear that it would simply not proceed to the vote of confidence until the candidate-commissioners had complied with its wishes. Indeed, Parliament postponed the vote scheduled for December 1994 to January 1995. This was to allow MEPs from the new member states (Sweden, Austria and Finland), who joined the EP in January, to take part in the vote on the Commission which would, after all, exercise its responsibilities over their countries as well. At the same time, it illustrated Parliament's power to delay its vote.

In the absence of an individual vote on each Commissioner, the hearings lacked the incisiveness sometimes obtained by confirmation hearings in the US Senate. None the less, they did provide a useful opportunity to scrutinise the views and qualifications of prospective Commissioners. Above all, they illustrated the fact that Commissioners are not unaccountable bureaucrats but members of a political executive accountable to the EP. Later in the 1994–99 Parliament, it was the EP's powers of scrutiny over the Commission which came to the fore. Parliament had never adopted a censure motion on a Commission, but in March 1999 the Commission resigned when faced with the near certainty of the adoption of a censure motion. The crisis arose in 1998–9 after the EP, at its December 1998 plenary, declined to give the Commission discharge on its 1996 accounts amid allegations of financial mismanagement. A censure motion was tabled and put to the vote during the January 1999 plenary. Although it obtained substantial support, a majority of MEPs, rather than move to an immediate censure, voted instead to set up a Committee of Independent Experts to investigate the actions of the Commission. The subsequent critical report of the Independent Experts in March 1999 led to the collective resignation of the Santer Commission, within hours of the reaction of EP political groups making it clear that there was a majority for censure.

These events highlighted the nature of the relationship between the Commission and the EP as one requiring the Commission to have MEPs' confidence. This was further underlined by the Amsterdam Treaty changes which recognised the binding nature of the EP's vote on the candidate for Commission President by turning it into a legal requirement, and at the same time giving the Commission President the power to decide on the remaining Commission members jointly with national governments and giving him/her the power to distribute and reshuffle Commission portfolios.

Development of the European Union

The EP was able to play a significant role in the IGC that led to the Treaty of Amsterdam. The member states' governments invited it to appoint two MEPs to the 'Reflection Group' which prepared the negotiations. The EP nominated Elisabeth Guigou (Socialist) and Elmar Brok (Christian Democrat) to join the personal representatives of the heads of government and Commissioner Oreja (himself the previous chair of EP's Committee on Institutional Affairs) in preparing a detailed blueprint for the negotiations. In the IGC itself, opposition from the then governments of France and the UK prevented them from participating in all the meetings, but they met the IGC negotiators for one of their three meetings each month, and were able to obtain and discuss all the proposals tabled and put forward their own texts. The resulting treaty met a number of the EP's wishes, not least as regards its own powers. As indicated, this included the extension of the scope of the co-decision procedure (from 15 to 38 subjects), changes to the procedure and a change in the nature of the vote on the Commission President.

Conclusions

The 1994–99 EP was largely successful in further developing the role of the European Parliament in the EU system. The presence of a body of full-time elected representatives at the heart of the Union's decision-taking system makes the whole EU more open, transparent and democratic. Without the EP, it would be dominated by bureaucrats, diplomats and technocrats, loosely supervised by ministers periodically flying into Brussels. Parliament's presence brings greater pluralism with members from both governing and opposition parties, from the whole political spectrum and from the capital city and the regions of each country.

Until 1994, the EP's lack of clout had undermined MEPs' credibility. The Maastricht Treaty gave them the opportunity to develop their role. The 1994–99 Parliament was successful in doing so, and at the same time in winning new powers entrenched in the Amsterdam Treaty. Its presence as a key participant in decision-taking in the EU was undoubtedly reinforced. Yet this enhanced role did not lead to a higher turnout in European elections. The reason lies partly in the peculiar nature of European elections: in national elections in European countries, most voters go out to vote to either change or keep a government since it is the government that is central to the election campaign. Every five years, they have to vote in European elections where no government is at stake and where no heads roll from the executive as a result of the elections. Electing the EP in isolation, with no consequential effects on any executive, is not what people are used to in EU member states. It is therefore not surprising that arguments in favour of the Commission emerging from a majority in Parliament have again come to the fore. Such a majoritarian system is, however, problematic in other ways at the current level of European integration. As a compromise, it has been suggested that just the Commission President be elected by the EP, with the rest put together in the usual way. The President is, after all, usually a relatively well-known figure. If he or she were to be elected by the EP, the main European political families would feel obliged to put up candidates and announce them prior to the elections. This might add something to the election campaign in that a visible effect on who heads the executive would be at stake. The prospects of such a reform at this stage seem slim. None the less, it is one to watch for the future.

# 3
The EP since 1994: Making its Mark on the World Stage

Christopher Piening

The EP has grown considerably in power since the first direct elections in 1979. Then it was a consultative body with a few budgetary and supervisory powers. The succession of landmark treaties that, in the 1980s and 1990s, transformed the European Community (EC) first into a single market and then into a European Union – the Single European Act (SEA), the Maastricht and Amsterdam treaties – has changed all that. The EP now has a major legislative role, and its supervisory powers, especially as they relate to the Commission, have also expanded. The EP, naturally enough, tends to be seen by most people, most of the time, as an institution that adds a measure of democratic control to the processes that determine the impact of the European Union on the lives of its own citizens. Here I touch on the EP's impact during this past legislative period since 1994 – and indeed more generally – *outside* Europe. This is, after all, part of the bigger picture, even though it may be of only marginal interest to the average voter at the European elections.

Foreign policy, or external relations, is a field in which the EP has only limited competence in a legal sense, but in which it nevertheless plays a considerable part. By external relations, I do not mean necessarily the Common Foreign and Security Policy CFSP. I believe the formal lines drawn between foreign policy as defined in the CFSP and external relations more generally to be misplaced: for example, the numerous agreements with third countries, usually in the form of co-operation and trade agreements, are products of treaty-based procedures, not CFSP deliberations, but they do have foreign policy as well as trade and economic implications. Development policy lies at the heart of the earliest EC endeavours to incorporate former colonies and dependencies in the policy framework, and includes a wide range of instruments, such as the Lomé Convention, the Generalised System of Preferences (GSP), emergency and disaster relief, EC Humanitarian Aid Office (ECHO) and so on. Yet development policy is also foreign policy. Other aid programmes include such things as disaster relief, food aid programmes (e.g., to Russia) and so on. Whom to help and with what resources is a foreign policy matter. EU enlargement is an ultimate example of foreign policy, yet is provided for in the Rome Treaty; it is not a CFSP matter.

These are all examples of external relations activities that are dealt with through normal Community procedures and not through the intergovernmental process that is the CFSP. As such, they are subject to the full range of EC machinery, involving the Commission, Council and the EP. Yet while they are considered to be external relations matters, and are therefore subject to the 'Community method', they are all equally construable as foreign policy. So however this aspect of EU business is categorised – as foreign policy or external relations – the EP has a central role to play. It has powers in the field of external relations. Some of these are direct, others indirect. I differentiate between them by calling them statutory or non-statutory powers, and I will address this difference in a minute.

Parliaments in Europe generally are not major players in foreign relations. In most EU member states, the government of the day determines foreign policy, and the majority on which it rests in Parliament endorses that policy. It is rare in European national politics to find a parliament taking an independent line on such matters. But unlike most national parliaments, the EP can and does take an independent line towards external relations, since it is not beholden to a government. Its role is more akin to that of the US Congress than to any national parliamentary practice familiar in the Union's own member states.

Statutory versus non-statutory involvement

The EP's statutory involvement resides in the fact that the EP has to play a role in the EU's external relations where these are covered by the 'Community method' (i.e., as stipulated by the treaties). Responsibility for external relations is shared amongst the EU's institutions. The Commission is effectively diplomat in chief in the area of trade and commercial policy. It is even equipped with its own foreign service, consisting of Directorate-General I as its Brussels-based ministry of external relations, and over a hundred delegations – effectively, diplomatic missions, or embassies – in the capitals of countries around the world. The Commission enjoys wide-ranging power under the EC treaty both to make trade policy and to negotiate it. Increasingly, it is also involved in non-trade issues at international level, such as the law of the sea or environmental issues. The Council of the Union provides political cover to the Commission, in the form of negotiating mandates, but plays little or no active diplomatic role in trade or other negotiations. However, the Council is the central player in CFSP activities from which the Commission and Parliament are – officially, at least – largely excluded. Major external relations initiatives or policy in the real sense can also be taken at the level of the European Council. The EP's responsibilities are mostly set out in Article 300 of the EC Treaty. This grants the EP the right to give its assent to agreements between the EC and third countries or international organisations, as well as acts under such agreements (such as additional protocols) or amendments to them. Article 49 of the Treaty on European Union (TEU) provides for parliamentary assent in the case of enlargement of the EU, arguably an ultimate act of foreign policy. Finally, the implementation of some external trade acts may require further legislation which is subject to the standard procedures (Articles 251 and 252, the co-decision and co-operation procedures).

What does this mean? It means that before the EU can enter into key foreign agreements, the EP will have to ratify those agreements. Is this merely a rubber-stamp power? Far from it. There have been a number of occasions since 1994 on which the EP has used the assent procedure as a political lever in shifting the positions of third countries, and serving warning on the Community's other institutions that their negotiations must take account of the EP's views. They included sometimes lengthy delays in ratifying several partnership agreements with the countries of the Commonwealth of Independent States, including with Russia itself, where the latter's involvement in Chechnya led to strong opposition in Parliament. Concerned at the lack of democracy in Kazakhstan, the EP persistently held up its assent on the partnership accord with that country, only granting it in March 1997, almost two years after it had been signed. One of the most notorious examples concerned Turkey's customs union with the EU. The EP had long been critical of Turkish human rights violations, including the arrest and imprisonment of members of the Turkish Parliament. Indeed, the EP's relationship with the Turkish Grand National Assembly, with which it had had a formal link (in the shape of a joint parliamentary committee, or JPC) since the 1960s under the terms of the Ankara Agreement, had often been stormy; so much so in recent times that the usual twice-yearly meetings had been suspended altogether in 1994 and only resumed in June 1996. The EP announced even before the signature of the March 1995 decision on customs union that it would withhold its assent in the absence of clear commitments by the Turkish government on a number of questions. Against considerable internal opposition, the government in Ankara did push through a raft of reforms during 1995, including a series of constitutional amendments strengthening the democratic rights of citizens, trade unions and parliamentarians, and a relaxation of the anti-terrorism laws. As a result, two Kurdish MPs were released from jail and a promise was given that four others could have their cases heard by the European Court of Human Rights. A further 79 prisoners were set free by the end of the year as a result of the changes in Turkey's anti-terrorist laws. The EP was not overly impressed with these moves, which it felt did not go far enough. Nevertheless, it acknowledged that the Turkish action represented progress and, under considerable political pressure from national EU capitals, approved the customs union in December 1995, just days before the 1 January 1996 deadline.

This is a real power. It can be used as a threat as well as in practice. Now that the Commission is aware that any agreement that it negotiates will have to be 'got through Parliament', its whole approach is inevitably modified. The Commission talks seriously to Parliament and its committees about what it intends to do, discussing in depth the negotiating mandate it has received from the Council. Armed with the EP's views, it conducts the actual negotiations accordingly. If it is clear that an agreement could only be reached on the basis of terms that would be unacceptable to a majority of MEPs, the Commission will have no interest in pursuing them. Such was the case in May 1996, when talks with Cuba on an association agreement were abruptly brought to a close by the Commission once it became clear that the Castro government would not accept the standard human rights clause in the accord. Indeed, though it has so far never come to it, it can be safely assumed that the Council would today be unlikely to give the Commission a negotiating mandate to begin talks on an agreement with a country on which the EP had expressed misgivings a priori.

In addition to the assent procedure, the EP can use its budgetary muscle to influence external policy. It has done this in relation to Turkey, blocking the use of MEDA (Assistance to Mediterranean countries) funds on the grounds of human rights concerns (1996). In 1998, the EP refused to release the money destined to allow EC participation in KEDO (Korean Peninsular Energy Development Organisation) unless it was consulted on how the money was to be used. The result was an agreement between Commission and Parliament that the latter will in future be consulted on international nuclear energy accords. More generally, the EP can set emphases in the way it assigns money under the non-compulsory expenditure sections of the budget. In this way, it has influenced the money spent on programmes to combat the destruction of tropical forests, to promote democracy and human rights in the Central and Eastern European countries, and to help the countries of Southern Africa following the end of apartheid in South Africa.

Turning to the EP's non-statutory role, it is clear that it can affect the EU's external relations through force of personality, as it were, rather than the exertion of its legal muscles. While the EP lacks a right of legislative initiative, it can ask the Commission to make proposals for new legislation. It has done this on numerous occasions. Some of the classic EU–third country trade spats had their origins in Strasbourg. If we look back over the years since the EP was first elected, we find a host of examples. The ban on the importation of seal products came about as the result of an EP initiative, with MEPs reacting to massive public dismay at footage of baby seals being clubbed to death in Canada. Similarly, bans on kangaroo products from Australia, of old-growth lumber from Canadian rainforests, and of furs from animals caught in leghold traps in Russia and North America all had their origins in parliamentary reports and resolutions which were subsequently taken up by the Commission. On the face of it, these are issue-specific, but they also have international repercussions, and may indeed have implications under General Agreement on Tariffs and Trade (GATT) or World Trade Organisation (WTO) rules. The countries concerned with these particular measures reacted strongly to the imposition of restrictions. To the extent that the EP may be instrumental in encouraging or initiating trade disputes, it is making an impact on the EU's external relations. To take the most topical, though also one of the longest running examples, that of growth hormones in beef: the ban on the import of US beef raised on such hormones would never have been implemented had it not been for EP pressure. That pressure still exists as it reflects public concern over the issues, and is heeded by the Commission.

Further, in what I term the human rights syndrome, the EP's preoccupation with human rights has had one of the most significant influences on the EU's relations with third countries. EP resolutions on human rights issues – usually adopted following 'urgency' debates under Rule 47 of the EP's Rules of Procedure – generally highlight specific cases and call on the authorities of the offending country to rectify the abuses in question. Often, in addition, they demand action from the EC's other institutions, usually through requests to the Commission and the Presidency to make representations to the target country. Sometimes, however, they go farther than this, proposing specific responses either by the other institutions, or by the EP itself.

References to human rights in resolutions adopted by the EP do not fall on deaf ears. States or regimes that are targets of such resolutions regularly react, often sending letters containing detailed rebuttals of the allegations (and often protesting that the resolution represents an interference in their internal affairs). For example, in June 1997 in a 'Solemn Statement', the Foreign Affairs Committee of the Chinese National People's Congress reacted with fury to the EP's resolution on the Commission's Communication on a long-term policy for China–European relations, declaring that the EP had 'poisoned the atmosphere of China–Europe relations and thus disrupted their normal development. The Standing Committee of the National People's Congress of China... expresses its utmost indignation at the anti-China clamour of this resolution.' I can also quote from my own experience of the first four years of the 1994–99 EP's term, during which it adopted no fewer than 38 resolutions on human rights abuses in China. As head of the interparliamentary relations department, I would receive the counsellor from the Chinese mission to the EU, accompanied by a first secretary, almost every month, to receive either a written or a verbal dressing down.

EP resolutions, while not in themselves necessarily resulting in any amelioration of a given abuse situation, none the less form part of wider campaigns directed at the perpetrators of human rights violations and thus serve to strengthen ongoing international efforts. In some cases, however, there is a clear linkage between alleged human rights violations in a third country and the European Union's formal relations with that country, in the form of an existing or pending agreement. Here, the EP may use its powers under the assent procedure to threaten to hold up or reject the accord subject to improvements in the human rights situation.

The culture evoked by the EP's long-standing concern with this issue has certainly permeated through to the other EU institutions. Nowhere is this clearer than in the practice in force since the early 1990s of including a 'human rights clause' in all agreements negotiated or re-negotiated with third countries, whether simple trade accords or complex association agreements. These clauses allow abrogation of the agreement, by either side, where human rights abuses are established. And they have come to have repercussions in the external relations field: Australia and New Zealand both refused to sign framework agreements with the EU in 1997 because of the EU's insistence that a human rights clause be included.

Related to human rights is democracy and good governance. During the course of the 1994–99 legislature, the EP devoted considerable attention to the imminent renewal of the Lomé Convention. Alleviating poverty in the African, Caribbean and Pacific countries will be at the cornerstone of the Convention. Of special significance, because it marks a new departure in the almost 40-year history of the EU's relations with the developing world, is the move towards recognising in the ACP states a political entity; and more than that, a political entity with certain responsibilities for upholding the principles of democratisation, respect for human rights and good governance. At EP insistence, the Commission accepted the principle that the new Lomé Convention should have a political dimension, incorporating the idea that long-term sustainable development in the Third World can only be achieved by governments applying the principles of good governance and the rule of law. Negotiations with the ACP on the new agreement thus centre not merely on the amounts of Overseas Development Assistance and the size of the Export Earnings Stabilisation Fund, as was usually the case in the past, but include for the first time a clearly political dimension. It would not be true to say that this new increase in conditionality necessarily appeals to all ACP members. But it is worth noting that the subject has been discussed and repeated in meetings of the EU–ACP Joint Parliamentary Assembly over the past few years. Assembly resolutions – which required majority support from both sides, EP and ACP, to be adopted – endorsed the new approach. The Joint Assembly, like various other interparliamentary bodies set up by the EP, thus influences policy.

The EP is unique among parliaments in having an institutionalised system of interparliamentary delegations. At present, there are 34 such bodies. Some are delegations to JPCs or parliamentary partnership committees (PCCs) established under association agreements between the EU and third countries; others are 'non-obligatory' bodies that exist by mutual agreement with the parliaments of partner countries such as the USA or Japan. But the work they do – creating a parliamentary dimension to the relationships that exist between the EU and third countries – has become of increasing importance in ensuring that the EP is able to exercise informed political control over the EU's external relations. Some of these delegations have grown interesting appendages that further enhance Parliament's role. The delegations covering Latin America hold an EU–Latin American Interparliamentary Conference every two years with members of the Latin American Parliament. In 1998, the various EP delegations involved with the countries of the Mediterranean Basin launched the Euro–Mediterranean Parliamentary

Forum, a body designed to monitor implementation of actions under the Barcelona Process and give the process a parliamentary dimension. The EP's delegation for relations with the USA and its counterpart in the US House of Representatives have agreed to upgrade their interparliamentary relationship to what is being called the 'Transatlantic Legislators Dialogue': it is a response to the New Transatlantic Agenda, which sets out objectives for achievement at executive level between the USA and the EU.

Related to the activities of the standing delegations are those of other more *ad hoc* delegations, such as those sent as election observers or to participate in ministerial meetings between the EU and its partners, for instance, the Barcelona Conference on the Mediterranean, or the EU–Rio Group meetings, or meetings of the Baltic States. Parliament is increasingly insisting on playing a part in such international gatherings, refusing to be excluded from what used to be meetings exclusively reserved for representatives of the Commission and member states' governments. Having achieved this right to participate, the EP finds itself involved in international events from which even the member states' national parliaments are excluded (except, of course, to the extent that government ministers are usually also members of their national parliaments).

All these activities are about making it clear to the other EU institutions – Commission and Council – that external relations are not the sole preserve of the executive, but need democratic oversight. Space limitations do not allow me to detail the roles played by the EP's specialist committees (Foreign Affairs, Industry and Trade, and Development). Suffice it to say that it is in these committees that most of the work is done on the nitty-gritty of preparing Parliament's formal positions and responses to external relations matters. Finally, it is worth mentioning an odd aspect of the EP that relates to its international image. Parliament has become a sort of mecca, a place of pilgrimage, for international statesmen. Presidents, prime ministers, and other leaders ranging from the Pope through Yasser Arafat to the Dalai Lama, all put in their appearances. The EP offers a convenient platform for outsiders – supplicants or not – to reach a wider European audience. The appearances give added credibility to the EU's external relations efforts. By providing a forum, open to the press and public, for important international figures to address matters of foreign policy before a wide European audience, the EP can offer a stage that no national parliament can provide. In doing so, it underlines its own claim to a role as an international player. Rarely does a high-ranking foreign visitor leave the EP without extending an invitation to Parliament's President to pay a return visit. Indeed, along with the EP's delegations, its president tends to be a frequent traveller.

These varied activities are all 'non-statutory', in the sense that Parliament is not required to undertake them. It is not required to operate a system of interparliamentary delegations (and if the existence of the JPCs and PCCs is mandatory under the relevant agreements, then only because of parliamentary pressure on the Commission when the agreements were being drawn up). It is not required to adopt resolutions on human rights, or to send teams of observers to monitor elections. Its committees are not required to draft own-initiative reports on the EU's relations with third countries. Its President is not required to undertake international travel. Parliament is not required to offer a platform to statesmen and women from around the world, or to provide the facilities and organisational expertise for international conferences such as the Euro–Mediterranean Parliamentary Forum. Yet it does all this and more, for a variety of reasons, and with a variety of results.

In conclusion, the EP is an actor in the field of external relations both because it has to be and because it can be. As such, it is doing what any parliament does: making use of the powers and influence at its disposal. But unlike most parliaments in the EU's member states, it has the freedom to act independently in the use of its powers, and is not obliged to follow the dictates of government policy. In this sense, external relations is a domain in which the EP has carved out a niche for itself that is unusual in the national parliamentary context. It has an impact on the world stage that is both evident and unusual.

Part II
Country Case Studies

4
Austria

Josef Melchior

Introduction

The 1999 elections were the second European elections in which Austria participated after EU accession in 1995. After an intensive campaign in favour of joining the EU the governing coalition of the Social Democratic Party (SPOe) and the People's Party (OeVP), supported by almost all major institutions and the social partners, convinced two-thirds of the population to approve of Austria's EU membership. The leading right-wing opposition party, the Austrian Freedom Party (FPOe) of Jörg Haider, succeeded in turning the first direct elections to the EP in 1996 into a protest vote against the government and its austerity measures which were attributed to the need to qualify for EMU. Since then the public attitude concerning EU membership has slightly improved. By the end of 1998 approximately 38 per cent of Austrians believed that EU membership was 'a good thing', and 41 per cent believed that they had benefited from membership, in contrast to 19 per cent and 34 per cent respectively who believed the reverse (European Commission, 1999, 39). Nevertheless, because the public remains split over European politics and policies, opposition parties have plenty of space to mobilise the discontented. The EP elections were only one of a series of provincial and local elections in 1999 and were primarily perceived as a dry run for the national elections that were held in October 1999. Unlike in 1996, economic and employment indicators were good and the majority of the public was content with government policies around the time of the elections. The main question seemed to be whether the governing coalition would be able to fend off the attacks of FPOe and whether the small opposition parties – and particularly the Liberal Forum (LIF) – could keep their seats in the EP.

The players

Seven parties contested Austria's 21 EP seats: the two partners in the grand coalition, SPOe and OeVP, FPOe, LIF, the Greens (*Die Grünen*), the Communist Party (KPOe), and the Christian Social Alliance (CSA). The Social Democratic Party has been represented in government since 1970 and has been the leading party of the coalition government since 1987. In the first EP elections in 1996 it only came second with 29.2 per cent of the vote and six seats in the EP, compared to the 38 per cent it had won in the 1995 national elections. The junior coalition partner, the OeVP, had become the strongest party in the 1996 EP elections with 29.6 per cent of the vote and seven seats, compared to 28.3 per cent in 1995. This had been the only victory for the OeVP in nation-wide elections for decades. The Austrian Freedom Party, led by right-wing populist Jörg Haider, had been the main domestic opponent of the two governing parties since Haider became FPOe chairman in 1986, since when it has continually increased its share of the vote. Although many had expected the FPOe to beat the OeVP in the 1996 EP elections, it came third with 27.5 per cent of the vote and six seats, which nevertheless represented an increase over its 21.9 per cent share of in the 1995 national elections. The Greens and the Liberal Party are the other two opposition parties, located on the centre-left of the political spectrum. The Green Party is an established opposition party in Parliament which was critical of Austria joining the EU, but it changed this stance after the 1994 referendum. The Liberal Forum was formed in 1993 when a dissident group of MPs decided to split from the FPOe. Since then the Liberal Party has had to fight to survive. The Greens and the Liberals won one seat each in the 1996 EP elections. Karl Habsburg, son of Otto Habsburg, who had held a seat in the EP for the Bavarian Christian Social Union party since 1979, founded the CSA shortly before the 1999 EP elections. Previously, he had represented the OeVP in the EP but was not re-nominated after a financial scandal in a charity organisation which he headed. The CSA presented a strictly conservative programme emphasising Christian values, opposition to abortion, and a strong European security policy. The Communist Party, whose role in Austrian politics is negligible, vigorously opposed Austria's membership of the EU and has stuck to its pronounced anti-European course since then without electoral success.

The players in a domestic context

Party dynamics in Austria are determined by competition between the grand coalition on the one hand, and the opposition parties on the other hand, the FPOe being the main challenger. Both coalition partners steer a pro-European course. Soon after the coalition was formed in 1987, Austria applied for membership in the EU. OeVP was the first one to push in this direction but soon SPOe joined in. Since then the two partners have competed for recognition as the leading European party in Austria. Together they managed to enable Austria to qualify for joining the euro but not without hassles over how to achieve a more balanced budget. The People's Party even forced early elections in 1995 – after regular elections had taken place in 1994 – because the coalition partners could not agree on the budget in the first instance. In the second half of 1998 Austria held the presidency of the EU. In preparation for the presidency, however, the OeVP and SPOe failed to come to any agreement on Austria's future course in the field of security and defence policy. This turned out to be the major rift between the coalition partners. While the OeVP would like Austria to join the North Atlantic Treaty Organisation (NATO) and develop the Western European Union (WEU) into a strong military arm of the EU, the Social Democratic Party wants to secure Austria's neutrality, which it sees as compatible with a possible future European security framework. In most of the other areas of EU policy, the governing parties are quite close. This holds true for Eastern enlargement, which both parties favour providing that transition periods are implemented, particularly in relation to the free movement of persons. Similar positions were formulated on the reform of the institutions, and the fight against corruption and fraud (particularly as regards the Commission). Slightly different emphasis is put on the various means to combat unemployment: while the Social Democratic Party stresses the role of politics and national and European action plans, the People's Party stresses economic competitiveness and tax harmonisation as the best route to job creation.

The Freedom Party has steadily gained support since the end of the 1980s, winning over 22 per cent of the share of the vote in the national elections of 1994 and 1995. The FPOe holds pronounced counter-positions to those of the government and does not shy away from mobilising public resentment of foreigners and migrants. Although the party does not question Austria's EU membership any more – as it did before 1994 – it strongly criticises EU politics and government policies towards the EU. It engages in right-wing populism denouncing the European institutions, warning about centralisation, mismanagement and over-spending in the EU. It adopted the slogan that it is against the 'Maastricht-Europe' but in favour of a 'Europe of fatherlands' in which most powers remain with the member states. The FPOe opposes Eastern enlargement as long as those countries have not achieved similar standards in social, economic and environmental development, and it demands strict asylum and immigration rules.

The Green Party is one of the two smaller opposition parties in Parliament with 7.3 per cent and 4.5 per cent of the vote in the 1994 and 1995 national elections respectively. Its programme focuses on environmental issues but also stresses social policy. The emphasis on social issues aims at competing with the SPOe for leftist voters, but the Greens also appeal to OeVP voters on ecology. Concerning EU politics, the Greens campaign against the use of atomic energy, genetic engineering of food and the use of genetically manipulated products in agriculture. It favours public over private transport and demands the democratisation and constitutionalisation of the Treaties.

The Liberal Party competes for a similar segment of the voters. Although the party came into existence by splitting from the FPOe, it does not compete with it for voters. This is due to its ideological positioning left of the centre. It gained 6 per cent and 5.5 per cent of the vote respectively in the 1994 and 1995 national elections. It developed a programme combining free market rhetoric with social responsibility and tolerance for different lifestyles. The latter resulted in a public image that associates the party with being primarily concerned with minorities such as gay people, national minorities, foreigners and immigrants, and their respective rights. The Liberals always presented themselves as the most pro-European party favouring economic integration and political integration, together with a European security system including a European army which could act independently of NATO. They also argued for swift Eastern expansion of the EU.

The EP elections were part of a series of national and regional elections in 1999. The most important ones were the elections to the national Parliament in October 1999 and the regional elections in Carinthia, the southern part of Austria, an FPOe stronghold. The elections in Carinthia took place in March and resulted in a landslide victory for Haider's FPOe. He won 42 per cent of the vote and became provincial president, a prestigious and powerful position in a federal state such as Austria. His victory put paid to any considerations to link European with national elections with a drop in turnout being likely; indeed, in the event it fell sharply to 49 per cent. This was a highly significant figure given the traditionally high rates of 80–90 per cent turnout in national elections. In 1996 voter turnout had been comparatively high, reaching 68 per cent. This was primarily due to the fact that the 1996 EP elections were the first in which Austria had participated and local elections in Vienna and St Pölten had been scheduled for the same day. After the decision had been taken to hold national elections in October, the EP elections were seen as of minor importance. It soon became clear that the parties would use the EP elections to test: (i) the popularity of the government, and (ii) whether Haider's victory would have any impact on the FPOe's attractiveness for voters in nation-wide elections.

The campaign

Chancellor Viktor Klima, head of SPOe, came to power only $2\frac{1}{2}$ years before the EP elections and wanted to use them to test his popularity and campaign strategy. Therefore he was heavily involved in the campaign, in contrast to his opponent in the coalition, Vice-Chancellor and Foreign Minister Wolfgang Schüssel. The main themes of the SPOe's campaign were the successful representation of Austria's interests in the EU, employment policy and securing Austria's neutrality. The OeVP relied on the popularity of Mrs Ursula Stenzel, a well-known ex-journalist who had held a job in the public broadcasting corporation (ORF) and was a political outsider when nominated as candidate for the EP. She was largely responsible for the OeVP's victory in the 1996 EP elections owing to her popularity and credibility as a serious and engaged person. The FPOe's strategy was to mobilise anti-EU feelings among the population. In the EP elections of 1996, it had recorded its best results in nation-wide elections by directly asking the electorate to cast a protest vote against the austerity budget of the government, and against the measures introduced to qualify for the euro, and so on. The FPOe intended to repeat its successful 1996 strategy by revealing plans to introduce a 'European tax'; by criticising Austria's position as a net-contributor to the EU budget; by opposing Eastern enlargement; and by using the Commission's resignation to accuse EU institutions of fraud, waste and over-centralisation. The Greens tried to present themselves as credible and sincere defenders of Austria's neutrality, emphasising their competence in controlling EU politics and promoting ecological issues. The Liberal Forum tried to present itself as a pro-European alternative to the government parties by stressing its competence in economic matters.

While it was expected that the campaign would follow the government–opposition logic of the 1996 EP elections, the strategic setting had changed completely because of NATO's engagement in the war against Serbia. This had two effects. First, the security issue moved into the centre of public attention with the result that parties favouring NATO membership had a difficult case to make, and parties favouring maintaining Austria's neutrality status gained public support. The SPOe, and particularly the Green Party, gained from this. Second, the latent conflict between the SPOe and OeVP over Austria's future security policy came to the forefront, and allowed the two coalition partners to present themselves as distinctive and separate parties, having different views on important issues of the future. The Social Democratic Party in particular tried to emphasise the decisive character of the forthcoming elections. Focusing on the differences between the two major parties offered the voters a real choice and detracted attention from the competition between the government and opposition parties. As a result, both governing parties increased their share of the vote compared to 1996; the Greens won their highest score ever; FPOe dropped by more than 4 per cent; and the Liberal Forum did not pass the required 4 per cent threshold and is not represented in the new EP.

Mobilising the vote

The most alarming signal of the 1999 EP elections was the decline in turnout from 67.7 per cent in 1996 to 49.4 per cent. This represented a historic low in nation-wide elections. Voting in EP elections is not compulsory in Austria. The low turnout was due to the timing of the elections, and to the fact that a whole series of elections took place in Austria in 1999 and that voters perceived the EP elections as of minor importance. This holds true in spite of the state-backed financing of the election campaigns. The campaign ran for nearly six weeks. It was financed by the parties who get a certain amount refunded in proportion to the seats they win. All this contributed to a relatively high public profile of the EP elections, which were also closely covered by the print media and public broadcasting, at least in the run-up to the elections. Even though all leading candidates, the president and the social partners advertised to encourage turnout, relatively few people showed up at the ballot.

This meant that which parties were chosen by voters was as important as whose voters did not vote. Analysing the flow of votes between parties, and between voters and non-voters, underscores this (Institute for Social Research and Analysis, 1999). FPOe lost nearly 50 per cent (or 516 000) of its voters in 1996 to the non-voters, a figure which could not be compensated for by winning voters from others parties. Jörg Haider was not very visible in the campaign. Only two months earlier he had won elections in Carinthia where he was busy performing his re-won political job as provincial president. Instead of using him, the FPOe had nominated a young woman without political standing and not widely known to the voters as the leading candidate. Haider's presence in campaigns and his appearing in public was and is decisive for the FPOe, as this election proved once again.

The second biggest loser was the SPOe, 312 000 of whose voters abstained, and only 162 000 of whose voters defected to other parties. The SPOe's success was due therefore to it making good this loss by winning approximately 227 000 voters from the FPOe and OeVP. In contrast to the FPOe chancellor, Viktor Klima engaged heavily in the campaign. He tried to present the EP elections as decisive for the future course of Austrian security and employment policy. The SPOe hired a well-known journalist and author, Hans-Peter Martin, who wrote the best-seller, *Globalisation Trap*, to lead the team of candidates. Martin, who came from outside the party organisation, was supposed to make the SPOe credible and attractive in its fight against unemployment and for securing neutrality. In taking this step, the SPOe had followed the example of the OeVP who, in 1996, had made an independent journalist its leading figure in the EP elections and won. This time only 95 000 of its 1996 voters abstained, but more than twice as many defected to the SPOe and FPOe. By mobilising more than two-thirds of its 1996 voters this time the OeVP again succeeded in increasing its share of the vote, although it lost more voters to other parties than any other party.

The Greens also succeeded in mobilising their voters. A long-standing Green Party activist who held the Greens' only seat in the EP led the campaign. The voters took him seriously in defending Austria's neutrality and ecological standards and honoured his engagement in transit matters. He was also seen as a competent MEP. A newcomer led the Liberal Forum while the national political leader, Heide Schmidt, stood in the background. The Liberal Forum presented itself as the 'positive opposition' but failed to mobilise its followers. Only 25 per cent of their previous supporters participated in the election, with the result that the Austrian Liberals lost their only MEP.

The issues

Unexpectedly, the question of the future course of Austria's security policy dominated public controversies before the elections (see *Der Standard*, 11 June 1999, A10). NATO's intervention in the Kosovar conflict was a welcome opportunity for SPOe and the Greens to mobilise voters, who overwhelmingly approved of Austria's neutral status. In guaranteeing Austria's neutrality, the SPOe opposed OeVP, which had long argued for joining NATO. This controversy was not directly about a European issue but it was intended to determine the role Austria might play in developing a European security policy. The OeVP's position was geared to joining the majority of European countries who are both members of NATO and the EU, while recognising that such a move would incur considerable adaptation costs. The SPOe argued for a broad conception of security policy within Europe while keeping the status quo in Austria. Focusing on these questions did not allow the FPOe's populist themes to dominate public discourse.

A second major theme was the question of Eastern enlargement. The two governing parties held similar views. Both favoured enlargement but argued for particular arrangements for new entrants and for transition periods to soften possible negative effects on the economy, employment and security. Their position stood in stark contrast to FPOe's opposition to enlargement, which corresponded to widespread public concern about international crime, immigration and the loss of jobs. This second theme never came to dominate the controversies, since it seemed too remote to capture the emotions of the electorate and because the security issue was so dominant.

A third issue that arose in the campaign was the question of fraud and the working of the EC. The two governing parties competed over who did most to fight mismanagement in the EU. The SPOe pointed out that one of its MEPs had a record of fighting EU and had participated in designing the new anti-fraud agency, OLAF, and that the chancellor had devised an 'action plan for a clean EU'. The OeVP's leading candidate, Ursula Stenzel, emphasised the EP's role in forcing the Commission's resignation and pointed out her plans to reform the institutions and strengthen the EP's control powers *vis-à-vis* the Commission. Given this rivalry between the two biggest parties it was difficult for the opposition parties – and particularly the FPOe – to distinguish themselves as critics of the EU institutions.

While the above-mentioned items dominated public discourse, voters were also mobilised by specific issues and competences attributed to particular parties. The Social Democratic Party had chosen to promote the question of employment by demanding that the EU develop into a 'social and employment union' (SPOe, 1999). Its election manifesto was in line with that of the European Socialists but focused more on institutional reform to combat mismanagement and corruption and develop a European security architecture 'under the condition of Austria's neutrality'. While the transnational party manifesto and the Party of European Socialists (PSE) summit were hardly noticed in Austria, the SPOe tried to present chancellor Viktor Klima as working with other social democratic leaders in Europe on an equal footing for the advancement of Austria's interests. There was even wordless billboards showing Mr Klima, Mr Schröder and Mr Blair. The OeVP campaign focused almost exclusively on presenting Mrs Stenzel as the most competent, credible and serious candidate for the EP. Since Mrs Stenzel was an independent candidate who was even prepared to criticise some OeVP officials for not sticking to the OeVP undertaking to join NATO, the election manifesto as such was of minor importance (OeVP, 1999). While the programme and its themes should have been presented by the head of the campaign, Mrs Stenzel's allusions to transnational party manifestos were negligible.

Almost all parties included in their teams political newcomers or independent candidates in order either to convince voters that these candidates could be trusted or to win attention by presenting an 'attractive' or well-known person. While the SPOe's Chancellor Klima was engaged extensively in the campaign, the OeVP's Vice-Chancellor Schüssel and the FPOe's opposition leader, Haider, were hardly visible. In comparison to the 1996 EP election the dominant campaign issues had a clear European focus, although all parties competed on the basis of being best able to represent Austria's interests. While one SPOe poster said that it would devise 'the right way' for Austria in Europe, the subtitle of OeVP's party manifesto read 'the best way for Austria'. A second challenge lay in trying to mobilise the electorate given a high proportion of sceptical or uninterested voters.

The results

Surprisingly, both governing parties increased their share of the vote over the 1996 elections. The SPOe became the strongest party again, although its share was down by nearly 6 per cent compared to the national elections in 1995, closely followed by the OeVP, whose share of the vote rose. Haider's opposition party gained only 2 per cent and came third. The Greens achieved their highest result in nation-wide elections and won a second seat in the EP, while the Liberals failed to pass the 4 per cent threshold (see Table 4.1).

Three parties thought of themselves as winners (SPOe, OeVP, and the Greens) and two as losers (FPOe and the Liberals). The composition of the Austrian EP delegation changed slightly, adding a seventh member to the group of Social Democrats and a second member to the group of Greens from Austria at the expense of FPOe and LIF. So Austria's contribution towards the composition of the EP ran counter to the overall

Table 4.1 The EP elections in Austria (votes and seats)

	Votes (%)		Seats	
	1999	1996	1999	1996
SPOe	31.7	29.2	7	6
OeVP	30.6	29.6	7	7
FPOe	23.5	27.5	5	6
Grüne	9.3	6.8	2	1
LIF	2.6	4.3	0	1
others	2.2	2.6	0	0
turnout	49.4	67.7		

Source: Adapted from Plasser, Ulram and Sommer (1999), 11.

trend in the EP elections. While the balance between the SPOe and OeVP is likely to inhibit any change of policy priorities in the EP, the election is likely to have an impact on EU policy via domestic governmental policy. The SPOe is committed to defending Austrian neutrality and the OeVP to joining NATO. Given that the SPOe remained in power after the October 1999 national elections, this need not prevent Austria from participating in the further development of a common foreign and security policy, but it will cause some 'creative interpretation' of what neutrality means in the future. If the OeVP and FPOe form a government – a more realistic option after the FPOe became the second strongest party in the national elections of October 1999 with the OeVP in third place – Austria's stance towards the CFSP could change dramatically. If the coalition between the SPOe and OeVP continues, Austria's foreign and defence policy may become deadlocked.

The parties agree (with qualifications, however) that there is no alternative to the EU's eastern expansion. Both coalition parties stress the need for transition periods before the free movement of persons can be allowed. All parties seek extra funds for Austria's peripheral regions which will be affected by trans-border work migration. The FPOe even wants to delay enlargement until differences in social and economic standards between the EU's current members and their eastern partners level out. All parties signalled that in the enlargement negotiations they would insist on enhancing the security standards of atomic power plants in the new member states, including the closure of old ones. The 1999 EP also revealed some common ground among the main parties (with the exception of FPOe) on EU policies such as institutional reform, cutting Austria's position as a net payer, job creation (although they diverge over the means to this), and developing the economic and social dimensions.

Conclusion

In the 1999 EP elections, protest motives were not decisive. Instead, the logic of 'second-order national elections' prevailed in which situational factors played an important role. The Kosovo crises influenced the campaign to a high degree but less so the outcome. Instead, tradition, personalities and issues determined party choice (see Plasser, Ulram and Sommer, 1999, 30–5). The crucial question was whether the parties were able to mobilise their followers.

The FPOe's relatively poor result was mainly due to the fact that their voters could not be mobilised. Abstainers for the first time in nationwide elections were in the majority, which is a cause of concern particularly regarding the EU, because abstainers' attitudes towards European integration were disproportionately negative: 50 per cent of non-voters disapprove of Austria's membership, 63 per cent oppose eastern enlargement, 51 per cent feel ill-informed about the EU, 21 per cent are not interested in the EP, 21 per cent are sceptical of the EU in general, 14 per cent cannot make sense of the elections, and 13 per cent are discontented with particular EU policies (see Plasser, Ulram and Sommer, 1999, 16–17). Therefore, it would be misleading to read the FPOe's relative defeat in the national elections as a symptom of a lack of support for parties critical of major future European projects. The tough competition between the coalition partners over Austria's foreign and security policy helped both of them, but their success was short-lived. In the October 1999 national elections, the SPOe again became the strongest party (as indicated above). While the outcome of the EP elections will not make a big difference to Austria's behaviour in European politics, the composition of the national government will.

References

Der Standard, 11 June 1999, A10
European Commission (1999) Eurobarometer, 50 (Brussels).
Institute for Social Research and Analysis (1999) *Wählerstromanalyse EU-Wahl 1999* (Vienna).
OeVP (1999) *Besser für Österreich. Unser Europaprogramm* (Vienna).
Plasser, Fritz, Peter A. Ulram and Franz Sommer (14 June 1999) *Analyse der Europawahl 1999. Wahlverhalten und Entscheidungsmotive* (Vienna).
SPOe (1999) Wahlprogramm der SPÖ für die Wahlen zum Europäischen Parlament.

List of Austrian party abbreviations used

CSA   Christian Social Alliance
CSU   Christian Social Union
FPOe Austrian Freedom Party
KPOe Communist Party
LIF   Liberal Forum
OeVP People's Party
SPOe Social Democratic Party

5
Luxembourg and Belgium

Derek Hearl

Introduction

This chapter deals with the only two EU member states whose national general elections coincided with the European Parliament election of 1999. For Luxembourg, of course, this was nothing new, but for Belgium it was an innovation that is unlikely to be repeated.

In spite of the very real, if not always appreciated, differences between these two countries, on this occasion at least there were a number of coincidences between them. Both had outgoing Christian Democratic–Socialist governments led by popular and charismatic prime ministers needing to claim as much of the credit for the popular things that had been done while simultaneously off-loading the blame for those that had gone wrong on to their erstwhile socialist partners. The problem for the latter, of course, was to try to do precisely the opposite. Equally, both countries had increasingly confident and power-hungry Liberal opposition parties which needed to attack both governing parties as effectively as possible, while at the same time putting forward distinctive policies of their own.

On a less positive note, there had been a number of political scandals in both countries which could be expected to have important repercussions not only on the overall outcome of the national election, but even on the fate of the national government itself.

It is hardly surprising therefore that in neither country did European issues actually come to the fore at all, even though most parties did devote some literature and/or a couple of election broadcasts to Europe (although in both countries time for the latter, of course, is guaranteed by law). There is little evidence that these somewhat token efforts had any real impact on either journalists or on the voting public. It is obvious that when there is a national parliamentary election at which the existing government's continued existence is at stake, any simultaneous 'second-order' election, such as that for the European Parliament,[1] will inevitably take a back seat. Certainly, in both Belgium and Luxembourg, almost every aspect of the 1999 European election was obscured by the national election and its outcome. Consequently, it is not possible to separate the two campaigns in what follows and no attempt will be made to do so.

Luxembourg

Luxembourg is a unitary state with a uni-cameral *Châmber vun Députéirten* (although the appointed *Conseil d'Etat* does perform some of the functions of a second chamber). As already mentioned, European elections in the Grand Duchy have coincided with national ones ever since their introduction in 1979 and are best seen as an integral part of the national election campaign itself. Consequently, in a very real sense they should perhaps not be seen as 'second-order' at all.

The party system is of long standing and is relatively straightforward, based as it is upon three traditional parties, the Christian Democratic *Chrëstlech-Sozial Vollekspartei* (CSV), the socialist *Lëtzeburger Sozialistesch Arbechter Partei* (LSAP), and the liberal *Demokratesch Partei* (DP), which are full (and, in spite of their small size, often not uninfluential) members of the European People's Party (EPP), the Party of European Socialists (PES) and the European Liberal, Democrat and Reform Party (ELDR) respectively.

In recent years these three parties have been supplemented by GLEI-GAP, a merger between two formerly separate Green parties and running under the label of *Déi Gréng* (the Greens) as well as a pensioners' protest party, the *Aktiounskomitee vir Demokratie a Rentegerechtekeet* (Action Committee for Democracy and Pensions Justice), ADR. On the other hand, the once relatively strong *Kommunistesch Partei vu Lëtzeburg* lost its last remaining seat in the *Châmber* in 1994 and has now disappeared from the scene altogether.

The electoral system is also rather straightforward, being based on 'open' (i.e., unordered) lists in which candidates' final rankings are wholly dependent on the personal votes they receive. This means that every elected MEP or national *députéirt* owes his or her seat to the fact that he or she obtained more personal votes than other party colleagues running on the same list. Nevertheless, since the order of names as they appear on the ballot paper is a matter for each party or other group to decide for itself, it has become the practice in recent years for the party's or group's most prominent candidate to appear first on the list (the remainder usually being in strict alphabetical order). Inevitably therefore there is a great deal of interest in the personal votes obtained by these *Spitzenkandidaten*, both as compared to their individual performances at previous elections and *vis-à-vis* each other.

For European elections the country constitutes a single six-member constituency (as opposed to the four used in *Châmber* elections). However, the law allows parties (and, in theory, other groups) to run up to twelve candidates on each list in order to allow for attrition not only due to candidates' death or resignation but to their taking national parliamentary seats or even joining the government. Party leaders and other prominent candidates for national office now routinely run in the European election as well and this has led to the latter being seen primarily as a so-called 'Beauty Contest' in which personal vote scores can be directly compared nation-wide (as they cannot in national elections due to the varying numbers of votes voters have in the *Châmber* constituencies).

It is in this sense in particular that European elections in Luxembourg have now become an established part of the national campaign. No party can afford not to have its most prominent candidates for national office not also on its European list, even though they may have no intention of leaving national politics to take up the EP seats to which, of course, as prominent politicians, they have a high chance of being elected.

Background to the campaign

Five years previously, at the 1994 election, the then outgoing 'Santer–Poos' government, a coalition between the CSV and LSAP, had been re-elected. However, in January 1995, Jacques Santer left the government to take up the post of President of the European Commission and was replaced by his erstwhile Finance Minister, Jean-Claude Juncker. Since then Mr Juncker, who has for some years now been the single most popular politician in the Grand Duchy, is widely considered to have done a good job and is certainly his party's greatest electoral asset.

Nevertheless, in the months leading up to the election there had been a number of embarrassments for the government although, compared to the various scandals which had rocked the Belgian political establishment over the previous few years, they were very minor. At the same time, they could obviously be expected to have some impact on the government's overall reputation. The most important of these was the so-called Enregistrement Affair in which a newspaper had accused the Interior Minister, the CSV's Michel Wolter, of favouritism and maladministration in a case concerning his local tennis club. The Minister took legal action against the editor of the newspaper leading to a full parliamentary Commission of Inquiry. However, just one month before the election, the *Châmber* decided for the second and final time to reject the complaint and to exonerate the Minister.[2] Nevertheless, the affair had rumbled on for a long time and the ensuing publicity was certainly somewhat embarrassing.

No sooner had this had been weathered than the news broke that the President of the *Chambre des Comptes* (i.e., the state audit office) had been charged by a parliamentary committee with fraud and suspended from his post.[3] His immediate resignation while protesting his innocence, combined with the fact that constitutionally he is an officer of parliament and not the government, obviously helped take some of the heat off the latter but none the less the affair might have been expected, however unfairly, to have adversely affected the government's reputation as a whole.

However by far the biggest political problem the Christian Social Party had to face was the fall-out surrounding the resignation of the European Commission in March 1999 in the wake of charges of corruption, mismanagement and cronyism. As President of the Commission, Jacques Santer obviously had to bear a large measure of political responsibility for this *débâcle* and as a result not unnaturally became the target of considerable criticism and negative publicity throughout Europe. However, this was very far from being the general view in his native country where Mr Santer remained as popular as ever, being widely perceived as having been the victim of the affair and badly let down by his colleagues. It is in this light that his decision to accept the CSV's nomination as *Spitzenkandidat* in the European election, something else for which he was criticised abroad, must be viewed. (Indeed, all reports indicate that he came under considerable pressure not only from Prime Minister Juncker personally, but also from the party as a whole, to do so.)[4]

Mr Santer was of course not the only prominent candidate in the European election. The country's long-serving foreign Minister, Jacques Poos of the LSAP, having announced his intention to retire from ministerial office, was also running as his party's *Spitzenkandidat* alongside his party colleague, the outgoing Economics Minister, Robert Goebbels. For its part, the DP ran its three most recent party leaders, the current party president and Burgomaster of Luxembourg City, Lydie Polfer, Charel Goerens and Colette Flesch, herself Poos's predecessor as Foreign Minister. Finally, of course, the most popular politician in the country. Prime Minister Jean-Claude Juncker, was – at least nominally – a candidate on the CSV European list (while all the time letting it be known that it was Jacques Santer who was the party's real *Spitzenkandidat*).

The campaign

With few exceptions the main parties concentrated on pensions and education, the DP insisting in an echo of Tony Blair that 'education must be the first, second and third priority' for the new government. In attempt to appear part of the 'cyber revolution' the LSAP chose to run under the rubric 'LS@P' and it was this form which in due course actually appeared on the ballot papers. Among the minor parties, the Greens stressed the need for an integrated road, rail and tramway public transport system as well as the need for measures to improve traffic safety and a national system of cycle-ways. They also organised a Round Table minicruise for their sister Green parties on a Moselle boat sailing from the Luxembourg village of Schengen which has given its name to the EU's common travel area.

Of the smaller parties, only the Greens had held a seat in the outgoing European Parliament won in 1994 by Jup Weber. However, Mr Weber, who had subsequently defected from his party to the EP's ARC (Rainbow Alliance) Group, ran for re-election on the new *Gréng a Liberal Allianz* (GaL) list. The latter, as its name implies, is a coalition of dissident Greens and people pledged to fight what they see as discrimination against the liberal professions, including (inevitably) the pensions issue.

As is generally the case in the Grand Duchy, the election campaign was a relatively calm affair with the usual round of party meetings throughout the country and much low-key face-to-face electioneering on the part of individual candidates. However, there was one highly publicised incident which may have had the opposite effect from the one intended. The daily *tageblatt*, itself closely linked with the LSAP, reprinted an alleged interview with Jacques Santer's wife, Danièle, originally published in the British *Sunday Telegraph*, in which her husband had been described as 'feeble'.[5] Danièle Santer denied ever having given such an interview and threatened to take legal action against both newspapers. The ensuing row was one of the few enlivening, if not enlightening, events in the entire election campaign.

The result

In the event, the outcome was pretty much what seasoned observers have come to expect. The traditional 'pendulum' swung against the

Table 5.1 Luxembourg Chamber of Deputies election, 13 June 1999 (change from 1994 in brackets)

	CSV	LSAP	DP	Greens	ADR	Others
Votes (%)*	30.1	22.3	22.4	9.1	11.3	4.8
Seats	19 (−2)	13 (−4)	15 (+3)	5 (=)	7 (+2)	1 (+1)

* As weighted by STATEC Service Information et Presse, *Bulletin d'Information et de Documentation*, Edition Spéciale (Luxembourg, 1999).

Table 5.2 EP election in Luxembourg, 13 June 1999

	CSV	LSAP	DP	Greens	ADR	Others
Votes (%)	31.7	23.6	20.5	10.7	8.98	4.6
Seats	2	2	1	1	0	0

government as a whole which lost votes and seats to the opposition parties, with the LSAP and DP respectively losing and gaining the most. The pensioners' party, ADR, also gained two seats and the extreme left, *Déi Lénk*, one (see Table 5.1).

In the light of these results the outgoing Prime Minister, Jean-Claude Juncker, was asked by the Grand Duke to form a new government which he did very quickly by bringing the newly victorious Liberal DP into the government while the LSAP went into opposition (see Table 5.2).

The overall national result was of course closely mirrored in the European poll too although, given the much smaller number of seats to be filled, there was no change in any of the parties' EP representation. As usual, most interest centered on the outcome of the 'Beauty Contest' which was won by Mr Juncker with 29 715 personal votes, beating Jacques Santer into second place with 26 590. The next highest was the DP's Charel Goerens with 19 369 votes, closely followed by his colleague Lydie Polfer with 18 464. The best retiring Foreign Minister Jacques Poos could achieve was to come third on the LSAP's list with 10 798.

As now happens after every election in Luxembourg, the usual merry-go-round of candidates then followed as leading figures who had been elected to both parliaments were forced to choose between the two institutions and, in some cases, government office. So it was that Prime Minister Juncker resigned his EP seat to be replaced by Mme Viviane Reding who, however, was then almost immediately appointed by the new government to the European Commission. Consequently, she was in her turn replaced by outgoing MEP Astrid Lulling. Similarly, both the

DP's two leading European candidates, Charel Goerens and Lydie Polfer, resigned to take up posts in the new government, leaving the party's sole EP seat to the former Foreign Minister, Colette Flesch. The LSAP's leading candidate, Alex Bodry, also opted for the *Châmber* and did the same thing, although outgoing foreign minister Poos chose to go the EP as promised where he has been joined by his colleague in the last government, former Economics Minister Robert Goebbels. However, the most blatant example, for which it was heavily criticised in the press, was *Déi Gréng*'s decision to order all its first six EP candidates to opt for the Luxembourg *Châmber*, leaving the party's sole EP seat to its *seventh* most popular candidate, Claude Turmes. The fact that a party such as *Déi Gréng* can do this is another indication of how relatively unimportant European elections in Luxembourg have become.

Nevertheless, the overall outcome of these manoeuvrings has been that, with a former Prime Minister and Commission President, two former Foreign Ministers and a former Economics Minister to represent it, the Grand Duchy clearly now boasts its most high-powered European Parliament delegation ever (and indeed one of the most high-powered of any member state).

Belgium

Since the most recent set of constitutional reforms came into force in 1993, Belgium has been a fully-fledged federal state with what now may be the single most complex set of institutional arrangements anywhere in the world. At the cost of some over-simplification, the new system can be described as being essentially based on the three geographical regions of Flanders, Wallonia and Brussels, each of which now has its own directly-elected 'parliament' and 'government'[6] responsible for a range of functions including, *inter alia*, economic development, spatial planning and supervision of local government. This fairly straightforward territorial arrangement is then overlaid with *non*-territorial Flemish, Francophone and German-speaking 'Community' Councils having jurisdiction over such things as social security, education and other so-called 'personalisable' matters.

In principle, therefore, in addition to the national level which, with its bi-cameral Parliament and Government, retains all the usual functions of foreign policy and defence, currency, macro-economic policy and so on, the country now has six subordinate parliaments and executives although, since the Flemish arrangements are to all intents and purposes fused into a single set of institutions, there are only five in practice.

52 *Luxembourg and Belgium*

To go with this complex decentralised structure Belgium today has what may be described as a 'dual' or 'parallel' party system in which more or less the same set of 'sister' parties exists on each side of the linguistic 'frontier' between the Flemish and French-speaking communities.[7] In ideological family terms these are the Christian Democratic CVP and PSC, the socialist SP and PS and the liberal VLD and PRL respectively. In addition to these three 'traditional' party families, 'Agalev'[8] and 'Ecolo' represent the increasingly significant Green movement, while north of the linguistic 'frontier' moderate Flemish nationalism continues to be represented by the once significant but now fast-declining Volksunie.

In recent years these 'democratic' parties have been confronted by increasingly worrying electoral threats from the extreme right, principally on the Flemish side in the form of the hard-line separatist and anti-immigrant Vlaams Blok, but more recently also from the smaller and nominally bi-lingual National Front.

As will be seen from Table 5.3, with the exception of the Vlaams Blok the same party 'families' are present in all three communities. This allows the so-called 'traditional' parties as well as the Greens to retain strong cross-community ties, evidenced not only by a range of policy co-ordinating organs and in some cases shared headquarters buildings, but more importantly by the fact that parties of the same family always go into, or stay out of, government together.

More recently, and following the demise of its Walloon sister in the 1980s, the Brussels regionalist party, the Front démocratique des bruxellois francophones (FDF), has entered into an electoral alliance with the French-speaking Liberal PRL and is consequently now rapidly declining as a distinct political force. In a related development, the Francophone Parti social-chrétien (PSC) suffered a damaging split at the end of 1998 when its former President and MEP, Gérard Deprez, broke away to form his own *Mouvement des Citoyens pour le Changement* (MCC). The new

Table 5.3 Principal Belgian political parties by linguistic community

Language community	Christian Democrat	Socialist	Liberal	Regionalist	Extreme right	Green
Flemish	CVP	SP	VLD	VU	VB	AGALEV
French	PSC	PS	PRL	(FDF)*	FN	Ecolo
German	CSP	?	PFF	PDB	–	Ekolo

* Brussels party now in electoral alliance with PRL.

formation then also entered the PRL–FDF alliance, thereby enabling the latter to present much more of a 'broad front' to the Francophone electorate. Unfortunately from Mr Deprez's point of view, however, since Belgian electoral law does not allow party labels on ballot papers to exceed six letters, the alliance continued to run under the 'PRL–FDF' label.

Finally there is the small, but what might be termed 'classic', extreme right *Front National/Front voor de Natie* (FN). Officially bilingual, it in fact appeals more or less exclusively to the Francophone community, leaving the Vlaams Blok with a virtual monopoly of the extreme right, and especially the xenophobic, vote in Flanders. The FN made something of a breakthrough in 1994 when it won one seat in the European Parliament. Since then its support has waned and earlier this year it shattered into half a dozen competing factions.

For European purposes the Belgian electorate is divided into three 'colleges', one each for the Flemish-, French-and German-speaking communities. Fourteen of the 25 Belgian EP seats are allocated to the Flemish college, eleven to the Francophone and one to the German. In unilingual areas only one set of lists is printed on the ballot paper, that for the linguistic community concerned. However, in bi-lingual (i.e., Flemish/French and French/German) regions two separate sets of lists are printed on the same sheet, allowing voters in those areas to vote in whichever college they choose. Otherwise the electoral system is a straightforward one based on flexible lists and using the d'Hondt method of seat allocation. There is one aspect of the system which is unique to Belgium, however, and that is that casual vacancies occurring between elections are filled not by lower-placed candidates on the same list but from a subsidiary list of 'substitutes' (*suppléants/opvolgers*).

This means that in Belgium, unlike Luxembourg, list order *is* significant although there is still provision for voters to cast individual preference votes which, theoretically, have the effect of improving an individual candidate's chances of election (although this rarely happens in practice). This means that there is always stiff intra-party competition for 'electable' list positions and to be the first-placed candidate (known as *Tête de Liste* or *Lijsttrekker*).

Background to the campaign

A number of prominent outgoing MEPs did not stand for re-election in 1999. They included former Prime Ministers Leo Tindemans and Wilfried Martens, as well as the former Leader of the FDF, Antoniette Spaak, and the veteran PSC politician, Fernand Herman. Of these the surprise was Mr Martens who, as EPP Group Leader in the European Parliament, had expected to be chosen as his party's *Lijsttrekker*. When in the event he was only offered second position on the CVP list he refused to run in the election at all. Nevertheless, in his other capacity as the EPP's President, outside the Parliament he continued to direct the Party's Europe-wide campaign until the election was over.

The coincidence of the European, federal, regional and Community elections on the same day in June 1999 came to be known at any rate on the Flemish side as 'The Mother of all Elections'. Even more than was the case in Luxembourg, therefore, the Belgian European campaign was totally eclipsed by the national and regional ones: to such an extent, indeed, that (except in the most formal sense) there can hardly be said to have been a European election in Belgium at all. Consequently the following discussion necessarily has to be almost entirely about the domestic campaign since it was that, and only that, which governed the outcome of the European election itself.

The last five years or so have been among the most difficult in Belgian public life. There has been a series of political and criminal justice scandals which have shaken public confidence in the public authorities to an unprecedented degree. There had been the complete failure to solve the former PS leader's (André Cools) murder some years before the forced resignation of NATO Secretary-General and former SP Party Treasurer, Willy Claes, to face corruption charges in 1995, and the judicial and political controversy surrounding a former PS Vice-Prime Minister's alleged homosexual relations with minors in 1996.

On top of all this came the Dutroux affair in which not only did a parliamentary committee of inquiry condemn the judicial and police system for incompetence in the investigation of a series of child murders which had profoundly shocked public opinion, but the prime suspect, Marc Dutroux, had actually briefly escaped from police custody in April 1998 leading to the then Interior Minister's resignation. On a somewhat lighter note, but none the less still one hardly calculated to inspire public confidence in the dignity of the country's institutions, an Agalev senator had admitted to smoking cannabis during public sittings of that House.

The campaign

Initially, most observers seem to have expected the campaign to concentrate on a range of issues, none of which would particularly dominate, and initially this seemed to be the case. Obviously, the various scandals surrounding the outgoing government as well as its general record were expected to play a major part as, in particular, would law and order and the environment. The opposition parties generally, and on the law and order issue the Vlaams Blok in particular, could be expected to push these issues hard. Against this the government as a whole, and the CVP in particular, had the towering figure of Prime Minister Jean-Luc Dehaene who had become known, somewhat irreverently but none the less affectionately and with a good deal of respect, as the 'National Plumber' in acknowledgement of his legendary ability to fix any problem.

At the same time the Vlaams Blok was itself a live issue not only from the beginning of the election but right through even the dioxin crisis to the very end of the campaign. A particular concern among the Blok's opponents was that, owing to the complicated dual nature of the Brussels regional constitution, it might win enough seats in the regional parliament to be able to constitute a blocking minority and thereby prevent the formation of an executive.

Indeed, as the campaign wore on, this seemed to be an increasingly realistic possibility as the Blok openly tried to solicit votes from among those sections of the French-speaking Brussels population who might be attracted by its hard-line law and order, as well as its associated anti-immigrant, stance. This led many politicians from the other parties to call for a united front against the Vlaams Blok. The two socialist parties jointly declared the Vlaams Blok to be *the* enemy on both sides of the language frontier while Prime Minister Dehaene, supported by other party leaders, said that the extreme right would be the principal theme of the election. To all this, the Blok's President and outgoing MEP, Frank Vanhecke, retorted that he considered it a 'great honour' to be seen as the National Plumber's enemy number one.

The election was also marked by a number of unprecedented legal challenges, mostly stemming from new anti-racist legislation which both the Vlaams Blok and the FN were several times accused of breaking. For example, the Blok was forced by the courts to withdraw a (significantly bi-lingual) leaflet it had circulated as part of its law and order campaign in Brussels and purporting to show the faces of a number of 'typical' criminals, all of whom were non-white! For its part, the Vlaams Blok brought cases in both Brussels and Antwerp against a number of publicly funded community newspapers to force them to carry its advertisements and challenged the procedure for electronic voting in the capital which it claimed discriminated against it amongst French-speaking voters.

The dioxin crisis

It was in this already over-heating atmosphere that the dioxin crisis broke on 28 May, exactly two weeks before the election. A shocked public was told that due either to accident or fraud – no one knew – an unknown quantity of poultry feed had been contaminated by a chemical poison, dioxin, as a result of which chicken, eggs and related products had been declared unsafe and withdrawn from the market. Worse, it seemed that the government had known about the problem ever since February but only now, some five months later and under pressure from the European Commission, had it taken steps to protect public health by banning the suspect products.

At first the government, and especially the Prime Minister, tried to hold the line by defending the actions of the two ministers most under attack, the CVP's Karel Pinxten at Agriculture and Marcel Colla, the SP Minister of Health. In a high-profile press conference on 30 May, the Prime Minister rejected calls for the two ministers' resignations, insisting that he had every confidence in them. Yet the very next day, the two Ministers resigned anyway, Mr Dehaene apparently having come to the conclusion in the meantime that their positions had become untenable.

The impression of a government which had lost control heightened as the crisis deepened over the next few days. Following the withdrawal of Belgian-produced eggs and poultry from the domestic market and their export being forbidden by the EU, the ban was extended, first to pigs and pork products, and then to beef, milk and dairy produce.

The affair swept everything else, including the election, off the front pages as well as out of the rest of the media. It is a measure of its seriousness that it was not until 11 June, 13 days after the crisis had first hit the headlines, that the leading Flemish daily *De Standaard* did not devote its main front page headline to the dioxin scandal, and even then it was replaced by news of the events in Kosovo rather than the Belgian general election, which by then was only two days away.

Naturally the opposition Liberals, led by Guy Verhofstadt and Louis Michel, leaders of the Flemish and Francophone sister parties respectively, did their best to exploit the government's difficulties. In effect, Verhofstadt blamed the entire crisis on a culture of complacency and ostrich-like attitudes within the government as a whole, while Louis Michel attacked a situation in which the two Christian Democrat parties had monopolised the Agriculture Ministry for nearly 50 years.

The result

In the event in what was what was widely seen as a 'political earthquake' the governing parties suffered heavy losses throughout the country. (The electorate, however, did not turn against Prime Minister Dehaene personally, whose own preference vote rose from 484 000 in 1995 to 562, 239.) The greatest advances were made by the two Green parties, Agalev and especially its Francophone sister Ecolo, while both the Vlaams Blok and the Volksunie saw their votes rise slightly too, although the former failed to make the breakthrough it had expected, either in Brussels or elsewhere. However, the real winners of the election were the Liberal parties which, for the first time since 1882, ended up as the strongest political force in the country (see Table 5.4).

In due course the King called upon the Flemish Liberal leader, Guy Verhofstadt, to form a coalition government at federal level between the Liberal VLD and PRL on the one hand, and the Socialist SP and PS on the other, thereby making Mr Verhofstadt the first Liberal Prime Minister in Belgium since Paul-Emile Janson in 1937,[9] while the CVP and PSC have gone back into opposition for the first time since 1954.

As was only to be expected, the European result closely mirrored the national one (see Table 5.5 for details). Four leading candidates who had initially been declared elected to the European Parliament, Franck Vandenbroucke, Bert Anciaux, Annemie Neyts and Phillipe Busquin, *Lijsttrekkers* and *tête de liste* for the SP, the VU, PVV and PS respectively, did not in the event take up their seats. The first three became ministers in the new Federal, Flemish and Brussels regional governments while Mr Busquin, for his part, was nominated as Belgium's European Commissioner, all four being replaced by their substitutes. The new EP

Table 5.4 Belgian Chamber of Representatives election, 13 June 1999 (% national vote and seats)

Community	Liberals + Allies	Socialists	Christian Democrats	Greens	VU	Vlaams Blok	Others
Flemish	14.3	9.6	14.1	7.0	5.6	9.9	
	23	14	22	9	8	15	
Francophone	10.1	10.1	5.9	7.3			
	18	19	10	11			
Total	24.4	19.7	20.0	14.3	5.6	9.9	3.6
	41	33	32	20	8	15	1*

* 1 seat won by the extreme right *Front National/Front van de Natie* with 1.5% of votes.

Table 5.5 Results of EP election in Belgium, 13 June 1999
(vote in % by community and seats, with change from 1994 in brackets)

Community	Liberals + Allies	Socialists	Christian Democrats	Greens	VU	Vlaams Blok	Others
Flemish	21.9	14.2	21.2	12.0	12.2	15.1	3.3
	3 (=)	2 (−1)	3 (−1)	2 (+1)	2 (+1)	2 (=)	
Francophone	27.0	25.8	13.3	22.7			11.2
	3* (=)	3 (=)	1 (−1)	3 (+2)			0 (−1)
German	19.6	11.4	36.4	17.0			15.6
			1 (=)				
Total	23.7	18.5	18.6	16.0	7.6	9.4	6.2
	6*	5	5	5	2	2	

*Includes Gérard Deprez elected on the PRL–FDF–MCC (Liberal + Allies) list.

delegation consists therefore of five each for the Liberals, Socialists and Greens and six for the Christian Democrats of whom one, Gérard Deprez, was elected on the Liberal dominated PRL–FDF–MCC list. In addition, the Volksunie and the Vlaams Blok each won two seats. Of the 25 outgoing Belgian MEPs, only ten have been returned to sit in the new European Parliament.

Notes

1 Karlheinz Reif, 'Ten Second-Order National Elections', in K. Reif (ed.), *Ten European Elections: Campaigns and Results of the 1978/81 First Direct Elections to the European Parliament* (Aldershot: Gower, 1985).
2 *tageblatt*, 21 May 1999.
3 *tageblatt*, 27 May 1999.
4 *Luxemburger Wort*, 8 June 1999.
5 *tageblatt*, 3 June 1999.
6 Technically, they are 'Councils' and 'Executives' but the terms 'Parliament' and 'Government' are now in general use at both Regional and Community level.
7 There are separate German-speaking parties as well, of course, at any rate nominally (see Table 5.3).
8 Short for *Anders gaan leven* ('go and live differently').
9 Theo Luyckx, *Politieke Geschiedenis van België van 1789 tot heden* (Brussels: Elsevier, 1973).

List of Luxemburgeois and Belgian party abreviations used

Luxembourg

ADR	Aktiounskomitee vir Demokratie a Rentegerechtrekeet
CSV	Chrëstlich-Sozial Vollekspartei
DP	Demokratesch Partei
GaL	Gréng a Liberal Allianz
LSAP	Lëtzeberger Sozialistesch Arbechter Partei
PDB	Partei der deutsch sprächigen Belgier
PFF	Partei für Greiheit und Fortschritt
PSC	Parti Social-chretien

Belgium

Agalev	Anders Gaan Leven
CVP/PSC	Christian Democratic Party
Ecolo	Mouvement 'ecolo' – les verts (Greens)
FDF	Front democratique des bruxellois francophones
FN	Front National/Front voor de Natie
PRL/VLD	Liberal Party
SP/PS	Socialist Party
VU	Volksunie

6
Denmark

Hans Jorgen Nielsen

The Danish Euro-election of 1999 was almost a non-event, a repeat of 1994, 1989 and even earlier. Once again, the campaign focused on institutions and not on the policies of the EP; only half of the electorate was sufficiently interested in turning out to vote; and once again EU-sceptic parties and lists got around a third of the vote. This confirmed the picture of the Danes as reluctant Europeans. They favour EU membership but not joint decision-making and there is a constant suspicion that EU institutions encroach upon the sovereign domain of national institutions. Therefore, there is not much space left for proposals for new European initiatives and it is difficult to generate debate among parties that broadly concur on this. The European Parliament has come close to being a 'real Parliament' but the Euro-election did not produce 'a real election campaign' akin to those of national or local elections. Rather than focusing on issues – which policies to pursue in different political fields – the parties' campaigns focused on institutions and, as the moderate level of turnout indicates, institutional matters seldom inspire the public.

Danish attitudes towards the EU (and before it the European Economic Community, or EEC, and the EC) have been tested in several referendums. In 1972 a high 63 per cent voted *Yes* to membership and in 1986 a moderate 56 per cent accepted the Single European Act. Then came the shock when the Maastricht Treaty fell in June 1992, when only 49 per cent voted *Yes* against 51 per cent voting *No*. At the Edinburgh summit, Denmark got four exemptions from the Maastricht Treaty which in this revised form was then accepted by 57 per cent of the voters in a referendum in 1993. Finally, the Amsterdam Treaty was accepted in 1998 by 56 per cent of the voters. These referendums had several consequences, not least for the strategies of the political parties. First, while the Yes-majority always has been overwhelming in Parliament (80 per cent or higher), EU enthusiasm has always been much lower in the electorate, something politicians are careful to heed fearing being out of touch with the general public. Second, this elite–mass discrepancy is not easily discounted as turnout at the EEC/EC/EU referendums have been high, varying from 75 per cent to 90 per cent with an average 82 per cent turnout. Therefore, as the public seems to be involved, politicians fear that the elite–mass discrepancy in EU policies may have consequences for them in terms of the public trust invested in them. Third, apart from the first referendum the majorities have always been quite slender. Therefore, there is always a fear that a narrow lead for a *Yes* in the polls may end up with a narrow lead for a *No* at the ballot box, as happened at the first Maastricht referendum in 1992. For the pro-EU parties this boils down to a basic strategy: caution is essential. It is easy both to insist that the parties should present general perspectives for Danish EU policies and it is easy to understand why pro-EU parties are reluctant to do so as the risks may be too high. EU-sceptics, on the other hand, have an easy job. They can always build on widespread scepticism in the public.

This picture may easily be taken too far. First, membership is no longer the issue it was until the end of the 1980s. According to Table 6.1, a clear majority now see Danish membership of the EU as 'a good thing'. Eurobarometer also shows Denmark generally to be more positively inclined towards the EU than average. Instead of membership itself, the real issue is the degree of integration to be allowed: an absolute majority favours national decision-making over both withdrawal from the EU and further integration within the EU (see the second question in Table 6.1). The Danes may either be called 'Gaullist' as they prefer a 'Europe des Patries', or they may be seen as state-righters in the American sense. In any case, the Danes want to be members of the club even if they prefer to do as much as possible on their own.

Specific policies are not judged solely on the criterion of whether they imply more integration. The subject matter is important, and on several occasions opinion polls, including Eurobarometer, have shown large Danish majorities in favour of joint European action. For example, according to the polls almost everybody is for strong joint action against organised international crime, and not because the action will be joint but rather because the Danes are just as much against mafia-gangs and heroin-rings as everybody else. In the field of environment protection this pattern is especially pronounced: the very parties and lists who fight any tendency towards more integration make an exception and favour *stronger* – not weaker – joint intervention against pollution.

Table 6.1 The Danes and the EU

Basic Attitude towards Membership	
Danish membership of the European Union is	
a good thing	49
neither good nor bad	26
a bad thing	21
don't know	4
Basic Policy Alternatives	
Denmark should leave the European Union	19
In the EU all member states should retain full sovereignty and have the right to veto EU decisions	55
All important decisions concerning the EU should be made jointly	12
Eventually the EU should become the United States of Europe	7
None of the above + don't know	8

Disagreement between parties and voters over EU policies may only have weak consequences for the general standing of the parties in the public. Surveys show that only a minority consider European policies important and the referendums have shown that voters vote against party advice without serious consequences for the strength of the parties in the opinion polls. In sum, there is room for some policy initiatives but the basic rule remains that it is wise for the parties to be cautious.

The actors

Two kinds of actors ran for election to the EP: parties and political lists, and individual candidates within these parties and lists. It may appear superfluous to mention the latter as parties always have candidates. However, at Danish Euro-elections candidates can play a strong role, independent of their parties, and the parties may depend more on the candidates than the candidates on the parties. Apart from the small left-wing Unity List all parties in the Danish Parliament participated in the EP election. Two 'movements', the People's Movement against the EC and the June Movement, also run for seats. The basic cleavage is between pro-EU and Euro-sceptic parties and lists. The pro-EU lists are primarily mainstream parties which on the left are flanked by three EU-sceptic lists, namely the two 'movements' plus the Socialist People's Party (SPP), and on the right by the Progress Party and the Danish People's Party. EU membership itself is not the issue. Only the People's Movement reject membership, while the other EU-sceptic actors take Danish EU membership as a given fact and concentrate on blocking any move towards more integration. Among the pro-EU parties only the Liberal Party and the small Centre Democrats are in favour of the principle of federalism, while the other parties are much more reluctant. They are willing to accept joint decisions and majority decisions when it is necessary to reach an agreement with the EU, but they feel they would be better off without. Traditionally, EU-scepticism was primarily a left-wing phenomenon. The parties to the left of the Social Democrats were united, at first against membership itself and later against further integration. The EEC/EC/EU was seen as a block of rich capitalist nations confronting the Third World and which was also a danger to the Nordic welfare-state model. In addition there was a scattering of opponents in the political centre but they did not really count. This location of opposition to the EU makes the discussion of Danish EU policies somewhat dull. On the left, EU-scepticism neutralises the general wish for more interventionism in fields such as employment, environment protection and so on. Consequently, there is little difference between the policies of the parties who generally refrain from saying very much. It is only a guess, but a fair guess, that the situation would have been very different if opposition to the EU had been a mainly right-wing phenomenon. In that case, there would have been a stark contrast between proposals for intervention from the left, not blocked by any EU-scepticism, and the rejection of the same proposals from the right, both because the proposals meant more intervention and because they meant more EU. As it is now, it is often difficult for voters and even for politicians to tell the difference between the parties, apart from often minor differences in attitude towards the EU.

In the 1990s there were signs that this traditional pattern would break up. In the early 1990s in the light of the Maastricht referendum, it seemed that the Progress Party was becoming more Euro-sceptic, as was the new Danish People's Party. This was not surprising as the party was founded by former Progress Party leader Pia Kjærsgaard, who tired of their chaotic internal fights. EU-scepticism on the left was joined by EU-scepticism on the right with a different agenda. In addition to national independence and self-determination, right-wing EU-sceptics focused on open borders and the assumed risk of a flood of immigrants and criminals. While left-wing EU-sceptics opposed what they saw as a 'Fortress Europe', their right-wing counterparts wanted almost the opposite, namely a 'Fortress Denmark'. At the same time, there was some erosion of left-wing opposition towards the EU, especially within the Socialist Party. So, for example, the party organisation advocated a *No* vote at the 1998 referendum on the Amsterdam Treaty while the majority of the (national) parliament members advocated a *Yes* vote. The same split was seen at the Euro-election. The SPP's top candidate was an ardent Eurosceptic but number two was quite the opposite, strongly in favour of further integration. The agenda had changed, at least for some of the left: the EU is no longer seen as a bloc against poor nations but in positive terms as an instrument for international co-operation, not least in the field of environment protection.

However, all this only hinted at future developments. Until the Euro-election nothing had really materialised. On the right, the Progress Party remained Euro-sceptic in line with its stance in 1994, but the party only got a third of the votes needed for a single EP seat. The same was true on the left. Against the advice of the parliamentarians, the vast majority of SPP voters followed the advice of the party organisation and voted no in the Amsterdam treaty referendum.

The electoral system

The importance of the individual candidates is due to the electoral system. For EP elections Denmark is one single constituency. The voter can vote for the party or one of the candidates within a party (a 'personal vote') and which candidates are elected depends exclusively on the number of 'personal votes' received. Being placed at the head of the list is of symbolic value because if the candidate in sixteenth position on the list gets more votes he or she will be elected ahead of the top candidate. In short, voters can decide everything, with the further consequence that candidates from the same party compete among themselves. With minor modifications the same system of 'personal voting' is used in all Danish elections but voters behave differently. At national elections less than half give a 'personal vote' and there are solid indications that most of those who vote 'personally' first choose a party and then find a candidate within the party. The situation is very different at both municipal and EP elections. Three-quarters of all voters vote personally and popular candidates can make a lot of difference. For example, at the last EP election the Conservatives had the former prime minister, Poul Schlüter, on their slate and the Conservatives rose well above their standing in the national opinion polls, while the Social Liberals had a prominent female politician, Lone Dybkjær (married to the Social Democratic prime minister). She got 90 per cent of all Social Liberal votes as 'personal votes' and doubled support for the Social Liberals, compared with the support at national elections and in the opinion polls. Poul Schlüter did not run for re-election. However, the Social Liberal Lone Dybkjær did, and the

Social Liberals tried to capitalise further on the personality factor by further nominating popular and successful immigrant authors. The Centre Democrats tried the same tactic by nominating the party leader, Mimi Jakobsen, who had the reputation of being able to mobilise undecided voters to her side by her excellent performance on television. Finally, the Danish People's Party approached Social Democratic voters by putting a former social democratic MP, Mogens Camre, on the top of the list.

The issues and the campaign

The European Parliament is no longer a 'Micky Mouse Parliament' (if it ever was). After all, its powers have increased and its actions led the Santer Commission to resign. Therefore, it was widely expected that the public would take the European Parliament and the election campaign seriously. There was a change among the politicians. There were almost no suggestions that the EP was superfluous, and there was also a greater appreciation that being an MEP was a full-time job. There was close to a general consensus that those elected to the EP would have to resign from national political posts – if any – and only the Centre Democrat Mimi Jakobsen insisted that she would remain a member of the Danish Parliament (and continue as party leader).[1] However, in the campaign the EP's powers featured less prominently than its failure to reform its own practices. Once again, much attention was paid to travel allowances, per diems, absenteeism and irregularities in the salaries for personal assistants. Most of the themes were classical but there was a variation in the targets as both the Social Liberal, Lone Dybkjær, and the People's Movement came under attack. Finally, the Social Democratic candidates had their own internal war, criticising each other and criticising the party leadership. For some of the candidates the evident goal was to better their own chances for election, even at the cost of other candidates from their own party. After the election comments were that the campaign had been dirty and it was right that mud-throwing played an important role: more important than in previous EP elections and more important than is normally the case in Danish politics. The campaign hardly increased the EP's prestige.

The more policy-oriented part of the debate focused on Danish exemptions from the Maastricht Treaty. The Liberals insisted that Denmark would do better without them and that they should be dropped; Denmark should join the monetary agreements and introduce the single currency. Therefore, a new referendum should be held as soon as possible. Basically the other pro-EU parties adopted a cautious wait-and-see line and the Social Democratic prime minister warned that an early referendum might produce a *No* majority and endanger Danish membership of the EU. Finally, EU-sceptics insisted that the Danish exemptions should stand. There was hardly anything new in this discussion. The same applied to the general debate about European Union. Euro-sceptics warned against further centralisation while the pro-EU parties warned that Euro-sceptics over-stated the trend. In sum, the parties played their old roles.

There were signs of reorientation among the Euro-sceptics: within the SPP a faction had become pro-EU while Euro-scepticism grew on the right. Only the latter became visible through the campaign. The Danish People's Party and Mogens Camre managed to get some attention for their agitation against immigrants. Also, a discussion between Mogens Camre and Naser Kader from the Social Liberals was one of the few campaign meetings to attract a big audience. It is unclear whether the campaign changed much. The pollsters disagree. According to the old and well-respected Danish Gallup the two Euro-sceptic 'movements' were below a 10 per cent level of support at the beginning of the campaign but then slowly started to increase their standing, ending up close to their position at the last election. This leads to the straightforward and almost classical conclusion that the campaign is extremely important by bringing the movements and their arguments to the attention of the voters. However, Megaphone – a new-comer among the pollsters – had the two 'movements' at their old level of support right from the beginning of the campaign, and for them the campaign had little impact on the outcome. Megaphone's exit-poll was more reliable than Gallup's. Megaphone presented a list with the party names to the respondents while Gallup only asked the respondents which party they would vote for. Therefore, Megaphone's questioning was much closer to the situation in the election booth.[2]

The results

Compared with the 1994 EP election, the results were remarkably stable. From 1989 to 1994 net volatility (sum of changes divided by two) was as high as 28 per cent. Part of this was due to the June Movement splitting from the old People's Movement, and it can be suggested that those changes in election results due to party splits should not be counted in the calculation of net volatility. However, even when the two movements are considered as one calculation, this was up at 20 per cent. Even by the standard of often unpredictable Danish national elections, the 1994 election scored highly. By comparison net volatility at the 1999 election was down to 16 per cent and, once again, discounting party split to 14 per cent. This was much closer to normal volatility at national elections. Furthermore, the strength of the main groups of contenders was almost unchanged. The two Euro-sceptic lists lost 2.1 per cent of the total vote but this was compensated for by a 2.1 per cent gain by the Euro-sceptic parties (see Table 6.2), leaving nothing either to gain or lose for the third category, namely the pro-European parties. However, there were some changes within the main categories. 'Soft EU-sceptics' now have a clear lead, well ahead of the hard-liners in the People's Movement. However, the pro-EU faction within the SPP did not succeed: the party got approximately the same support as in 1994 and the party's seat was won by the top Euro-sceptic candidate. More important was the gain by the Danish People's Party. In 1994 the Progress Party, standing for right-wing Euro-scepticism, got a poor 2.9 per cent of the vote; hardly enough to be taken seriously. In 1999 the Progress Party lost most of its votes but this was more than offset by the gain of the Danish People's Party. Yet, the old picture stands: again in 1999, left-wing Euro-sceptic parties and lists got a much larger share of the vote than right-wing Euro-sceptic parties. Now, however, right-wing Euro-scepticism must be taken seriously.

Among the pro-EU parties the results did not signify any major change with respect to European politics. The Liberal Party made a clear gain and the Conservatives had an even greater one, but both changes were most probably linked to changes in national politics. The figures for the Liberal Party brought the strength of the party in line with its strength at the last 1998 national election but not in line with the party's present political strength, according to the polls. This was almost a replication of the 1994 result which was also below the strength of the parties at that time. Further, in the 1990s the same pattern has applied to national elections in which the Liberals have made smaller gains than predicted by the polls. Either the polls over-rate the Liberals – maybe due to high profile of the Liberals in public debate – or the Liberals lack the ability to translate high popularity into votes. The Conservative loss was heavy, but not as heavy as could have been expected. In 1994 the party capitalised on the popularity of its leading candidate, former prime minister (1982–93) Poul Schlüter, and did better in the polls. Since then the Conservatives had been in big trouble, and since February 1997 they have even become a laughing stock, due to an everlasting series of bitter fights within the party leadership. Against this background, the party did reasonably well. Finally, there was a significant non-event: the Social

Table 6.2 The result of the European election: vote (%)

	1994	1999	Change	National election 1998	Present political strength
EU-sceptical lists					
June Movement	15.2	16.1	+0.9	–	–
People's Movement	10.3	7.3	–3.0	–	–
Subtotal	25.5	23.4	–2.1	–	–
EU-sceptical parties					
Socialist People's Party	8.6	7.1	–1.5	7.6	11.0
Danish People's Party	–	5.8	+5.8	7.4	9.2
Progress Party	2.9	0.7	–2.2	2.4	1.4
Unity List	–	–	–	2.7	2.4
Subtotal	11.5	13.6	+2.1	20.1	24.0
EU-positive parties					
Liberal Party	18.9	23.4	+4.5	24.0	34.3
Social Democratic Party	15.8	16.5	+0.7	35.9	26.3
Social Liberal Party	8.5	9.1	+0.6	3.9	4.6
Conservative Party	17.7	8.5	–9.2	8.9	6.3
Christian People's Party	1.1	2.0	+0.9	2.5	2.1
Centre Democrats	0.9	3.5	+2.6	4.3	2.4
Subtotal	62.9	63.0	+0.1	79.5	76.0
Electoral turnout	52.5	50.4			

Democrats got close to the same share of the vote as the party got in 1994, in both cases extremely low compared to the party's strength at national elections and in the opinion polls. It did not help that the party had experienced serious difficulties for almost a year, due to an unpopular tax reform and also to even more unpopular cuts in early retirement legislation. Yet the result first of all shows that the Social Democrats once again faced serious difficulties in Euro-elections.

There is a straightforward explanation for the discrepancies between the strength of the parties at Euro-elections and the strength of parties at national elections: Euro-elections are Euro-elections and not just another replication of a national election. Therefore, basic attitudes towards the European Union play an important role for the voters' choice of party. Those opposed to membership overwhelmingly supported Euro-sceptic parties and lists, while those in favour of further integration overwhelmingly supported pro-EU parties (see Table 6.3).[3] The electoral system had one characteristic which should have facilitated an almost perfect translation of votes into seats: seats were distributed proportionally and the different parties and lists could form electoral alliances in order to avoid spoiled ballots (seats were first distributed among alliances and lists outside the alliances, and in a second step among lists within alliances). The system could hardly be more open. However, another fact could easily have the opposite effect. There were only 16 seats to distribute, amounting on average to an electoral threshold of 6.25 per cent of the vote. Overall, the result was as proportional as it could be. The share of the seats for the broad categories of Euro-sceptic lists, Euro-sceptic parties and pro-EU parties corresponded almost exactly to share of the vote for each broad category. Therefore, there was only a minor change in the distribution of seats among the major categories: the EU-sceptic Danish People's Party got one seat by a narrow margin. Among pro-EU parties there was a substantial disproportionality: the Liberal Party was relatively over-represented and both the Centre Democrats and the Conservatives were under-represented. The three parties had formed an electoral alliance but the Centre Democrats would have needed a few more votes for a single seat and the Conservatives a few votes more for a second seat. In both cases, the Liberals capitalised on surplus votes from the partners in the alliance and thereby got five seats (see Table 6.4).

Only slightly more than half of the electorate turned out to vote. This was in line with previous results but was seen as a negative verdict on the European Project. However, this interpretation seems improbable as among abstainers 43 per cent saw EU membership as a good thing while only 22 per cent found it bad. This was clearly a positive balance, even when it was less positive than among those who turned out to vote (54 per cent against 20 per cent). Rather, the low turnout compared to national elections was probably due to lack of interest. Interest in European politics is clearly lower than interest in 'politics in general'. While 63 per cent of all voters declare that they have at least some interest in 'politics in general', only 47 per cent say the same with respect to European politics. Furthermore, participation in EP elections is more dependent on interest than participation in national elections. According to respondents' own reports, only 23 per cent of those who were not much interested or not interested at all in politics in general abstained at the 1998 general election, while the number of abstainers at the EP election was as high as 60 among those who were not much interested or

70   Denmark

Table 6.3   Attitudes towards the EU and vote at the European election

	Voted for EU-sceptical list	Voted for EU-sceptical party (Row %)	Voted for EU-positive party
Basic Attitudes towards Membership			
Membership of the European Union is a bad thing	61	24	15
Membership of the European Union is neither good nor bad	35	18	46
Membership of the European Union is a good thing	6	7	87
Basic Policy Alternatives			
Denmark should leave the European Union	60	20	20
EU-States should keep full sovereignty and the right to veto all decisions	21	14	65
Joint decisions	4	2	94
EU should become the United States of Europe	4	0	96
Vote at the Referendum on the Amsterdam Treaty, 1998			
Voted No	54	22	24
Voted Yes	4	7	89

Table 6.4   Distribution of seats

	Seats 1994	Seats 1998	Change	Disporportionality Share of seats (%) minus share of votes (%)
EU-sceptical lists				
June Movement	2	3	+1	+2.7
People's Movement	2	1	−1	−1.1
Subtotal	4	4	0	+1.6
EU-sceptical parties				
Socialist People's Party	1	1	0	−0.9
Danish People's Party	0	1	+1	+0.5
Progress Party	0	0	0	−0.7
Subtotal	1	2	+1	−1.1
EU-positive parties				
Liberal Party	4	5	+1	+7.9

Social Democratic Party	3	3	0	+2.3
Social Liberal Party	1	1	0	−2.9
Conservative Party	3	1	−2	−2.3
Christian People's Party	0	0	0	−2.0
Centre Democrats	0	0	0	−3.5
Subtotal				−0.5

not interested at all in European politics. Voting as a 'citizen duty' seems to be a reality at national elections but hardly at all at European elections.

Notes

1 This was a miscalculation. The argument that EP membership could be combined with other jobs did not lend credence to the Centre Democrats' strong assertion of the importance of EU institutions. Furthermore, it was visible to all that she had far from fully recovered after a serious illness, and it seemed highly unlikely that she would able to handle more than one job. The main explanation probably is that she was needed in two places – both to get the party into the EP and as party leader at home – and therefore had no choice between the two parliaments.
2 Of course, it could be argued that the two series are on common ground as they both indicate the importance for the two 'movements' to be brought to the attention of the voters, either by the campaign or by presenting a list. However, without a list the name of the movement may simply fail to pop up in the minds of respondents who right from the beginning would be inclined to support one of the movements. In that case, the low standing of the 'movements' are due to the methodological difference between an oral interview and a cross on a ballot paper with party names.
3 Based on the question of whether EU membership is a good or a bad thing, two groups are found to be highly likely not to vote for their national party at Euroelection: first, a majority of EU-sceptic voters within EU-sceptic parties vote for one of the two 'movements' and this is only partly explained by the fact that the Unity List do not run for Euro-elections. Next, a majority of those who disagree with their national party on the EU question (Euro-sceptics within pro-EU parties and pro-EU voters within Euro-sceptic parties) likewise vote for another party or list at Euro-elections. By contrast, three-quarters of all pro-EU voters within pro-EU parties stick to their national parties at Euro-elections.

List of Danish party abbreviations used

SPP   Socialist People's Party

7
Germany

William E. Paterson and Simon Green

Introduction

In common with many other member states of the EU, European Parliament (EP) elections in Germany have traditionally been non-events, which have usually failed to generate any meaningful enthusiasm among the electorate. Within the German context, turnout over the years has been correspondingly lower than at other domestic elections, and those who do vote tend to do so out of civic duty rather than as a result of any serious engagement with the problems of European integration. As a result, the main German political parties (the Christian Democratic CDU and its Bavarian sister party CSU, the social democratic SPD, the ecologist Greens, the liberal FDP and the post-communist PDS) have all followed the pattern evident in other member states of concentrating on domestic issues in an attempt to mobilise their supporters.[1]

However, European Parliament elections in Germany are also unusual compared with other member states in that one party, the CDU/CSU, has won each of the five elections, irrespective of whether it is in government or not. The 1984, 1989 and 1994 EP elections were all fought during the lifetime of Helmut Kohl's CDU/CSU–FDP Government, and the SPD only ever came close to beating the CDU/CSU in 1989. Indeed, it was the CDU/CSU's strong showing at the 1994 EP election that created the momentum which was to carry Kohl to his fourth federal election victory later that year (Paterson, Lees and Green, 1996). This fact stands in contrast to the conventional wisdom that the low profile of EP elections increases the likelihood of a protest vote, either in favour of the main opposition party, or even an extremist party. In Germany, EP elections only tend to benefit the opposition party if it is the CDU/CSU, and thus it comes as no surprise that its best two EP results, in

1979 and 1999, have come when the party has been in opposition. Moreover, only once, in 1989, has an extremist party (the *Republikaner*) gained representation to the EP, which could largely be attributed to specific public concern over immigration (Kolinsky, 1990, 78–9). It is against this background that the 1999 European Parliament election in Germany must be considered. During the course of this chapter, we will therefore analyse the domestic context, discussing in particular the period after the last federal election on 27 September 1998, when an SPD–Green Government under Gerhard Schröder replaced Helmut Kohl's administration in the first full change of government in the history of the Federal Republic (Green, 1999). However, this only tells part of the story. In Germany, EP elections, and indeed European integration more generally, must be understood within the framework of the remarkable cross-party consensus that has characterised policy over the years. Our discussion of the election campaign is therefore preceded by an examination of this unique feature of German politics.

European integration and the German consensus

In the context of a divided Germany, European integration initially divided rather than united the political parties. The CDU/CSU had from the beginning stressed the centrality of European integration but this priority was contested initially by the SPD and then in the late 1950s by the FDP who asserted the absolute primacy of national unity. The CDU/CSU's position was underpinned by business and labour who shared its view that, by opening up markets, European integration could provide the basis for export-led growth. Politically, European integration allowed Germany to overcome its 'semi-sovereign' status (Katzenstein, 1987) and, together with France, to play an influential role in the construction of Europe. These views gradually gained wide acceptance and a broad consensus on the desirability of European integration was established by the early 1960s which has persisted ever since. Although the Greens, who entered the party system in the early 1980s, were initially sceptical of many aspects of the EU, they supported all of its key elements with the possible exception of the CFSP by the mid-1990s. The PDS, by contrast, has retained its critical view of the EU and is now the only mainstream party to do so.

Some cracks in this consensus have begun to appear post-Maastricht. The CSU, the Bavarian sister party of the CDU, has developed a discourse of populist resentment about Brussels interventionism very similar to the discourses of the British Conservative Party. This discourse is most evident in *Land* elections and is much less apparent in federal or European elections. Support for the EU is also less apparent in the five new *Länder* in the former East, and the population as a whole has had serious misgivings about the loss of Germany's traditionally stable currency. Indeed, party political divisions over EMU threatened to erupt at various points in the run up to the third stage of EMU. For instance, Oskar Lafontaine, then Minister-President of the Saar, raised EMU as an issue in the last weeks of the 1994 EP campaign. In the wake of a disastrous SPD result in the Berlin *Land* election of 22 October 1995, Rudolf Scharping, the then party leader, and Gerhard Schröder also expressed strong doubts about the viability of the single currency and clarified their intention of making it a central issue in the 1998 federal election campaign. However, on those occasions when the SPD did attempt to use EMU to gain political advantage in *Land* elections, the tactic invariably backfired, as it did in Baden-Württemberg in 1996 and Hamburg in 1997. Within the CSU, Edmund Stoiber, the Minister-President of Bavaria, was an acerbic critic of EMU for a time but the CSU still voted for German entry into the third stage on 23 April 1998 with only the PDS voting against. In the *Bundesrat*, where state interests are represented, only Saxony (under Minister-President Kurt Biedenkopf) abstained. This position reflected public opinion in the eastern *Länder*, which is much less enthusiastic on the question of the European Union and displays especially strong reservations about EMU and enlargement.

Although Kohl was able to push through EMU, the erosion of Germany's traditional commitment to a maximalist position on European integration was apparent at the IGC preceding the Amsterdam Treaty in 1996–7. This IGC failed to address the deeper issues of institutional reform which continued to pose a question mark about the practicalities of enlargement. Kohl's declining personal standing and the reluctance of federal ministries and *Länder* governments to cede competencies resulted in even the UK being less often opposed to the extension of qualified majority voting (QMV) than Germany was. Despite the false dawn of the Schäuble and Lamers paper of 1994, which had identified political union as a desirable goal for Germany, there was simply not enough party political support to make significant progress.

The European policy of the SPD–Green Government

The award of the Honorary Citizenship of Europe to Helmut Kohl during the Austrian Presidency was a striking tribute to his generally heroic and visionary leadership style on European policy, especially in overcoming opposition to EMU (Paterson, 1998). However, the successor SPD–Green

Government was expected to be different in a number of respects. The most obvious defining factor was generational change. The new Government was largely made up of post-war baby boomers who had not shared the traumatic experience of Germany's shame and defeat in 1945. This point was made with some force by Gerhard Schröder:

> My generation and those following are Europeans because we want to be not because we have to be. That makes us freer in dealing with others. I am convinced that our European partners want to have a self-confident German partner which is more calculable than a German partner with an inferiority complex. Germany standing up for its national interests will be just as natural as France or Britain standing up for theirs.
>
> (Quoted in *Financial Times*, 16 November 1998, 6)

This generational change was reflected in a much greater readiness to talk the language of national interest which had hitherto been something of a taboo in Germany where policy was normally made 'in Europe's name' (Garton Ash, 1993). Chancellor Schröder was also determined to tackle the situation in which 'the Germans pay more than half the contributions which are frittered away in Europe' (Bulmer, Jeffery and Paterson, 2000). Schröder's rhetoric was counter-balanced by the more obviously *communautaire* discourse of Foreign Minister Joschka Fischer of the Greens, who shared many of Helmut Kohl's positions.

The new Government took over the Presidency of the EU on 1 January 1999, with the Commission's Agenda 2000 proposal as its top priority. However, the Berlin summit of the European Council on 24–6 March 1999 turned out to be concerned with providing an answer to the crisis of the Santer Commission and preparing for action in Kosovo. The Schröder Government received much praise for the way in dealt with both issues. Even so, the resolution of Agenda 2000 did not have to wait for the subsequent Finnish Presidency, as many had anticipated. While Germany was only able to secure a marginal reduction in its contributions, the agreement at Berlin of the EU's financial framework from 2000 to 2006 removed a crucial road block to enlargement. Agreement at Berlin was also helped by the resignation of Oskar Lafontaine on 11 March 1999 (see below). In his brief incumbency as Finance Minister since the election, Lafontaine had brought incoherence into the centre of government. Committed to a full-blooded Keynesian position, Lafontaine was soon in conflict with the fledgeling European Central Bank over interest rates and with the UK on tax harmonisation. The latter dispute even earned him the honour of an epithet as 'the most dangerous man in Europe' from the *Sun* (25 November 1998). His challenging personality and lack of flexibility would have greatly complicated the Federal Government's negotiating position at Berlin had he still been in office.

The Cologne Summit of the European Council

The Cologne summit on 4–5 June 1999 was the first unequivocal success of the SPD/Green government. During the conference, Schröder, together with the Finnish President Martii Ahtisaari, was able to announce that Serbia had agreed to accept the conditions put forward by NATO and Russia to end the war in Kosovo. The final proposals relied very heavily on the input of Foreign Minister Fischer, who had become by that time by far the most popular politician in Germany, and who enjoyed wide respect for his 'Stability Pact South Eastern Europe'. The Summit was particularly notable for the gains made in the area of security policy, with Javier Solana, the former NATO General Secretary, being appointed to the new post of 'Mr CFSP'. In its final resolutions, the Summit not only built on the St Malo declaration in its proposals for incorporating the WEU into the EU but went beyond the rhetoric of a European defence and security identity to start the process of creating a European security and defence capacity. The relative success of the German Presidency and of German efforts in relation to Kosovo should, therefore, have positioned the Government well for a reasonable result in the imminent EP elections. Unfortunately for it, the domestic policy position was far from propitious.

The domestic context in 1999

Despite these achievements, and the momentous developments in European integration in recent years, the European Union played only a very minor role in determining the outcome of the 1999 EP election in Germany. Instead, the roots of the German result lie in the 1998 federal election, which had constituted a watershed in modern German politics. Chancellor Helmut Kohl's crushing defeat not only ended one of the greatest careers in post-war European politics, but also ushered in the EU's first Social Democratic/Green Government under Gerhard Schröder.

In addition, the 1998 *Bundestag* election confirmed long-term underlying changes in the party system which have been becoming ever more apparent in German politics. Already, the two traditional societal cleavages, religion and class, which have in the past explained most election results, have been losing relevance since the 1980s as the Western electorate has undergone a process of 'de-alignment' (Dalton, 1996). This was given an extra dimension by unification in 1990, which added an essentially non-partisan, 'non-aligned' electorate in the East to the already de-aligning voters in the West. The result has been that the electorate as a whole has progressively become much more willing to switch parties from one election to the next, a phenomenon known as 'volatility'. It is probably unsurprising that such electoral behaviour has been particularly common in the East, given the weakness of partisan loyalty and the absence of traditional voting patterns there.

In addition, the essence of party competition remains highly different between East and West, which is largely due to the divergent values of the two populations (Veen and Zelle, 1995). Beyond the fact that unemployment is the top policy priority among the population in both East and West, considerable differences are in evidence. The concern for social justice is generally higher in the East, while neo-liberal values and environmental concerns find greater resonance in the West. In addition, as we have argued above, the population in the new *Länder* is also less enthusiastic about Europe than its Western counterpart. These differences have effectively created a dual party system in Germany: the West is dominated by the CDU/CSU, SPD, FDP and Greens, while in the East, competition is restricted to CDU, SPD and PDS. Indeed, although the PDS is only marginal in West, in the East it is a considerable political force, which by early 1999 was threatening to relegate the SPD into third place.

By producing the first complete election-induced change of Government in the history of the Federal Republic, this long-term transformation of the party system reached its preliminary apotheosis at the 1998 election. The SPD's victory was primarily crafted in the East, where the former citizens of the GDR deserted the 'Chancellor of Unification' (*Kanzler der Einheit*) in droves (Green, 1999; Weins, 1999). With substantial variation in voting patterns from election to election most likely in the East, it was there that both the large parties would have the greatest potential for winning (or losing) votes in the 1999 EP election.

Yet despite its historic victory at the *Bundestag* election, the left's initial euphoria had dissipated by mid-1999. When confronted with the realities of government, the SPD's central election promises of lower unemployment and greater social justice revealed themselves to be mutually exclusive policy goals. It was simply impossible to cut industry's high tax burden to encourage job creation while simultaneously financing the redistributive measures demanded by unions and the party's grassroots to increase social equality. These political problems were compounded by the 'semi-sovereign' institutional nature of German domestic policy-making (Katzenstein, 1987), under which governments are restricted in their actions by multiple veto points in the policy process. In particular, the negotiations for Schröder's flagship policy, a tripartite 'Alliance for Jobs' to tackle unemployment, rapidly degenerated into stalemate, in which no side was able to strike through the Gordian knot that emerged. Ironically, it was the Kohl Government's failure to make any headway with precisely these same problems which had been a major factor in its downfall just a few months earlier.

With little tangible policy progress, the SPD itself rapidly became embroiled in damaging squabbles between the modernisers, led by Chancellor Schröder, and the traditionalists, represented by the party leader, Finance Minister and darling of the rank and file, Oskar Lafontaine. Lafontaine's Keynesian policies stood in stark contrast to Schröder's more neo-liberal emphasis and his much-publicised camaraderie with industry bosses, which seemed to symbolise his commitment to the 'New Centre' (*Neue Mitte*) of German politics. The long-standing personal rivalry between the two leaders, which dates back to the mid-1980s, thus came to personify the struggle for ideological supremacy within the party. On 11 March 1999, shortly after the outbreak of hostilities in Kosovo, the simmering tensions between the two suddenly erupted, when Lafontaine abruptly resigned from all government and party positions and returned in self-imposed exile to his native Saarland.

The resignation rocked the Government and although Schröder was confirmed as the new party leader on 12 April 1999, the underlying problems of the Government, and especially the controversy over the future direction of the SPD, did not abate. For instance, whereas the imposition of social insurance taxes on previously exempt low-paid jobs (so-called 'DM630 jobs') was hailed by the unions and SPD as an improvement in employee protection, employers balked at the prospect of higher insurance contributions on their part. Even many employees in such jobs were unhappy, as any benefits of social protection were immediately offset by a reduction in their net income. The main sticking point, though, has been unemployment, which has remained persistently high, especially in the East. Within this context, the new Finance Minister Hans Eichel's plan to address Germany's substantial budget deficit by slashing DM30 billion (£10 billion) off public expenditure caused an outrage, with the left and the unions rejecting outright the need for substantial cuts in welfare expenditure. Furthermore, the joint British–German paper on the 'Third Way' for social democracy, published shortly before the EP election on 8 June 1999, received a decidedly lukewarm reception among the SPD grassroots and served only to highlight the unease of its core, unionised electorate with Schröder's course. Opinion within the party, which had never really warmed to Schröder's style or politics, was thus further polarised.

To make matters worse, the Government's malaise also encompassed the SPD's junior coalition partner, the Greens, who had found the transition from the anti-establishment protest party of the 1980s to a partner in national government particularly difficult to manage. In the short time that the Government had held office, the Greens had already been forced to accept unpalatable dilutions to its flagship policies, the liberalisation of Germany's Wilhelmine citizenship laws (see below), the withdrawal from nuclear power and the introduction of energy taxes. The NATO intervention in Kosovo, in which Germany played a key role, was the last straw for many party activists, for whom the war contradicted all of the pacifist principles that have long lain at the heart of the ecologist movement. Indeed, this almost brought down the Government: only after a special congress of the Greens in Bielefeld in early May was the crisis temporarily defused. However, this did not stop the disillusionment translating into poor results at the polls. At state elections in two of their strongest states, Hesse (February 1999) and Bremen (June 1999), the Greens' share of the vote dropped by four percentage points to 7.2 per cent and 9 per cent respectively. The situation in the all-important East was no better, where the party has practically disappeared, both in terms of membership (under 500 in most Eastern states) and in terms of results.

With the 1999 EP election offering the first opportunity for the electorate as a whole to pass judgement on this hitherto unique experiment in coalition government, the SPD–Green Government thus entered the campaign with a sense of foreboding. By contrast, the opposition CDU/CSU was able to approach the election secure in the knowledge that public disenchantment with the new Government's record of achievements was widespread. It also benefited from the fact that it had regrouped quickly after its loss of power the previous year, even though the scale of its defeat in 1998 was comparable to the collapse of the Conservative Party in the 1997 UK general election. The leadership of the CDU was seamlessly passed to Helmut Kohl's preferred successor, Wolfgang Schäuble, who enjoyed a formidable reputation as a pragmatic and skilful politician. In addition, the popular Bavarian Minister-President, Edmund Stoiber, was installed as CSU leader in December 1998, having previously made his mark as a vigorous defender of his state's interests at national and European level (Paterson, Lees and Green, 1996, 77–8; see above).

The CDU/CSU was handed an unexpected opportunity to regain the political initiative in early 1999, when the Government published its proposals for amending Germany's citizenship legislation to ease naturalisation for its large immigrant population. The plans included the *de jure* acceptance of dual citizenship for the first time in Germany's history and represented the radical change for which the SPD and Greens had long campaigned. However, they were resolutely opposed by the CDU/CSU, which launched a public petition campaign against dual citizenship in January 1999. This campaign was unexpectedly successful in capturing the public mood, and had amassed some 5 million signatures by May. Politically, it created momentum in favour of the CDU at the crucial Hesse state election on 7 February 1999. Indeed, it was largely because of public opposition to dual citizenship that the incumbent SPD-Green state Government in Hesse suffered a heavy defeat and was replaced by a CDU–FDP coalition.

This shock result had significant implications for national politics, because it changed the balance of power in the upper house of parliament, the *Bundesrat*, in favour of the CDU/CSU opposition, which now enjoyed a blocking minority. As well as requiring the Government to amend its original proposal substantially, including the restriction of general dual citizenship to a temporally limited period for foreigners born in Germany, it served further to frustrate the efforts of the SPD–Green Government in other policy areas, thereby contributing decisively to the gridlock which prevented any reforms being enacted.

Yet while the CDU/CSU was able to bounce back quickly, its former coalition partner, the FDP, was finding the adjustment to life in opposition much more difficult. In post-unification Germany, the FDP has struggled to find a *raison d'être* in the party political landscape, a process which was cushioned up to 1998 by its participation in federal government. However, the return to parliamentary opposition for the first time since 1969 has cruelly exposed the party's structural weaknesses. Its leader, Wolfgang Gerhardt, has lacked the charisma and gravitas of its elder statesman, the former Foreign Minister Hans-Dietrich Genscher, and its policy portfolio has been dominated by electorally unappealing neo-liberal measures. With the redundancy of its traditional role as a centripetal force on either a CDU/CSU or SPD-led federal government (Padgett, 1999, 103–4), the FDP has come to be seen increasingly as a party representing the interests of the rich, which has further contributed to its political marginalisation. This danger was epitomised by its repeatedly poor election results at regional and national level, which by June 1999 saw the party represented in only four out of sixteen state legislatures. Like the Greens, the Party's results in the East have also fallen to an almost insignificant level: in the Mecklenburg Western Pommerania state election, which was held simultaneously with the federal poll in 1998, the Party could only muster a meagre 1.6 per cent of the vote. Despite all public protestations to the contrary, the FDP could entertain few hopes of passing the 5 per cent hurdle necessary for representation at the EP election.

By contrast, for the fifth main party in German politics, the ex-communist PDS, the EP election offered a golden opportunity for extending its growing position in the East. Following unification, the PDS, which had emerged out of the GDR's ruling Socialist Unity Party (SED), appeared to have little chance of long-term survival. However, by rebranding itself as defender of Eastern regional interests, it has successfully tapped into the latent Eastern suspicion of *Wessis* that persists in German society. Electorally, it has attracted an eclectic mix of voters, ranging from former SED apparatchiks to the masses of unemployed 'losers of unification' and even those seized by GDR nostalgia. Backed up by the highest membership levels for any party in the East, its strategy has been highly effective: in the 1998 *Bundestag* election, the PDS scored almost 20 per cent of the vote in the East, and even cleared the 5 per cent hurdle nationally for the first time.[2] Given the SPD–Green Government's problems and the continued economic weakness of the East, the Party could look forward to the EP election with justified confidence.

The campaign

The EP campaign, which began formally with an SPD rally in Dortmund on 8 May 1999, bore the hallmarks of previous elections, with an apathetic electorate and a broad cross-party consensus on the issues which were formally at stake. Inevitably, then, the parties chose instead to attack each other on domestic issues, such as the 'DM630 jobs' and to limit their debates over Europe to relatively vacuous platitudes. The SPD's bland main slogan was 'Good for you and good for Europe' (*Gut für Sie, gut für Europa*), while the CDU's catchphrase 'Europe must be done properly' (*Europa muss man richtig machen*) was similarly uninspiring. The preponderance of domestic issues was reinforced by the dominance of national political figures in media coverage, with the parties' actual candidates for the EP itself virtually unknown among most of the electorate.

The profile of the campaign was not helped by exogenous factors in the form of the Kosovo crisis, which helped to keep the EP election out of the media spotlight, especially as the end of the military phase of the war coincided with the final days before the election. Moreover, the war itself critically weakened the Government's capacity for electioneering, as its most popular ministers, Chancellor Schröder, Foreign Minister Joschka Fischer of the Greens and Defence Minister Rudolf Scharping (SPD) were simply unavailable for most of the campaign. More seriously for the Greens, many of its pacifist grassroots members actually refused to campaign in protest against Germany's participation in the conflict. Indeed, a number of local party organisations (for instance, in Duisburg) chose to spend their EP election budgets on anti-war demonstrations, thereby causing considerable damage to the Greens' public perception. The PDS sought to capitalise on this by constructing its entire EP campaign around the fact that it was the only main German party in outright opposition to the war.

The results

Despite the fact that the broad outcome of the EP election was in the main predictable, the final results did hold some surprises. Certainly, the scale of the CDU/CSU's victory exceeded its own expectations, and the scale of the SPD's defeat confirmed its worst fears. In common with other EU member states, turnout fell from 60 per cent in 1994 to a record low for any national election of 45.2 per cent. Although in some states where local elections were being held simultaneously it remained above 50 per cent, the turnout in the Eastern state of Brandenburg (29.5 per cent) constitutes the lowest ever experienced at any state or national election (Table 7.1).

Among the individual parties, the SPD's result represents its worst at any national election since 1949. It lost almost 12 million votes compared to its 1998 *Bundestag* result, and its share of the vote dropped in every state except Hamburg and Bremen; only in Bremen and Brandenburg did it emerge as strongest party. In the Eastern states of Saxony and Mecklenburg–Western Pommerania, the SPD, with 19.6% and 20.3% respectively, even came an embarrassing third to the CDU and PDS. Most worryingly for party strategists, the SPD failed to make an impact in its heartlands, the big industrial states of Northrhine–Westphalia, Hesse and Lower Saxony. The population's dissatisfaction with the Government's performance, and to a lesser extent the party's failure to mobilise its core voters adequately, were the principal reasons behind

Table 7.1 Results of the German EP election, 13 June 1999

	1999		1994	
	Vote share (%)	Seats	Vote share (%)	Seats
CDU/CSU	48.7	53	38.8	47
SPD	30.7	33	32.2	40
Greens	6.4	7	10.1	12
PDS	5.8	6	4.7	–
FDP	3.0	–	4.1	–
Republikaner	1.7	–	3.9	–
Others	3.7	–	6.2	–
Total seats		99		99
Turnout (%)	45.2		60.0	

Source: Forschungsgruppe Wahlen (1999).

the SPD's poor result (Forschungsgruppe Wahlen, 1999). Crucially, the SPD could not reap any electoral rewards from its successful handling of Germany's military involvement in the Kosovo war, which enjoyed broad public support. Moreover, the timing of the publication of the British–German paper on the 'Third Way' on 8 June 1999 proved to be ill-judged, as the remaining five days to polling day were dominated by negative reactions to the ideas contained therein. The fact that criticism came not only from the opposition but also from within the SPD and its unionised supporters was all the more damaging, a fact acknowledged later by Chancellor Schröder (*Financial Times*, 5 October 1999, 8).

As expected, the principal beneficiary of the SPD's troubles was the CDU/CSU, which polled its best result at national level since 1983, even though the very low turnout levels actually meant that it attracted fewer votes in absolute terms than in 1994. None the less, its strength was evenly distributed throughout the country: the party took over 40 per cent of the votes in 12 of the 16 states, and over 50 per cent of the votes in the key states of Baden-Württemberg, Rhineland-Palatinate and Schleswig-Holstein. The CSU even scored an astonishing 64 per cent in Bavaria, thereby becoming third largest party nationally in its own right. Indeed, the CSU was the only main party to actually increase the number of votes it received.

For the FDP and Greens, the results were altogether depressing. The FDP fell well short of the 5 per cent hurdle for representation again, and even failed to rise above this relatively modest level (for a national party with aspirations of government) in any of the 16 states. The Greens paid dearly for their failure to deliver in Government, as well as for their support of the war in Kosovo, as their normally loyal and motivated electorate stayed away from the polls. As a result, the Greens lost over half their voters compared to the 1994 EP election, with their heaviest losses in previously strong states such as Hamburg, Hesse and Schleswig-Holstein. The mere fact that the Party remains represented in Strasbourg was thus considered a success by its leadership. By contrast, the election represented a resounding success for the other small party, the PDS: it polled 23% of the vote in East and even 40.9% in East Berlin, which equated to 5.8% nationally. This not only ensured its first representation in the European Parliament, but also its best ever result in an election at federal level. With the exception of the extreme right-wing *Republikaner*, no other party even scored over 1% of the vote.

The election was over-shadowed by the negligible turnout among citizens from other EU member states, who enjoy the right to vote at local and EP elections since the Treaty on European Union came into force in 1993. Whereas German nationals, once included on the electoral register, remain on it from year to year, EU citizens are required to re-register for each election at which they can vote. Many missed the poorly publicised deadline and, as a result, only around 2% of the 1.6 million theoretically eligible EU citizens received ballot papers. In the wake of the election, the European Commission threatened Germany with legal action unless it addressed these procedural differences between its own nationals and citizens of other member states.

The results also provide further evidence of the difficulties of party competition in Germany ten years after unification. As Table 7.2 shows, the differences in voting between East and West affected all main parties except the FDP, which just did badly everywhere. But Table 7.2 also shows how difficult it is for parties to hold on to voters. When compared with the previous election at federal level, the 1998 *Bundestag* election, all main parties have experienced considerable fluctuations in their vote share, especially in the East. In many respects, the challenge for the parties now is not to win voters, but rather to keep hold of them at the next election.

Conclusion

Overall, the SPD–Green Government was punished at the 1999 EP election for its failure to make greater progress towards improving the domestic economy. Its lack of internal discipline, especially among the

Table 7.2 Results of German EP election, 13 June 1999, by West and East and compared to 1998 federal election

	All Germany (%)		West (%)		East (%)	
	1999 EP	1998 BTE	1999 EP	1998 BTE	1999 EP	1998 BTE
CDU/CSU	48.7	+13.5	50.7	+13.5	40.6	+13.0
SPD	30.7	−10.2	32.6	−9.8	23.6	−12.0
Greens	6.4	−0.3	7.4	+0.3	2.9	−2.3
PDS	5.8	+0.7	1.3	+0.2	23.0	+3.5
FDP	3.0	−3.2	3.3	−3.7	2.2	−1.4
Others	5.4	−0.5	4.7	−0.5	7.7	−0.8
Turnout	45.2	−37.0	44.5	−38.3	47.9	−32.4

Note: 'West' includes western Berlin, 'East' includes eastern Berlin.
EP = European Parliament election.
BTE = *Bundestag* election.
Source: Forschungsgruppe Wahlen (1999).

SPD, has on occasions contributed to an unco-ordinated and dilettantish presentation of its policies. For instance, the SPD paid dearly for suggesting in the run-up to the election that, as part of its austerity measures, rises in the state pension should be inflation-only; it subsequently lost support heavily among the large numbers of pensioners in the electorate (Forschungsgruppe Wahlen, 1999).

Significantly, the CDU/CSU was able to provide the electoral outlet for public frustration, rather than it finding release via extremist parties, as had been the case in 1989. The immediate political effect of the 1999 EP election was similar to 1994, in that it created a momentum in favour of the CDU, which helped the Party achieve impressive victories at four of the five state elections which were held in September 1999. Within the domestic context, the EP election has further shifted the balance of power in the *Bundesrat* in favour of the CDU/CSU, which can now veto most of the Government's important policies. This in turn raises the prospect of a *de facto* Grand Coalition of the kind that dogged Helmut Kohl from 1994 onwards. It remains to be seen whether the SPD can avoid being blamed for any resulting policy immobilism.

The 1999 EP election was also the last election to be held at federal level in the Bonn Republic. It was a considerable disappointment to the Bonn political class, though this disappointment was obviously less in the CDU/CSU. The relatively poor turnout at European elections has long been a concern in Germany (Bulmer and Paterson, 1987, 120) but this time round it was especially low, despite Germany holding the

Presidency. The extraordinarily low electoral participation rates in the eastern *Länder*, the lowest in the history of the Federal Republic, underline the continuing divergence between West and East almost ten years after unification. It also gives some indication of the difficulties that Germany would face in the enlargement process on issues such as free movement of labour, and their implications for the East. The lesson that any government was likely to draw from a humiliating defeat in the EP election was that elections are won on domestic issues and the necessary focus on European and security policy in the first six months of 1999 had to be replaced by more popular domestic policies. There is therefore a direct connection between Schröder Government's poor performance at the EP election and its decision in December 1999 to pump support into the ailing Holzmann construction firm.

German European policy is now more visibly concerned with securing German interests. This imperative was very visible in the negotiations for the new Commission where Günter Verheugen (SPD) became the new Commissioner for Enlargement and Michaele Schreyer of the Greens took over the budgetary portfolio. An attempt by the CDU to build on its electoral success in claiming one of the two Commission posts was never likely to succeed, given that Chancellor Kohl had followed a policy of giving both posts to members of the governing coalition. Overall, the EP elections were something of a disaster for European socialists and social democrats, as the EPP became the largest grouping in the European Parliament. Within that grouping the CDU/CSU looms very large and German priorities will consequently play a significant role.

The dramatic fall in turnout at the EP election also underscores a declining interest on Germany at a mass level in European integration, which has become essentially an elite concern. Although there was some erosion of the elite consensus on EMU, the indications are that it is still largely intact save for the PDS. Plebiscitary procedures do not exist at a federal level in Germany and the scope therefore for European integration to proceed on the basis of elite consensus is thus broader than in a number of other member states. What has narrowed in Germany is the budgetary space. Unification has had a corrosive impact on public finances and Germany's ability to play the mediator/paymaster role, while still present, is more constrained than in the past. It can therefore be concluded that EMU and enlargement will continue to be supported but that the bottom line will be more visible and that the benefits to Germany will need to be more immediately demonstrable than on some occasions in the past. None the less, there will still be a place for abstract values in German European policy and policies, such as the suggestion for a charter of basic rights at an EU level, which do not reflect a purely bottom-line approach.

Notes

1 Although formally two separate parties, the CDU and CSU have formed an electoral alliance at national elections since 1949 and are usually considered together. The formal title of the Greens is Bündnis 90/Die Grünen, but the shorter anglicised version will be preferred here.
2 In 1994, the PDS had gained representation in the *Bundestag* by winning three constituency mandates.

References

Bulmer, Simon, Charlie Jeffery and William E. Paterson (2000, forthcoming), *Germany's European Diplomacy* (Manchester: Manchester University Press).
Bulmer, Simon and William E. Paterson (1987), *The Federal Republic of Germany and the European Community* (London: Allen & Unwin).
Dalton, Russell (1996), 'A Divided Electorate?', in Gordon Smith, William E. Paterson and Stephen Padgett (eds), *Developments in German Politics 2* (London: Macmillan).
Financial Times, 16 November 1998, 'Germany Annual Country Review', 6.
Financial Times, 5 October 1999, 'Schröder admits new centre blunder', 8.
Forschungsgruppe Wahlen (1999), *Europawahl. Eine Analyse der Wahl vom 13. Juni 1999*, Bericht 95, 16 June 1999 (Mannheim: Forschungsgruppe Wahlen).
Garton Ash, Timothy (1993), *In Europe's Name: Germany and the Divided Continent* (London: Jonathan Cape).
Green, Simon (1999), 'The 1998 German Bundestag Election: The End of Era', in *Parliamentary Affairs*, 52(2), 306–20.
Katzenstein, Peter (1987), *Policy and Politics in West Germany: The Growth of a Semisovereign State* (Philadelphia: Temple University Press).
Kolinsky, Eva (1990), 'The Federal Republic of Germany', in Juliet Lodge (ed.), *The 1989 Election to the European Parliament* (London: Macmillan).
Padgett, Stephen (1999), 'The Boundaries of Stability: The Party System before and after the 1998 *Bundestagswahl*', in *German Politics*, 8(2), 88–107.
Paterson, William E. (1998), 'Helmut Kohl, the "Vision Thing" and Escaping the Semi-Sovereignty Trap', in *German Politics*, 7(1), 17–36.
Paterson, William E., Charles Lees and Simon Green (1996), 'The Federal Republic of Germany', in Juliet Lodge (ed.), *The 1994 Elections to the European Parliament* (London: Pinter).
Sun, 25 November 1998.
Veen, Hans-Joachim and Carsten Zelle (1995), 'National Identity and Political Priorities in Eastern and Western Germany', in *German Politics*, 4(1), 1–26.
Weins, Cornelia (1999), 'The East German Vote in the 1998 General Election', in *German Politics*, 8(2), 48–71.

List of German party abbreviations used

CDU   Christian Democrat Union
CSU   Christian Social Union
FDP   Free Democratic Party
PDS   Party of Democratic Socialists
SED   Socialist Unity Party
SPD   Social Democratic Party

8
Greece

George Kazamias and Kevin Featherstone

Introduction

The 1999 European elections in Greece took place during a significant period of transition on several fronts. This transitional quality was also reflected in the election results. No party could claim a major victory, and neither did the campaign have a deep impact on any matter of long-term consequence. The major backdrop of the campaign – NATO's actions in Kosovo – was, essentially, a short-term distraction. The elections came well before turning points that are much more important for Greece: the decision on entry into Stage 3 of Economic and Monetary Union (EMU) and national parliamentary elections, both expected in 2000. Nevertheless, the elections provided a means of gauging the process of transition. The defining elements of the transition can be summarised as follows.

1 Greece is no longer the 'black sheep' of EU politics. For much of the period when Andreas Papandreou was prime minister (1981–9; 1993–6), Greece was a dissenter from foreign policy declarations issued under European Political Co-operation and was often seen by its partners as wasteful of the aid granted under the Structural Funds. Its reputation in Brussels and beyond suffered as a result: it was seen as a liability. Under Costas Simitis, however, Greece has returned to the EU's political mainstream. On the vast majority of issues, the Simitis Government has been part of the prevailing EU consensus.
2 Simitis has thus brought stability to Greece's EU policy and has brought it into line with the high long-term public support for 'Europe'. The steer for Simitis's European policy has undoubtedly been his resolute commitment to obtain EMU entry. There is widespread

political support for this objective and it has been raised to a level of national pride and identity. The near-desperation of Simitis to enter is somewhat ironic, however: Panellinio Socialistiko Kinima (PASOK) regained power in 1993 promising a 'softer' path to EMU, but Simitis is now attacked for his exclusive concern for EMU 'accountancy'. His government secured the entry of the drachma into the European Exchange Rate Mechanism (ERM) on 14 March 1998. At the June 2000 EU summit, Greece was judged to have fulfilled the Maastricht criteria and therefore was accepted into EMU in January 2001.

3 The two major contenders for government – PASOK, the socialists, and New Democracy (ND), of the centre-right – were themselves in the throes of their own transitions as political parties in 1999, and this had implications for the wider domestic system.

Within PASOK, the difference of style between the highly charismatic Andreas Papandreou and the worthy, but much less colourful, Simitis represented a change of greater policy substance. Simitis's succession was based on him leading a body of 'modernisers' (*eksynchronistes*) within the party, urging internal party reforms and a set of policy shifts (loosely, pro-West; pro-EMU; and pro-market economics). The modernisers defined themselves by contrast to what they saw as the populism of Papandreou and his supporters (the loyal *proedrikoi*). The latter were portrayed as too prone to demagoguery, clientelism via the party, an excessive reliance on the personality of the leader, and to making unrealisable policy declarations. In addition to Simitis, the most prominent 'modernisers' were Vasso Papandreou (now Interior Minister), Theodoros Pangalos and Georgios Papandreou (the ex-Prime Minister's son, now Foreign Minister). Following Andreas Papandreou's resignation, his 'populist' legacy was most closely associated with Akis Tsohatzopoulos (the current defence minister) and, more latterly, with Pantelis Oikonomou. Simitis's succession to the leadership had been difficult: he had defeated Tsohatzopoulos for the 'soul' of the party at its fourth congress in June 1996, but his own aura of calm competence consistently failed to ignite his party.

Party tensions came to a head at the fifth congress in March 1999, in the lead up to the EP elections. Simitis faced no rival in his bid to be re-elected as party leader, but his endorsement was not unequivocal. He gained 66 per cent of the votes, but the remainder either returned blank or invalid ballot papers. Critics of the leadership charged that it was pursuing EMU too strictly and with disregard for the party's natural lower-income constituency. Polls suggested that PASOK's electoral base was, indeed, shifting towards the more educated, more affluent voter. Tsohatzopoulos indicated that he would challenge both policies and individuals if the party failed to obtain more than 30 per cent of the vote in the forthcoming European contest. Simitis's hold on his party thus faced a notable electoral test.

New Democracy had undergone a prolonged period of fratricidal conflict after it left government in 1993. Again, the struggle was for the very soul of the party: its ideology and electoral identity. Miltiades Evert wrested the party leadership from former premier, Constantine Mitsotakis, in November 1993, and endeavoured to ditch the heavy emphasis on neo-liberalism in favour of a more consensual populism. Evert never managed to establish coherence and unity, however, as party factionalism veered almost out of control. The turmoil after the party's defeat in the 1996 elections saw Evert initially re-elected as party leader, but his weakening hold led to his replacement at a party congress in March 1997. The new leader, Costas Karamanlis, is the nephew and namesake of one of the most towering figures of post-war Greek politics. His elevation owes much to his name: he was just 41 years old when he became leader and he had not served in any of the previous ND governments. As the former Foreign Minister, the ebullient Pangalos, once remarked, the young Karamanlis had never even run a *periptero* (street kiosk). He does have charisma, however – unlike the somewhat desiccated Simitis – and he has used his 'newness' to assuage the conflicting tensions and egos of the remaining party barons. Karamanlis has shifted the party to be somewhat more populist and closer in orientation to the Popular Party of the 1950s, appealing to an 'underdog' culture.

The fundamental aspect of these transitional elements is the extent to which they reflect a gradual process of 'Europeanisation' in the domestic political system. In place of the highly charged clashes of personalities and subcultures, the party contest has emphasised issues of competence, technocratic reform, and 'de-ideologicisation', each tied in some way to 'Europe' (Featherstone and Kazamias, 1997). On EU issues, there is very little of policy substance that divides PASOK and ND. The Amsterdam Treaty, for example, was ratified in Parliament in February 1999 with an overwhelming majority. The Kosovo conflict temporarily re-ignited old passions of anti-Westernism and asserting national independence, giving campaign advantages to ND and the smaller parties of both right and left. But the deft handling of the issue by Simitis and Georgios Papandreou avoided it becoming the cause of a deeper and a more general political cleavage.

The campaign

The campaign was, generally speaking, lacklustre; it was on the pattern of the 1994 EP elections and once more in sharp contrast to the colourful campaigns of the Papandreou years. Both PASOK and ND organised mass rallies in the two largest cities of Greece, Athens and Thessaloniki (and smaller versions elsewhere), though on a lesser scale than in the past. Both PASOK and ND gave particular attention to television advertising. On the left, the Kommounistiko Komma tis Ellades (KKE) and DIKKI chose to focus on a traditional 'soap-box' style campaign; while Stefanos Manos's new party, the Liberals, chose a coach-tour campaign, not dissimilar to the one conducted by John Prescott for the UK's national elections of 1997.

The question of funding of the elections remained unchanged: the state contributed sizeable sums, though the parties remained free to find alternative sources for topping-up their campaign funds. The state funded a non-partisan television campaign to publicise the EP elections and encourage voting in the elections. Political parties received airtime on state television to present their views. In addition, the largest parties also paid for their own television campaign; this was not uncontroversial, leading to allegations of 'grey' advertising early in the campaign. Unsurprisingly, the PASOK television advertising campaign focused on the progress of major public works (road building, railway modernisation, the Athens underground and new airport, and telecommunications) but also on the government successes in securing continued EU funding for the period 2000–6. ND's television campaign focused on the shortcomings of PASOK, particularly on foreign policy, tax reform, the rising crime wave and inadequate privatisation; PASOK and ND each spent in the region of 1.5 billion drachmas (approximately £3 million) for television advertising. No leaders' television debate took place due to disagreements between PASOK and ND, mainly over the rules to be applied. The campaign load was shouldered mainly by national party leaders and their close collaborators, though MEPs and other senior personnel also participated to varying degrees. Most government ministers pitched into the battle alongside Kostas Simitis, the Prime Minister. The leader of New Democracy (the largest opposition party), chose to carry most of the load in a 'one man show' campaign. This may have been partly the result of intra-party tensions. In any event, Karamanlis tried to capitalise on his youthful image and his undoubted ability as a speaker, though he may also have used the campaign to build up his own standing with the electorate; that was probably also reflected in the party election slogan, 'New Start'. Overall, the flavour of the elections was clearly and openly national: even the party slogans reflected this: only PASOK's slogan had a clear European component. PASOK's slogan was 'For the Greece we want to the Europe we aspire to'. On the other hand, DIKKI's party slogans were unashamedly Greek-centred: 'Greeks are worth it, Greece can do it' and 'Resist'. The Liberals' slogan was 'Whoever cares, dares'. In the course of a press conference three weeks before the elections, Simitis accused the other parties of focusing on non-European issues and thus obliging PASOK to respond and shift away from European issues.

European personalities as well as European issues (with two exceptions) were noticeably absent from the campaign; developments such as the Commission's resignation played no role, though some attention was paid to issues such as democratising the EU and an EU common foreign policy and defence. Participation by Greek political parties in Europe-wide activities was relegated to a third or even fourth place. Practically the only trans-national party activity that received any media attention was the participation of five PASOK delegates (including the Prime Minister) in the PES Congress in Milan, in early May 1999. The Prime Minister addressed the congress on development and employment (both issues close to his own heart in a domestic context); and Akis Tsohatzopoulos, his erstwhile domestic rival, was elected to one of the 7 vice-president posts of the PES. Kostas Karamanlis, the leader of ND, also participated in the EPP Conference in Brussels in early February 1999, where he was also elected to one of the 7 vice-presidencies of the party. Trans-national party manifestos were practically unknown in Greece, as was any significant participation of Greek parties in their drafting. Indeed, it seems that in view of diverging views in the two major non-national issues that dominated the campaign (see below), it was in the interest of the major Greek parties to distance themselves from their European Parliament partners.

In the midst of national issues, there were two 'external' issues that received very considerable attention from all sides. These were the question of Greece's participation in EMU and the war in Kosovo. On the first, the major parties exhibited a considerable degree of agreement; Greece's entry to the EMU is actively supported by both PASOK and ND, with only relatively minor differences on strategy noticeable between them. The main opponent of EMU is the unreformed Communist Party, the KKE. It has continued to denounce EMU as a target chosen by 'the plutocracy' (local, European and global), which thus seeks to maintain its hold on the peasants and workers. The KKE denounced

EMU, declaring that it fails to safeguard national interests and protect Greece's borders and claiming it cannot contribute to the shift of the EU's character in a more 'people-friendly' direction. In this the KKE made clear its differences from both DIKKI and Synaspismos, their more moderate left-wing rivals. The two of them had at different times stated their support for a 'reformed' Europe with a human face, governed by broadly socialist, centre-left governments.

The second non-national issue was Kosovo, and on this matter more divergence was noticeable. The ruling PASOK government followed a line of moderate support for intervention, at least once operations had begun. It allowed NATO forces to use both the port and the airport of Thessaloniki and granted them transit rights to the northern border of Greece. However, its own contribution was very low key and it made clear it would neither contribute forces to an invasion force, nor permit the use of Greek airports for bombing missions. ND followed a more ambiguous line, denouncing atrocities and pointing to the potential dangers for Greek interests from developments in the territorial status of the Balkans, but stopping well short of outright opposition to the actual operations. Once more, the KKE took the more extreme line and denounced foreign intervention in the Former Yugoslavia as an imperialist plot. Party activists took the lead in demonstrations, demanding that no Greek soldier should participate in the multi-national force preparing to invade the rump of Yugoslavia; supporting a handful of conscripts who publicly declared their opposition to the war and refused to join their units; and joining in 'practical jokes' such as the instance when military signposts were 'rearranged', diverting a disembarking NATO column to the wholesale fruit and vegetable market of Thessaloniki (instead of their destination, the Greek border).

Where ideology is concerned, it becomes clear that the EP elections gave the smaller parties (Synaspismos, KKE, DIKKI, Liberals) a chance to reaffirm or even redefine their ideological positions. The KKE presented itself as strident and militant, and it probably benefited from the clarity of its views. Synaspismos, among the losers of the campaign, showed again its pro-European face. DIKKI consolidated its position left of PASOK, using populist rhetoric. The Liberals tried (relatively successfully, if we take the very short life of the party into consideration), to capture the centre ground between PASOK and ND. Politiki Anoixi (Pol. An.) used the same, rather tired nationalist rhetoric as in the past and saw its following shrink even further. However, for the two large parties (capturing between them just under 70 per cent of the vote), the contest was de-ideologised. Kostas Karamanlis's remark on this issue was that the ideological convergence of the large parties (or rather PASOK's move to the centre) is a victory for ND's views. A noticeable innovation in these elections was the creation of an 'abroad region' (overseas constituency). This gave the right to the numerous Greeks living outside the country to vote in the EP elections whilst remaining in their place of residence. In the event, over 42 000 voters registered to vote in this 'region'. Interestingly, PASOK captured over 44 per cent of the vote, to just over 38 per cent for ND, an outcome very different from that registered in the homeland.

The results

The EP election results reinforced some earlier trends. Turnout continued to fall, down to the lowest level recorded for any European election in Greece. Abstention was a little under 30 per cent of registered voters, while blank and spoilt ballots between them accounted for a little over 4 per cent of the votes cast. The decline may be attributable to a recent legislative change. Voting has long been compulsory in Greece, but new legislation removed the administrative sanctions for not voting (even though it reaffirmed that voting is an obligation of citizens). Of those who did go to the polling stations, more Greeks than ever before took the opportunity to vote for one of the minor parties. The fact that the overall election results failed to produce an outright winner further reinforced the sense that these were 'second-order' elections, of questionable significance for future trends in the party system. PASOK lost ground, but not as much as it had feared. It recorded its lowest ever vote in a EP election and its lowest vote in a nation-wide election since 1977. Its share of the vote was down 4.7 per cent on the 1994 EP elections, but down by a much larger margin (8.5 per cent) on its 1996 national election triumph. The fall in its vote since the previous national election (8.5 per cent) was much greater than that which had occurred between 1993 and 1994. (1.7 per cent), though less than that which it had suffered between 1985 and 1989 (9.8 per cent). On perhaps the most crucial test of all, PASOK was just 3.1 per cent behind ND and it had retained about a third of the total vote. There could be no major threat to Simitis's leadership on the basis of these results. The real electoral battle would commence in 2000.

ND could claim victory, though it was more modest than it must have hoped given the early campaign polls. Its share of the vote was up 3.3 per cent on the previous EP elections, but that had been a particularly poor performance. ND also suffered from voters backing the minor parties: its vote was 2.1 per cent down on its score in the 1996 national contest. Its young leader had invested much in these elections, but the results were not a very strong springboard for the looming national election.

The parties that had achieved the clearest success were those to the left of PASOK. KKE's share of the vote was 3.1 per cent up on its 1996 result and it gained an extra EP seat.[1] The vote of Synaspismos remained more or less as it had been in 1996, though far below the expectations the party had built up. The combined vote for the two parties was now back to the level they had achieved in 1989. DIKKI entered the EP for the first time, with a share of the vote 2.5 per cent higher than in 1996. Both KKE and DIKKI appear to have benefited from their populist-nationalist campaigns on the severity of PASOK's EMU strategy and its timidity in opposing NATO over Kosovo. In the new EP, DIKKI chose join Synaspismos and the KKE in the Confederal Group of the European United Left/ Nordic Green Left. Amongst the minor parties of the right, there was a contrast between the 'old' and the 'new'. Pol.An.'s fortunes continued to decline, down 6.4 per cent on the 1994 EP elections and losing its two seats in Strasbourg, making its future more uncertain than ever. By contrast, the Liberals, the new party of Stefanos Manos, a former National Economy minister who was expelled from ND in 1998, recorded its first EP vote, but won no seat. The party had been founded less than two months before the elections. Its very low vote seemed to pose little threat to Manos's former party.

The vote for the other parties was thinly dispersed. Fifteen parties shared just 5.4 per cent of the vote. The 'green' vote was again divided between two entities, but failed to reach 0.5 per cent in total. The largest of the minor parties was the Party of Greek Hunters (a countryside alliance) which saw its vote move up to 1 per cent. The remaining parties were fringe and difficult to identify. The television personality, Vassilis Leventis, a regular election 'no hoper', saw his party's vote almost halved to 0.8 per cent, whilst another maverick, Dimitris Kollatos, recorded an increase to 0.6 per cent. The plethora of small parties seemed again to be symptomatic of the 'second-order' nature of the election (see Table 8.1).

Table 8.1 The Greek elections to the EP, 13 June 1999

Registered voters	9 555 326	
Voted	6 712 684	(70.25%)
Valid votes	6 428 696	(95.77%)
Spoilt votes	150 537	(2.24%)
Blank votes	133 451	(1.99%)

Party	Votes	%	Seats	Votes in the 1996 national elections	% in 1996	Votes in the 1994 European elections	% in 1994
PASOK	2 115 844	32.91	9	2 813 245	41.49	2 458 619	37.64
ND	2 314 371	36.00	9	2 584 765	38.12	2 133 372	32.66
KKE	557 365	8.67	3	380 167	5.61	410 741	6.29
Synaspismos	331 928	5.16	2	347 051	5.12	408 066	6.25
DIKKI	440 191	6.85	2	300 671	4.43	–	–
Pol.An.	146 512	2.28		199 463	2.94	564 778	8.65
Liberals	103 962	1.62		–	–	–	–
Others		5.62			2.28		3.65

Conclusion

The 1999 European elections in Greece did not produce any significant change. They were fought as a scaled-down national contest. Indeed, in this respect, the character of the elections was not wholly dissimilar from that of the local government elections of autumn 1998. European issues were largely absent; indeed, in view of the war in Kosovo, maintaining distances from EU parties as well as the policy of most EU states was a safe strategy for the big players. No party could claim a major victory in the elections.

The 'winners' of the elections were the smaller parties and even the 'micro' parties, who are usually lost in national parliamentary contests. KKE and DIKKI enjoyed the most noticeable success. But, like the Eurovision Song Contest, the EP elections provide opportunities to 'mavericks'. Three of the lesser leftist parties (Organisation for the Reconstruction of the KKE, Marxist–Leninist KKE and ASKE) saw the number of people voting for them double from the national elections of 1996. The two Green parties enjoyed an increase in votes almost as large. Yet the percentage is negligible: the three leftist parties captured 0.48 per cent and the combined Green vote was 0.7 per cent. The political spectrum continues to exhibit a high degree of cohesion: the two largest parties combined account for just under 70 per cent of the vote; and the five parties represented in the European Parliament account for just under 90 per cent of the total vote. In short, the proliferation of parties offering themselves to the electors in the European contests does not seem to be translating into party fragmentation.

For the major parties, the EP elections were a beauty contest to boost or lower party morale. Kostas Simitis's loss was not great enough to put his leadership in jeopardy. Kostas Karamanlis's gain was not big enough to silence all opposition. Stefanos Manos's showing probably went some way towards establishing him and his party's future. Antonis Samaras (Pol. An.) suffered a bad defeat, putting his party's survival in doubt. Nikos Konstantopoulos (Synaspismos) also fared poorly, but appeared to weather the storm. More substantively, the importance of the EP contest lies more in the direction of the forthcoming national elections. These are scheduled for autumn 2000, but may well take place in the spring if no agreement on the election of the president of the Republic (due in March 2000) is reached among the parliamentary parties, notably PASOK and ND. These national elections will be greatly influenced by the EU decision on the question of Greece's participation in EMU, due in March 2000. This is the most serious hurdle Greece has had to pass since accession to the EU in 1981 and it has been elevated to a matter of great national pride. Success in the EMU target will probably bolster PASOK's chances of winning the elections; failure will most probably spell its doom. The electoral prospects for PASOK seem mixed. On the one hand, the fall in its vote across the country and the suggestion of voter fatigue with the party are worrying signs for the future. On the other hand, continued economic improvement and EMU entry should bolster the party's increasing reputation for stability and security. The outcome may hinge on the choice between the images of dull technocratic competence and safety, and that of a youthful fresh start with uncertainty. The political pendulum has moved in an ironic fashion. Simitis's main foe claims he is the natural leader of 'change' (reminiscent of Papandreou) and that he leads a party that appeals to all electors[2] regardless of past divisions. The swapping of party clothes is taking on confusing dimensions. The EP elections of 1999 should be a prelude to a much more interesting drama.

Notes

1 A comparison with 1994 is difficult due to the electoral collaboration at that time between KKE and Synaspismos.
2 Throughout the campaign Karamanlis repeatedly stressed that past divisions (of the civil war and subsequent eras) mean little to more and more people of 'his generation'.

References

Featherstone, K. (1994) 'Political Parties', in P. Kazakos and P. Ioakimidis (eds), *Greek Membership of the EC Evaluated* (London: Pinter).

Featherstone, K. and G. Kazamias (1997) 'In the Absence of Charisma: The Greek Elections of September 1996', *West European Politics*, 20, 2 (April), 157–64.

List of Greek party abbreviations used

ASKE   Agonistiko Sosialistiko Komma Elladas (Fighting Socialist Party of Greece)
DIKKI   Dimokratiko/Koinoniko Kinima (Democratic Social Movement)
KKE   Kommounistiko Komma-tis Elladas
ND   New Democracy
PASOK   Panellinio Sosialistiko Kinima
Pol.An. Politiki Anoixi

9
Finland

Tapio Raunio

Introduction

The EP elections in Finland were largely treated by the parties more as a nuisance forced upon them than as an opportunity either to discuss European matters or to engage in domestic party competition.[1] Turnout was extremely low at 31.4 per cent. Apart from the cross-national explanation of the distance between citizens and the EU, several country-specific factors contributed to the low turnout: lack of interest shown by parties and especially party leaders, very unfortunate timing of the elections, and the rhetoric of the Finnish politicians, who downplayed the legislative powers of the Parliament, and portrayed the Union first and foremost as an intergovernmental organisation. The concluding section will discuss the highly individualistic features of the candidate-centred electoral system, which are not conducive to generating constructive debates on integration.

The players

According to the Finnish law on EP elections (714/1998) candidates can be nominated by registered parties and voters' associations. Parties can form electoral alliances with one another, and voters' associations can set up joint lists. No voters' associations presented candidates. The maximum number of candidates per party or per electoral alliance is 20. The whole country forms one single constituency. Voters choose between individual candidates from non-ordered party lists. Altogether 140 candidates were nominated by eleven parties. This was less than in the first EP elections of October 1996, when fourteen parties and one voters' association fielded a total of 207 candidates.[2] The Social Democratic Party (SDP), National Coalition (KOK), Left Alliance (VAS), the Green League

(VIHR), and the minuscule Communist Party (SKP) fielded the maximum number allowed by the electoral law. Two electoral alliances were formed, one between the Centre Party (KESK), Swedish People's Party (RKP), and the Finnish Christian Union (SKL), the other between three marginal parties, the True Finns (PS), Ecological Party (KIPU), and Pensioners for the People (EKA). Of the candidates, 55 were women (39 per cent) and 85 men (61 per cent). The average age of the candidates was 46.8 years.

The players in a domestic context

Recent developments in the Finnish party system are dominated by two main features: the bargaining and ideological moderation necessitated by building broad, over-sized coalition governments, and the increasing significance of the rural–urban/centre–periphery cleavage resulting primarily from EU membership.[3] Table 9.1 shows the distribution of votes in national parliamentary and EP elections in the 1990s.

Table 9.1 Distribution of votes in Finnish national parliamentary elections in the 1990s and in the 1996 EP elections (%)

Party	1991	1995	1996 EP	1999
Centre Party of Finland (KESK)	24.8	19.8	24.4	22.4
Social Democratic Party (SDP)	22.1	28.3	21.5	22.9
National Coalition (KOK)	19.3	17.9	20.2	21.0
Left Alliance (VAS)	10.1	11.2	10.5	10.9
Swedish People's Party (RKP)	5.5	5.1	5.8	5.1
Green League of Finland (VIHR)	6.8	6.5	7.6	7.3
Finnish Christian Union (SKL)	3.1	3.0	2.8	4.2
Finnish Rural Party (SMP)	4.8	1.3		
Liberal People's Party (LKP)	0.8	0.6	0.4	0.2
Young Finns (NUORS)		2.8	3.0	1.0
True Finns (PS)			0.7	1.0
Åland Island*	0.3	0.4	0.4	0.4
Others	2.4	3.1	2.7	3.6
Total	100	100	100	100

* The autonomous Åland region has one MP who is not formally a representative of the Swedish People's Party, but in practice sits with the RKP group.
Note: Three parliamentary groups were formed during the 1995–99 legislative term: the Leftist Faction (3 MPs), a splinter group of the Left Alliance, and two independent members, Mr Pertti Virtanen and Mr Risto Kuisma. Only the Kuisma retained his seat in the 1999 elections as the Reform Group. The right-wing Young Finns were established in 1994. The party failed to gain seats in the 1999 elections and was subsequently disbanded.
Source: Statistics Finland.

The 1994 membership referendum (56.9 per cent for membership, 43.1 per cent against, on a 74 per cent turnout) indicated that European questions would prove to be problematic for the parties. The majority of parties were divided over the issue, particularly the Centre Party, Left Alliance and the Green League.[4] The large majority of the political, administrative and business elite has been consistently pro-integrationist before and after the referendum. The two rainbow governments – bringing together SDP, KOK, VAS, RKP and VIHR – headed by Prime Minister Paavo Lipponen (SDP) since 1995 have been steadily committed to integration and have emphasised that only a strong and efficient EU can guarantee the interests of smaller member states. Similar arguments were used to defend the decision to join the third stage of the EMU.[5] Lipponen was one of the first politicians to speak in favour of membership, and has shown determination ever since to lead Finland into the inner core of the Union. While the supporters and a section of MPs in his party have been more critical, Lipponen has not met any strong criticism from the social democrats. The National Coalition and the Swedish People's Party have pursued broadly similar policies throughout the 1990s. The Green League and the Left Alliance have found themselves in a rather awkward situation. As junior partners in a government committed to deeper integration, the party leaders have needed to strike a delicate balance between the pro-European views of the cabinet and the party followers who remain divided over Europe. The leadership of the main opposition Centre Party has likewise faced a tough challenge in pacifying their often strongly Euro-sceptic, primarily rural, voters while remaining cautiously pro-integrationist in order to maintain the party's credibility as a potential future governing party.[6] The parties which resisted membership before the referendum, SKL and PS (then as its predecessor, the Rural Party), have accepted membership but were against EMU and do not support deeper integration (Raunio, 1999). President Martti Ahtisaari (SDP), elected in 1994, has likewise given his support to the government policy.[7] The national media, together with the majority of provincial newspapers, can also be classified as broadly pro-integrationist.

This relatively coherent elite consensus is not replicated among the voters. For example, Eurobarometers polls show that the share of respondents who perceive that their country has benefited from EU membership is lower in Finland than in the EU as a whole. Particularly noteworthy has been the increasing salience of the rural–urban cleavage. While the left–right dimension remains the most important structure of competition in domestic politics, the rural–urban/centre–periphery cleavage is not far behind. Disaffection, and outright hostility, towards the EU is widespread among the rural population, a development explained primarily by the destructive impact of the Common Agricultural Policy (CAP) on the farming sector. The sparsely populated rural provinces are the natural stronghold of the centre, SKL, and to a lesser extent the SDP and the Left Alliance. The Lipponen governments have been accused of ignoring public opinion, particularly so in the context of EMU when the Government was blamed for its lack of willingness to engage in a wider debate about the pros and cons of single currency.[8] The saliency of the centre–periphery cleavage is linked to the party politicization of foreign policy matters. Previously the domain of the President, recent constitutional amendments, together with EU membership, have brought foreign policy and European questions into internal party debates.

The timing of the election could hardly have been worse. First, the electoral calendar was already full. National parliamentary elections were held on 21 March 1999. Presidential elections were scheduled for January 2000, and speculation about possible candidates and their respective chances of winning office received wide coverage in the media. SDP held its own primary in April, with the Foreign Minister, Tarja Halonen, becoming the party's presidential candidate. The Centre leader, Esko Aho, used the launch of his party's Euro campaign to announce his own presidential candidacy. The various party congresses held in May and June focused primarily on presidential elections, with EP elections relegated to a clear second position. Second, June is in any case a bad month for organising elections in Finland. With warm weather and bright days after the long, dark winter, Finns usually dedicate their summer to more pleasant things in life than party politics, such as relaxing at their summer cottages. Unfortunately for the elections, 13 June was an uncharacteristically sunny and beautiful day.

The campaign: mobilising the vote

Before and after the elections the media lamented widespread voter fatigue among the citizens. However, despite the national parliamentary elections held less than three months earlier in March, it is better to speak of party fatigue. The collective absence of party leaders during the campaign left the field open for individual candidates, whose most efficient electoral strategy in the Finnish electoral system is to focus on their personal qualities.

The EP information office located in Helsinki should not be blamed for the low turnout. Its funds for distributing information were doubled for the election year to 850 000 FIM. The Helsinki office began its information campaign in the autumn of 1998 by targeting key groups such as provincial and national media and civil servants, distributing a wealth of material and organising seminars. In the spring, the EP office undertook various activities to reach the wider public. A poster campaign, with 9000 posters in 72 municipalities, was targeted to reach 2.2 million citizens. No fewer than 400 000 copies of a free election special issue of *Euroopan ääni*, a monthly newsletter with information on the EP, was made available through six provincial newspapers. A letter was sent to all those entitled to vote for the first time (168 000 citizens), and 75 000 free post cards were distributed to restaurants and cafeterias. The office took part in organising a Euro-quiz for about 5000 upper secondary school students. Finally, about 20 different brochures were distributed, and the EP was present at a host of market events, fairs, election debates, sports events and seminars. The provincial media, including newspapers from the more Euro-sceptic provinces, cannot be blamed either, for most of them ran a series of informative stories on the EP and the candidates. Similarly the largest quality nation-wide daily, *Helsingin Sanomat*, provided fairly comprehensive coverage of the elections. The four national television channels, including the two state-owned ones, and the radio fared much worse. However, the parties were responsible for campaigning, not the media or the EP.

Absence best describes the behaviour of party leaders during the campaign. Party leaders did not take part in a single televised debate, with only the leading individual candidates representing their parties.[9] This was in stark contrast to national parliamentary elections, where party chairmen participated in a series of long televised debates. Some party leaders, notably Lipponen, did embark on the campaign trail during the final stages of the campaign, but by then it was already too late. The collective absence of party leaders gave the citizens the impression that the elections did not matter. If party leaders do not bother, why should voters? Given the strong open list electoral system, the input of party leaders is also crucial in terms of facilitating constructive debates. Since the most efficient electoral strategy of the overwhelming majority of individual candidates is to focus on personal qualities and to downplay 'real' issues, the party leaders could have counter-balanced this through articulating their parties' views to the voters, but in fact the parties were almost completely in the background.

The political elite discouraged turnout by projecting to the citizens an image of the EU as little more than an intergovernmental organisation. Such 'nation-state logic' has dominated Finnish integration rhetoric throughout the 1990s. Finland is a unitary, centralised state, and federalism has never been a part of Finnish political vocabulary. The majority of politicians emphasise the intergovernmental features of EU decision-making without giving adequate weight to the supranational institutions, particularly the EP. Democratic control of EU decision-making through the national parliament, the Eduskunta, and its efficient EU Affairs Committee, the Grand Committee, is underlined while the democratic credentials and legislative powers of the EP remain neglected. Thus citizens learn to view the Parliament as a weak institution. According to a survey just before the elections, 78 per cent against 10 per cent of the respondents respectively saw the Eduskunta elections as more important than EP elections.[10]

The issues

No single issue dominated the campaign. While individual candidates did, to a varying extent, try to stimulate debate on European matters, the individualistic nature of competition more or less forced most candidates instead to advertise their personal qualities. The majority of candidates with any realistic chance of gaining seats relied mainly on highlighting their previous political and international experience which would make them particularly efficient national ambassadors to the EP. The focus was therefore on looking after national interests, not on Europe. The Commission's resignation was exploited by most parties and candidates as an example of the urgent need to 'clean up' the institutions and to increase transparency and openness. In particular, the Euro-critical candidates saw the Commission scandal as further proof of the moral decay of the EU administration, the EP and its well-paid MEPs included. The Kosovo crises diverted attention from the elections, with the majority of parties merely stating that the EU should assume more responsibility for handling such conflicts in Europe. Interestingly, the two main themes concerned the election itself. The party elites and parts of the media had already argued for several months before the EP elections that turnout would be low and that it would be difficult to achieve visible and interesting campaigns. Thus the relevant actors themselves contributed to apathy and pessimism. The media criticised the composition of the lists. Compared to the October 1996 elections, the parties had substantial difficulties in finding electorally attractive candidates. Several MEPs elected in 1996 had either won a seat in the Eduskunta in the March 1999 elections, or had for other reasons decided not to seek re-election. Even the largest parties (SDP, KESK, KOK) could only muster a handful of nationally known candidates. The presence of a high number of candidates lacking previous political experience was also pointed out as evidence of the low importance of the EP.[11] The approaching Finnish Council Presidency – starting 1 July 1999 – could have offered the parties a chance to debate both the agenda of the Finnish Presidency and EU issues in general, but even the opposition parties failed to play this card.

Reflecting the overall ideological moderation necessitated by forming multi-party governments, and the candidate-centered electoral system, differences between parties were equally difficult to identify.[12] All Eduskunta parties support Finland's EU membership. There was substantial agreement among the main parties about increasing openness and transparency, enlargement, the Northern Dimension, CAP, regional policy, and maintaining the Council as the primary legislative organ of the Union. With the partial exception of the Centre Party, all the established parties are moderately pro-European. One issue which separated the parties was employment policy. Parties of the left saw the EU as a potential bulwark against the dominance of uncontrolled market forces. The SDP emphasised in its election manifesto the need to fight unemployment through European-level action and closer co-ordination of national economic policies. The Left Alliance wanted the EU to set minimum environmental, taxation and social security standards. However, the party's list contained also several strong Euro-critics, including the leading candidate, MEP Esko Seppänen.[13] The Greens argued that environmental and social policies should become central areas of EU action. Parties of the right argued instead against EU-level employment policies. The National Coalition stressed that the elections constituted a clear choice between a bourgeois and a socialist Europe, and emphasised that the Union did not need any substantial new powers, but should instead focus on making more efficient use of its existing competencies. Several KOK candidates stressed that the EU should do 'less but better'. RKP also opposed European employment agreements. The main opposition party, the Centre, stressed that the EU should be maintained primarily as an intergovernmental organisation. Along with SKL, the party also favoured continuing Finland's traditional foreign policy of military non-alignment. The government parties (SDP, KOK, VAS, VIHR, RKP) argued for more qualified majority voting, while the opposition parties resisted such moves towards more supranational decision-making. In the 1996, elections two anti-EU lists also put forward their own candidates, the Alliance for Free Finland and the Alternative to EU, receiving a combined share of 2.7 per cent of the votes. However, this time the various, very small, specifically anti-EU organisations failed to present any lists. Of the parties contesting the elections, only the marginal communists, SKP, demanded that Finland should leave the Union following a referendum. According to a candidate survey, 21 per cent wanted Finland to leave the Union, while 65 per cent thought that EU membership had been beneficial for Finland, including all SDP, KOK and RKP candidates. Nearly two-thirds supported EMU membership, but 71 per cent of respondents opposed the EU becoming a defence alliance. The majority of KOK and RKP candidates were in favour of a defence alliance. Two-thirds of respondents wanted the EP's powers increased.[14]

The most notable difference between a Finnish party and its EP group is illustrated by the Centre Party, which sits in the ELDR group. The intergovernmentalist ideology of the Centre is at odds with the federalist approach of the Euro-liberals. Such differences manifested themselves, for example, in voting behaviour, with Centre MEPs voting against their group more often than Finnish MEPs in general (Rauramo, 1999). Also the European policy of National Coalition is not as pro-integrationist, especially regarding institutional matters, as that of EPP. As in the 1996 elections, the transnational manifestos of the Euro-parties were completely absent from the campaigns. They were available via the Internet at the relevant parties' home pages, but not really used in the actual election campaigns by the candidates or the parties. Euro-parties as a whole remained firmly in the background, with Finnish parties not receiving any visible campaign help from abroad.[15]

The results

The partisan implications of the electoral results were overshadowed by the turnout of 31.4 per cent. Turnout was 28.9 per cent below the previous 1996 EP elections, and was 37 per cent below that of the March Eduskunta elections. Turnout was highest in the Helsinki constituency (40.5 per cent) and lowest in Åland (21.8 per cent) and in the Kuopio constituency (25.7 per cent).[16] Only twice in national elections since independence had turnout fallen below 50 per cent, in 1925 and 1931 when choosing the electoral college in the presidential elections. However, turnout in national parliamentary elections has also declined in recent elections, falling to 68.4 per cent in the March 1999 elections. Coming so soon after the March national elections, the elections had hardly any impact on domestic politics.

Low turnout undermines left strength

With the exception of the Green League, the record low turnout hit parties of the left especially hard. Turnout in traditional leftist strongholds, such as the working-class suburbs around Helsinki and other urban, industrialised municipalities, remained low. SDP lost one seat, winning only 17.8 per cent of the votes. Also the Left Alliance lost its second seat. The Green League won two seats, and was the only leftist party to increase its share of the vote (by an impressive 6.1 per cent compared with the March 1999 national elections). The Centre Party retained its four seats. The National Coalition became the largest party, receiving 25.3 per cent of the votes. However, its seat share remained the same as in the previous Parliament. The Swedish People's Party focused its efforts on getting its sole MEP, Astrid Thors, re-elected, and the party achieved its goal by a comfortable margin. The Finnish Christian Union won their first ever MEP, thanks in part to the electoral alliance formed with the Centre and RKP. Of the government parties, two saw their vote share decline (SDP, VAS), while three increased their support (KOK, VIHR, RKP). Despite the impressive gains make by the Green League and the National Coalition, it would be wrong to make any longer-term predictions based on the results. Gains and losses experienced by parties owe at least as much to the individual candidates as to the parties. In terms of seat distribution in the EP, the EPP and the Greens gained one seat each, while the PES and Confederal Group of the European United Left (EUL-NGL) both lost one. Table 9.2 shows the distribution of votes and seats between the parties.

Table 9.2 Results of the 1999 EP elections in Finland

Party	Candidates	Votes (%)	Seats	Seat change
KOK	20	25.3	4	
KESK	15	21.3	4	
SDP	20	17.8	3	−1
VIHR	20	13.4	2	+1
VAS	20	9.1	1	−1
RKP	3	6.8	1	
SKL	2	2.4	1	+1
KIPU	4	2.3		
PS	11	0.8		
SKP	20	0.6		
EKA	5	0.2		
Total	140	100	16	

Source: Statistics Finland.

Turning to individual candidates, the list of the new MEPs contained hardly any major surprises. The undisputed vote queen was the Green League MEP, Heidi Hautala, who received a total of 115 502 votes, 9.3 per cent of all votes cast. Over half of her votes came from Helsinki and the Uusimaa constituency surrounding the capital. Hautala's popularity is primarily based on her reputation as an active and influential representative, and on her skills in dealing with the media. Following the elections, Lipponen accused Hautala of populist campaigning as the latter has constantly spoken against the financial malpractices of the EP, and the secrecy surrounding the Council. However, Lipponen's anger is at least partly explained by the fact that Hautala won a substantial share of her votes in areas where SDP has traditionally enjoyed strong support. Hautala's success helped the Green League to secure a second seat. Matti Wuori, a lawyer with considerable international experience and known best for his previous stint as the leader of Greenpeace, also won the majority of his votes from Helsinki and Uusimaa.

Marjo Matikainen-Kallström of the National Coalition well too, winning 107 444 votes. Her support was fairly evenly spread across the country. The three other National Coalition MEPs are all new: the former rally world champion, Ari Vatanen; former party chairman, Ilkka Suominen; and 24 year old Piia-Noora Kauppi. The support of the four elected Centre MEPs was geographically concentrated. Paavo Väyrynen, a former minister and a strong Euro-critic who has constantly attacked the policies of the party chairman, Esko Aho, won strong support in the northern Lapland and Oulu constituencies. Kyösti Virrankoski was effectively a one-constituency man, receiving considerable backing from the provincial and local media and thereby winning 67.7 per cent of his votes in the western Vaasa constituency. The same applied to Samuli Pohjamo whose support came mainly from the Oulu region. The fourth Centre MEP is the former vice-speaker of Eduskunta, Mikko Pesälä. The three SDP representatives all have previous experience from EP. Reino Paasilinna and Riitta Myller, with the support of the latter centring in North Karelia constituency, were re-elected, while Ulpu Iivari sat in the Parliament in 1995–96, failing to maintain her seat in the October 1996 elections. Apart from the three elected candidates, the SDP list consisted of people lacking nation-wide electoral appeal. Considering the rules of the electoral game, it is no surprise that the party experienced a heavy defeat in terms of its vote share. The Swedish People's Party focused its efforts in getting Astrid Thors re-elected. Thors won the third largest number of votes, 81 092, doubling her vote tally from the previous elections. The sole remaining MEP of Left Alliance is the Euro-critic

Esko Seppänen, who, like Väyrynen, consistently criticised his party for too prointegrationist positions. Finally, the SKL member, Eija-Riitta Korhola, ran a moderate and 'secular' campaign compared to opinion among party activists. Korhola joined the EPP group, even though her party favours intergovernmentalist decision-making and is against the deepening of integration.[17] Both Thors and Korhola benefited from an effective electoral strategy whereby the party campaigns and votes were heavily centralised in their favour. Thors won 96.4 per cent of all RKP votes, and Korhola 94.8 per cent of all SKL votes.

Table 9.3 presents information on the 16 elected MEPs. Just under half, 44 per cent, were women, including the three most popular candidates. Nine of 12 outgoing MEPs won re-election. Of the unsuccessful ones, Jyrki Otila (KOK) had won a seat in the October 1996 elections, while Inna Ilivitzky (VAS) and Riitta Laurila (KOK) had been MEPs only since the March Eduskunta elections.[18] The average age of the elected MEPs was 47.6 years. Three have previous ministerial experience (Pesälä, Suominen, Väyrynen) and nine have experience of Eduskunta. The last column shows the estimated costs of their campaigns. The average cost of the 16 campaigns was about 61 000 euros (363 000 FIM), which is approximately twice the amount spent by elected MPs in national Eduskunta elections. Twelve elected MEPs reported that their own finances constituted their main source of funding. Korhola's campaign was fully funded by her party (SKL). Several candidates complained about the difficulties in raising money for their campaigns. The proximity of Eduskunta elections and the allegations concerning electoral funding that plagued those elections, drained funding sources, and further undetermined the importance of the EP for companies. While there are no reliable data available, it appears that only a few unsuccessful candidates spent over 100 000 FIM on their campaigns, with only a selected few among them reaching the average campaign cost of elected MEPs. The largest parties spent between 1 and $1\frac{1}{2}$ million FIM on the elections, excluding money spent by individual candidates. The majority of parties' campaign funds came from the state, which allocated the parties a total of 6 million FIM for informing the citizens about the EU.

Conclusion

In the immediate aftermath of the elections the discussion focused on ways to improve turnout. Not surprisingly, holding EP elections simultaneously with national elections received support. Prime Minister Lipponen, Minister of Justice Johannes Koskinen (SDP), and the SKL

Table 9.3 The 16 Finnish MEPs elected to the EP

Candidate*	Party	EP party group	Age	Votes	Campaign expenditure (euros)[†]
Mrs Heidi HAUTALA*	VIHR	Greens	43	115 502	26 000
Mrs Marjo MATIKAINEN-KALLSTRÖM*	KOK	EPP	34	107 444	50 000
Ms Astrid THORS*	RKP	ELDR	41	81 092	76 000
Mr Reino PAASILINNA*	SDP	PES	59	64 204	64 000
Mr Paavo VÄYRYNEN*	KESK	ELDR	52	64 009	34 000
Mr Esko SEPPÄNEN*	VAS	EUL–NGL	53	59 954	37 000
Mr Ari VATANEN	KOK	EPP	47	58 836	63 000
Mr Kyösti VIRRANKOSKI*	KESK	ELDR	55	50 075	144 000
Ms Riitta MYLLER*	SDP	PES	42	47 939	67 000
Mr Ilkka SUOMINEN	KOK	EPP	60	38 364	50 000
Mrs Eija-Riitta KORHOLA	SKL	EPP	39	28 095	92 000
Mr Matti WUORI	VIHR	Greens	53	26 846	9 000
Mr Samuli POHJAMO*	KESK	ELDR	49	25 333	81 000
Mr Mikko PESÄLÄ	KESK	ELDR	60	24 281	67 000
Mrs Ulpu IIVARI	SDP	PES	51	24 091	47 000
Ms Piia-Noora KAUPPI	KOK	EPP	24	18 221	67 000

* Re-elected MEP.
† Based on MEPs' own estimates. The figures report the cost of personal campaigns, and thus exclude the funds used by the party organisation.
Sources: Panu Uotila, 'Europarlamenttin päässeiden vaalikassa oli Keskimäärin 400 000 markkaa', Helsingin Sanomat, 23 July 1999; Statistics Finland.

parliamentary group proposed that municipal elections and EP elections should be held at the same time. Centre Party vice-chairman Sirkka-Liisa Anttila, herself a former MEP, suggested simultaneous EP and Eduskunta elections, while the Left Alliance MP, Esko Helle, called for simultaneous presidential elections and EP elections. The Centre Party opposed holding two elections simultaneously.[19] Adjoining municipal and EP elections would mean lengthening the electoral term in municipal elections from four to five years. However, combining the elections is bound to marginalise one of the two elections and confuse the voters as the agendas of the two campaigns will get mixed. The October 1996 EP elections were held simultaneously with municipal elections, with the result that municipal elections remained in the shadow of the EP elections. This was particularly problematic as recent changes have strengthened the position of the municipalities.

The second reform proposal concerns the number of constituencies. Lipponen and the leading opposition party, the Centre, both supported the idea of dividing Finland into four constituencies.[20] Smaller parties (VAS, RKP, VIHR) resisted this as being likely to damage their prospects of gaining seats unless there were to be some additional adjustment of seat allocation to distribute seats according to the share of votes gained throughout the country. Moreover, the establishment of four constituencies would mean once again changing the electoral system. In the 1996 elections the parties and voters' associations had the right to field both nation-wide candidates as well as regional candidates in the four constituencies, but all lists fielded only nation-wide candidates. The Eduskunta approved the existing law on EP elections, in October 1998 according to which the whole country operates as one single constituency.[21] Splitting the country into four constituencies has been proposed to ensure the representation of the regions in the Parliament and to stimulate debate on locally important issues such as CAP and regional policy. Geographical representativeness is a fundamental problem, and the division of the country into multiple constituencies would probably facilitate more intensive campaigns and boost turnout. Of the 16 new MEPs, ten live in the capital area, even though only around 20 per cent of the population live there. Several parts of the country, most notably the south-west Turku, the Tampere and the Jyväskylä regions, were left without their 'own' members. The reform would also most probably lead to more even competition among candidates, as the cost of the campaigns would be reduced.

The heavily candidate-centred, strong open list electoral system leads to more competition within than between parties. Individual candidates from the same party list pursue personal campaigns, with party programmes in a very minor role. When asked two weeks before the elections which factor most influences your voting decision, 48 per cent of the survey respondents said the candidate, 25 per cent the party, and 20 per cent the candidate's attitude towards EU, with 7 per cent replaying they did not know.[22] In two similar surveys carried out before the 1996 elections, 57 per cent and 63 per cent had agreed with the statement that the individual candidate is more important than the party when making the voting decision (Majonen, 1999, 76). Considering the potentially divisive impact of European integration on party unity, party leaders have good reason to support the existing rules of the electoral game. Protest or dissenting opinions get channelled through individual candidates, whereas in member states with closed lists organised factions often appear to contest the official party line. The idea of introducing four regional constituencies, together with a system of allocating seats according to the national vote share, deserve serious consideration. However, tinkering with the electoral rules can achieve relatively little without the positive input of parties, and especially of party leaders.

Notes

1 I am grateful to Tuomo Martikainen and Matti Wiberg for their insightful comments.
2 For detailed analysis of the 1996 elections, see Martikainen and Pekonen (1999).
3 During the past two decades governments have as a rule been formed around two of the three main parties: SDP, KOK, and the centre. With the exception of the 1991–95 bourgeois centre-led government, the cabinets have included parties across the ideological spectrum, most notably so the two Lipponen governments since 1995. For further information on the Finnish party system, see Borg and Sänkiaho (1995) and Sundberg (1994, 1996, 1997).
4 The percentage of party supporters voting 'yes' in the referendum by party was: KOK (89%), RKP (85%), SDP (75%), VIHR (55%), KESK (36%), and VAS (24%). Figures are from Sänkiaho (1994). The Left Alliance and the Green League did not adopt official positions prior to the referendum. The Centre Party adopted a pro-membership line, but only after the party chairman and then Prime Minister Esko Aho had threatened to resign were the party to oppose membership.
5 Among the policy objectives of the programme of the second Lipponen government, formed after the March 1999 parliamentary elections, are improving the efficiency of CFSP and of the EU as an international actor, and increasing QMV in the Council.
6 An illustrative example of such strategic behaviour was in the context of EMU. Accepting the Maastricht Treaty without any opt-out clauses as the leading government party when Finland entered the Union in 1995, the Centre nevertheless decided against EMU membership in the autumn of 1997. However, at the same time the party indicated that it would respect the outcome of the Eduskunta vote and would not seek exist from EMU in the future were Parliament to vote in favour of EMU membership.
7 See, for example, the speech given by President Ahtisaari in the EP on 17 June 1998, 'Presidenti Martti Ahtisaari: Suomi kannattaa vahvaa EU: ta.', *Helsingin Sanomat*, 18 June 1998.
8 See the editorials in *Helsingin Sanomat*, 'Hallituksen Emu-linjaa häiritsee ylimielisyys', 14 November 1997, and 'Kansalaismielipide haraa Emu-linjaa vastaan', 22 November 1997.
9 The television and the party chairmen blamed each other for not arranging any debates. The former defended themselves by claiming that the party leaders had been invited to participate in debates, and therefore the party chairmen are responsible. While the party chairmen, Paavo Lipponen (SDP), Esko Aho (KESK) and Suvi-Anne Simes (VAS), admitted that party leaders could have played a more active role, they also accused the media of neglecting the

114 Finland

elections. See Panu Uotila, 'Lipponen: Vahva EU turvaa suurten ylivallalta', *Helsingin Sanomat*, 12 June 1999; Esko Aho, 'Vaalikeskustelun sietämätön laimeus', *Turun Sanomat*, 11 June 1999; 'Puolueet laiminlöivät tehtävänsä EU-vaaleissa', *Helsingin Sanomat* 12 June 1999; Eija Kallioniemi, 'Siimes haastoi puoluejohtajat eurovaalitalkoisiin: Paavo Lipponen patistaa mediaa vaalityöhön', *Iltasanomat*, 21 May 1999.

10 See Kyösti Karvonen, 'Nuori "eurosukupolvi" arvostaa eurovaaleja muita enemmän', *Turun Sanomat*, 6 June 1999.

11 Among the candidates were an ice hockey coach, musicians, well-known television personalities, a former footballer who played for the national team, and most notably the former rally world champion, Ari Vatanen (KOK), who also got elected.

12 The European programmes of the parties winning seats are available at their websites: SDP (www.sdp.fi), KESK (www.keskusta.fi), KOK (www.kokoomus.fi), VAS (www.vasemmistoliitto.fi), VIHR (www.vihrealiitto.fi), RKP (www.rkp.fi), and SKL (www.skl.fi).

13 The party leadership and the trade unions closely attached to the party gave strong backing to the pro-European candidate, Matti Viialainen, wanting thereby to ensure that the party would also have a pro-integrationist MEP. However, Viialainen failed to gain a seat. See Paavo Rautio, 'Vasemmistoliiton johto työntää Viialaista EU-parlamenttiin', *Helsingin Sanomat*, 26 May 1999.

14 116 out of 140 candidates answered the survey. See Paavo Rautio, 'EU-ehdokkaat HS: and verkkolehden vaalikoneessa: Emu-jäsenyys saa kiitokset, puolustusliitto tyrmäyksen', *Helsingin Sanomat*, 12 June 1999.

15 The only exception was the visit of the PES General Secretary Jean-François Vallin to Finland during the week before the elections. See Jouni Mölsä, 'Eurodemarit odottavat Suomelta vahvaa panosta EU-puheenjohtajana', *Helsingin Sanomat*, 9 June 1999.

16 More detailed information on regional variation in turnout and party support is available at the website of Statistics Finland (www.stat.fi).

17 The SKL leadership has in recent years tried to change the party name to Christian Democrats in order to broaden the party's electoral appeal and to lessen its image as a religious movement. The party congresses held in Kajaani in 1997 and in Lappeenranta in 1999 voted against a name change.

18 Of the MEPs elected in 1996, five contested the March 1999 Eduskunta elections. Sirkka-Liisa Anttila (KESK), Outi Ojala (VAS), Kirsi Piha (KOK), Mirja Ryynänen (KESK), and Paavo Väyrynen (KESK) all won seats. Only Väyrynen chose not to take up his seat as his party remained in opposition.

19 See Pekka Ervasti, 'Oikeusministeri yhdistäisi euro-ja kunnallisvaalit', *Iltasanomat*, 15 June 1999; Panu Uotila, 'Keskustan Anttila: EU-vaalit liitettävä eduskuntavaaleihin', *Helsingin Sanomat*, 19 July 1999; 'Kristilliset: EU-vaalit liitettävä kunnallisvaalien yhteyteen', *Helsingin Sanomat*, 4 June 1999; Paavo Rautio, 'Keskusta ei yhdistäisi euro-ja kunnallisvaaleja', *Helsingin Sanomat*, 18 June 1999.

20 See Paavo Rautio, 'Puolueissa ei sopua eurovaalien asettelusta', *Helsingin Sanomat*, 17 June 1997; 'Virrankoski: Vaalipiirijako EU-vaaleissa uusiksi', *Helsingin Sanomat*, 11 June 1999; 'Aho: Suomi jaettava neljään vaalipiiriin eurovaaleissa', *Helsingin Sanomat*, 20 June 1999; Panu Uotila, 'Paavo Lipponen: Alueelliset EU-vaalit lääkkeeksi vaaliapatiaan', *Helsingin Sanomat*, 18 July

1999; 'Keskustelu eurovaaleista meni ylikierroksille', *Helsingin Sanomat*, 20 July 1999.
21 The law stimulated heated debate among MPs, with several members arguing on behalf of regional constituencies. A working party on electoral laws had proposed in its report in the autumn of 1997 that the whole country would function as one constituency. The government disagreed and proposed instead that parties could field candidates both nationally and in individual constituencies as in the 1996 elections. However, the Constitutional Committee of the Eduskunta decided in favour of a single nation-wide constituency. See Paavo Rautio, 'Vaalilait yhtenäistetään, piirit pysyvät ennallaan', *Helsingin Sanomat*, 18 November 1997; Juha Akkanen, 'Europarlamentaarikot valitaan vanhaan malliin', *Helsingin Sanomat*, 22 January 1998; Katarina Baer, 'Eduskunta kiisteli EU-vaalitavasta viimeiseen asti', *Helsingin Sanomat*, 23 September 1998.
22 See Kyösti Karvonen, 'Nuori "eurosukupolvi" arvostaa eurovaaleja muita enemmän', *Turun Sanomat*, 6 June 1999. Another survey undertaken during the same week reported that 55% of the respondents chose their candidate regardless of her or his party affiliation, while 40% said they chose the party first and then the candidate from the party list. See Jarmo Aaltonen, 'Kokoomus nousi eurovaaleissa suosituimmaksi puolueeksi', *Helsingin Sanomat*, 5 June 1999.

References

Borg, Sami and Risto Sänkiaho, (eds.) (1995) *The Finnish Voter* (Tampere: The Finnish Political Science Association)
Majonen, Pia (1999) 'Kauniita ja rohkeita vai aatteellisia ammattipoliitikkoja? Suomen eurovaalien vaaliteemat ja vaalikampanjointi 1996', in Tuomo Martikainen and Kyösti Pekonen (eds), *Eurovaalit Suomessa 1996: Vaalihumusta päätöksenteon arkeen*, Acta Politica No. 10, Yleisen valtio-opin laitos, Helsingin yliopisto, 70–129.
Martikainen, Tuomo and Kyösti Pekonen, (eds) (1999) *Eurovaalit Suomessa 1996: Vaalihumusta päätöksenteon arkeen*, Acta Politica No. 10, Yleisen valtio-opin laitos, Helsingin yliopisto.
Raunio, Tapio (1999) 'Facing the European Challenge. Finnish Parties Adjust to the Integration Process', *West European Politics*, 22(1), 138–59.
Rauramo, Juhana (1999) 'Euroopan parlamentin puolueryhmien koheesio ja koalitionmuodostus 1994–1998', in Tuomo Martikainen and Kyösti Pekonen (eds), *Eurovaalit Suomessa 1996: Vaalihumusta päätöksenteon arkeen*, Acta Politica No. 10, Yleisen valtio-opin laitos, Helsingin yliopisto, 281–97.
Sänkiaho, Risto (1994) 'Puoluesidonnaisuutta vai sitoutumattomuutta', in Pertti Pesonen (ed.), *Suomen EU-kansanäänestys 1994: Raportti äänestäjien kannanotoista* (Helsinki: Ulkoasiainministeriö, Eurooppatiedotus ja Painatuskeskus), 164–73.
Sundberg, Jan (1994) 'Finalnd: Nationalized Parties, Professionalized Organization', in Richard S. Katz and Peter Mair (eds), *How Parties Organize: Change and Adaptation in Party Organizations in Western Democracies* (London: Sage), 158–84.
Sundberg, Jan (1996) *Partier och partisystem i Finland* (Schildts: Esbo).

Sundberg, Jan (1997) 'Compulsory Party Democracy. Finland as a Deviant Case in Scandinavia', *Party Politics* 3(1), 97–117.

List of Finnish party abbreviations used

EKA	Pensioners for the People
KESK	Centre Party of Finland
KIPU	Ecological Party
KOK	National Coalition
LKP	Liberal People's Party
NUORS	Young Finns
PS	True Finns
RKP	Swedish People's Party
SDP	Social Democratic Party
SKL	Finnish Christian Union
SKP	Communist Party
VAS	Left Alliance
VIHR	Green League of Finland

10
France

David Howarth

Introduction

EP elections have become an important element of French political life. Despite widespread public apathy and high abstention rates, the elections involve a high level of party activity, a lengthy unofficial campaign, considerable debate and a great deal of attention from the national press. They are a test of the popularity of the government in power. The country-wide proportional representation-based system with a 5 per cent threshold provides an opportunity for smaller parties to win seats beyond the regional level. The elections thus encourage a multiplicity of lists and the tendency towards party system fragmentation, with the three traditionally largest parties – Socialist (PS), neo-Gaullist Rally for the Republic (RPR) and the centrist Union for French Democracy (UDF, a confederation of parties) – winning a smaller percentage of seats than in the legislative elections.[1] Previous elections gave seats to parties unrepresented in the National Assembly, notably the Far Right National Front (FN) and the Greens, as well as to lists created specifically for the European elections such as the one led by the discredited Socialist politician, Bernard Tapie, in 1994. European elections equally provide an opportunity for factions within the parties (or parties within the UDF) to run their own lists, thus allowing the expression of diversity while permitting unity at the time of the national elections. This was the case with the UDF politicians Simone Veil, who ran a strongly pro-federalist list in 1989, and Philippe de Villiers, who in 1994 formed a Euro-sceptic list with the Anglo-French financier, Sir James Goldsmith. The elections also provide an opportunity to focus debate on European issues to a greater extent than national elections, with some lists formed principally around the issue of European integration, although domestic concerns normally predominate.

The lists

As in 1994, twenty lists were organised (see Appendix 1 for the complete list of names and leaders). However, there was considerable difference in the component parties and candidates, list names and policies. On the centre right, for the first time since 1979 the RPR and UDF did not run a joint list. The RPR joined with Liberal Democracy (DL), a former component of the UDF, to form a joint list, the Union for Europe (UFE), led by the presidents of the respective parties Philippe Séguin and Alain Madelin. On 16 April, Séguin resigned from both his leadership of the list and the RPR, citing the deliberate efforts on the part of his rival, President Chirac, to undermine his leadership.[2] Chirac sought to prevent Séguin from becoming a credible neo-Gaullist challenger in the 2002 Presidential elections. Séguin was replaced by Nicholas Sarkozy, the former Minister of Budget, who had fallen foul of Chirac supporters in the RPR following his decision to run Edouard Balladur's campaign for the 1995 presidential elections. With the real threat of a poor result at the polls, the young and ambitious Sarkozy was the only remaining RPR politician of national stature willing to lead the list.

There was a debate within the UDF on the desirability of a joint centre-right list. The former President Giscard d'Estaing spurned the confederation that he himself had created to support the RPR–DL list. UDF leaders insisted that a joint list present a 'pro-federalist' position in order to distinguish itself clearly from the Euro-sceptic lists on the right.[3] The RPR could not accept such a condition for fear that it would lose the support of moderate Euro-sceptics. The UDF list (With Europe, Let's Move France Forward) was led by the confederation and Democratic Force (FD) president François Bayrou.

On New Year's Day, Charles Pasqua, the former RPR Minister of the Interior, announced the organisation of his own Euro-sceptic list, initially labelled Tomorrow France, which further undermined the electoral chances of a separate de Villiers list. Pasqua had left the RPR party executive in December 1998 because of his opposition to the Amsterdam Treaty. He sought to form a cross-partisan list of parties hostile to further integration.[4] However, his failure to attract left-wing leaders forced him to form a joint list alone with the traditional catholic aristocrat, de Villiers, known as Rally for France/Independence from Europe (RPFIE).[5] In 1994, de Villiers's Euro-sceptic list, Movement for France, won an impressive 12.3 per cent of the vote and 13 seats. Since then, he has failed to transform his list into a real party and he performed badly in the 1995 presidential elections.

On the extreme right, as in previous EP elections, the FN leader, Jean Marie Le Pen, led the FN list. For the first time, the party's position on the far right was seriously undermined by the organisation of a rival list by the recent offshoot National Front–National Movement (FN–MN) led by Le Pen's former lieutenant, Bruno Mégret. On the left, the Socialist-dominated list (Let's Build Our Europe), led by party president François Hollande, remained unified and dominant, unlike in 1994. It was joined by the Citizens' Movement (MDC), the left-wing republican nationalist party led by Jean-Pierre Chevènement, the Minister of the Interior. Chevènement had been discouraged by his separate list's poor showing in 1994 (when it scored only 2.5 per cent and won no seats). The Left Radicals also participated in the Socialist-dominated list, accepting only two of the top twenty places, including Catherine Lalumière, a former Minister and head of the Radical Alliance group, in fourth place. This would necessarily result in a substantial drop in the number of their EP seats. However, their presentation of a separate list in 1994 owed entirely to Bernard Tapie's leadership.

The Greens found a popular list leader in Daniel Cohn-Bendit, the student revolutionary leader of May 1968 who had been exiled to Germany by de Gaulle. After the great success of 1989, the Greens had performed poorly in 1994 due in part to division. In 1999, they were largely united except for the Euro-sceptic Green candidate, Antoine Waechter. The Communist Party (PCF) list, led by party leader Robert Hue, was determined to perform better than its 1994 vote of 6.8 per cent. While in the previous EP elections it had sought to represent a diversity of party opinion, this time it opened its list, Move Europe (BE), to non-Communists of the left (most crucially places two and five). On the extreme left, Arlette Laguiller, the leader of the Trotskyite Workers' Struggle (LO), agreed for the first time to form a joint list with Alain Krivine and his Revolutionary Communist league (LRC), hoping to attract those on the left disenchanted by the Socialist-led Government and to win seats for the first time in the European Parliament. The most significant of the eleven other lists was the Hunting, Fishing, Nature, Traditions (CPNT) party list, led by J. Saint-Josse, which sought to mobilise the public against recent European and national laws and court decisions which restricted hunting rights. This list had won 4.1 per cent and 3.9 per cent of the vote in the two previous European elections.

With the initial drafts of the bill on parity for regional and European electoral lists being discussed in the National Assembly, this was a major issue during the first half of 1999. Even without the law, all the major lists on the left were organised on the basis of gender parity, now considered a *sine qua non* of EP elections, and the centre right placed more women higher up the lists.[6] Such positive discrimination was not extended to ethnic minorities. None the less, minority candidates were more in evidence than in previous EP elections. The outgoing EP had no non-European ethnic minorities among its 87 French MEPs. Now a handful were placed in high places on several lists (notably, the Move Europe and Socialist-led lists) and even the FN placed a candidate of Algerian (*harki*) origin in tenth place on its list. The major rift between the lists was on the general issue of further European integration. The divisions within several parties on integration have, in most cases, been held in check. However, in the past two EP elections, new lists have been created to express minority views. Thus Veil led her own pro-federalist list distinct from the joint RPR–UDF list, while de Villiers ran a distinct Euro-sceptic list. The Maastricht Treaty campaign demonstrated the depth of division within the RPR which none the less remained united in the 1994 EP elections. The creation of the RPFIE released a large number of neo-Gaullist activists to present a more overtly hostile line on integration. The label 'sovereigntist', borrowed inappropriately from Quebec, was coined to describe the RPFIE, which argued that it was the true defender of de Gaulle's vision of a 'Europe of Nations', as opposed to the RPR which, it was argued, had become too pragmatic on European integration. The FN and FN–MN, whose positions on Europe and most other policies could not be distinguished, were even more hostile to integration, calling for the renegotiation – even abolition – of the Maastricht Treaty, as did the LO–LRC. The RPR–DL list presented a cautious pro-Europeanism: Madelin argued for a 'federation of nation-states', although Séguin presented a vague pragmatism on integration. In previous elections, the Communists were resolutely hostile to what they saw as capitalist-inspired integration. Now, due in part to the opening of their list to non-Communists and attempts to maximise votes for the list, they presented a more cautiously positive line, emphasising the need to focus on social rights and employment. The opposition of the MDC to a capitalist-inspired integration was muted in the context of their participation in the Socialist-led list, which argued officially in favour of a 'federation of nation-states': an entity that was more than a 'Europe of Nations' but less than a federal state. It accepted federal elements (notably the euro) but refused to recognise Europe as a true federation, insisting that the nation-state would remain the major political unit. The UDF ran the most pro-European list. Free of coalition constraints, it could wave the federalist banner. It is important to note that the most convinced Europeans in France would only support a very decentralised federalism in which the member states would remain the principal actors. The mainstream Greens, likewise, were strongly positive on further European integration, but one which placed less emphasis upon liberalisation, competitiveness and technocratic control and more emphasis on social rights, democracy and citizen participation.

In 1994, the presidential elections of the following year were of some relevance to party divisions. In 1999, no elections loomed on the immediate horizon. Moreover, as the previous elections had taken place in March 1998, the risk of voter fatigue seemed relatively limited. Despite the distance of the next presidential elections in 2002, the EP election campaign was largely about different party leaders positioning themselves in relation to other party leaders and leading members of their own party in preparation for the 2002 presidential elections. President Chirac had sought to weaken all potential centre right opponents to his second bid for the presidency. He thus preferred a joint UDF–RPR list run by, ideally, a less important political figure from the UDF, as in 1994. As a second best, he hoped that an RPR list would do well in the elections but not so well as to boost the list leader. In a deliberate effort to undermine Séguin and appeal to the more nationalist neo-Gaullists, Chirac claimed that both the RPR–DL and RPFIE lists represented presidential opinion. France was in a period of co-habitation, with a president and prime minister of opposing parties. Being responsible for European and defence policy, President Chirac insisted upon the compliance of his own party, the RPR, which was an important factor leading to disgruntlement in the party. On the left, the Socialist prime minister, Lionel Jospin, was in an unassailable position as the leading presidential candidate for 2002.

The campaign

Abstentionism in EP elections has always been higher than in nearly all other nation-wide French elections (legislative, regional, departmental or cantonal, and municipal), the second ballot of 1988 cantonal elections being an exception due principally to voter fatigue following the presidential and legislative elections of the same year. The rate of abstention reached a record 51.2 per cent in 1989, although it declined to 47.2 per cent in 1994.[7] Only a handful of French referendums have exceeded this percentage. The rise of abstention parallels that for other elections, although is generally higher. The various reasons for the increased abstention rates are magnified for the EP elections.[8]

In an effort to combat this apathy, the EP organised the most expensive European-wide campaign yet (5.5 million euros) to bring out the vote. The official campaign slogan of the French office of the EP was borrowed from Descartes: 'I think, therefore I vote.' Its campaign targeted specifically the young (18–24 year olds) and women, who in the past have been less likely to vote. The EP office ran film trailers, distributed brochures and placed posters in train stations throughout the country.[9] It also organised a petition – signed by leading French intellectuals and well-known personalities – published in national newspapers (11 June) inciting the population to vote in the forthcoming elections. The Ministry of European Affairs also organised a campaign to encourage participation – themed 'To vote is to exist' – using television and radio advertisements. For the first time, the French office of the EP placed a campaign website on the Internet, providing detailed information about the EP and the organisation of the elections.

Public apathy cannot be blamed on the party lists which campaigned tirelessly. Financing relatively large campaigns is not a problem for the larger lists, as the state provides a refund for up to 58.8 million francs for lists winning 5 per cent of the vote. For the first time the French elections were fought in the realm of cyberspace, as most of the lists invested a great deal in websites and many hoped and argued that the increased use of the Internet would help mobilise the population. The French public also had numerous opportunities to follow the campaign through interviews and debates in newspapers, on radio and television, including a televised debate on 8 June involving the participation of most of the major list leaders. There was an attempt to attract voter attention by the Socialists who kicked off their campaign on 27 May with a high-profile Paris rally of European Socialist prime ministers and party leaders including Tony Blair, Gerhard Schröder and Massimo d'Alema. None of the other lists organised such a rally, in large part due to a lack of unity with other European leaders and the comparative paucity of well-known party leaders. Party infighting may also have attracted the public's attention. However, it also disillusioned erstwhile supporters and discouraged them from voting. This was notably the case for the FN and FN–MN, which spent more time attacking each other rather than the other lists.

The campaign was noteworthy for the relative absence of the President. Although Chirac unofficially supported the RPR–DL list and was omnipresent in his backstage manipulations, he announced officially that he would campaign for no party. Prime Minister Jospin allowed the Socialist-led list to be run by the PS president, François Hollande. Nevertheless, Jospin was actively involved in election campaigning. The former Prime Minister, Michel Rocard, whose Socialist list had performed badly in 1994, remained a prominent figure in these elections. Former RPR prime ministers Balladur and Juppé largely avoided the campaign, initially because of their poor relations with Séguin and then because they recognised the real threat to the RPR and preferred to watch their younger ambitious colleague, Sarkozy, damage himself. Pasqua excited some interest in the campaign, particularly among the police and security forces with whom he was popular, and encouraged the anti-Europeans among RPR supporters to vote. Because of the permitted accumulation of electoral mandates (*cumul des mandats*), EP election lists in France tend to distinguish themselves from those in other EU countries by the inclusion of a large number of party leaders except for government ministers, who often none the less play an active role in the campaign. The Socialists – sensitive to complaints about the high level of French absenteeism in the EP and demands from party activists – adopted a rule against holding National Assembly and EP seats simultaneously. However, given the paucity of suitable high profile list leaders they were forced to break this rule. Providing considerable inspiration in second place on the Socialist-led list was Pervenche Berès, an incumbent PS MEP who had been severely disfigured in a scooter accident in 1995.

Numerous non-party political stars were added to lists to attract voter attention and votes. The most flamboyant and eloquent campaigner proved to be foreign: Daniel Cohn-Bendit. However, he was forced to modify his style somewhat as the campaign progressed because the Greens started dropping in the polls as his penchant for provocation began to annoy some potential voters. The UDF list relied to a great extent upon the self-satisfied General Philippe Morillon, who had led NATO troops in Bosnia and thus attracted ex-soldiers wherever he campaigned. Fodé Sylla, the former president of SOS-Racism, campaigned with Robert Hue's list, which sought to attract more of the ethnic minority and non-Communist vote. The star of the FN list, other than the indefatigable Le Pen, was Charles de Gaulle, an incumbent Other Europe (de Villiers's list) MEP who shocked his family and the neo-Gaullists by his move. However, other than his family name, de Gaulle possessed little else to win votes.

The issues

The debate on European integration rarely entered into any details about the precise shape of a future EU. Séguin and Sarkozy argued in favour of a pragmatic 'Euro-realist' approach to future developments rather than any grand vision. Séguin went so far as to claim that the centre right should not go to battle over European issues in the EP elections because this was a matter for legislative and presidential elections given that national governments, not the EP, had power over the evolution of the EU (*Le Nouvel Observateur*, 22–28 April 99). Séguin, the Euro-sceptic in charge of a moderately pro-European list, had good political reasons to make such a claim. Three leading lists argued in favour of a European constitution (UDF, Socialist-led and Green). The Socialists had already presented the idea of a constitution in the 1994 election campaign as part of their stated goal to strengthen the democratic legitimacy of Europe. However, no votes were to be won by entering into details: these lists simply argued in favour of a clarification of the powers of the EU institutions, the member states and the regions, and the inclusion of the European Convention of Human Rights. On institutional reforms, they tended to remain equally vague, focusing upon improved democratic legitimacy.[10]

Apart from the more general issue of European integration, only one major EU issue had considerable political saliency in France: Kosovo and the need to reinforce EU defence policy to handle such matters independent from the Americans. There is widespread agreement in France on the need to decrease American dominance in defence matters. However, there remains considerable disagreement as to the reinforcement of a European defence policy. All the lists sought to manipulate the public's perception of the Kosovo crisis to their advantage. The UDF, Socialists and Greens argued strongly in favour of NATO bombardment and claimed that the crisis demonstrated the need to reinforce the European defence pillar and improve diplomatic co-ordination. Daniel Cohn-Bendit was the only major list leader to argue in favour of the early deployment of NATO ground troops to prevent ethnic cleansing. However, his own party – which consisted of a large number of pacifists – was intensely divided on the issue. These lists claimed that it was contradictory for the Euro-sceptic lists to argue against American dominance in NATO and European defence matters, yet not challenge this dominance effectively through the establishment of an effective European defence (*Le Nouvel Observateur*, 22–28 April 99). The government was potentially riven with division on the issue of Kosovo, with the MDC and Communists strongly opposed to intervention. However, this division was contained successfully by Jospin who allowed a public internal government debate on appropriate French action. The RPR was in a difficult position over Kosovo, as Séguin was publicly opposed to President Chirac's position on French participation in the NATO bombardment.

Another issue of contention, upon which the CPNT focused its campaign, concerned a series of European and French laws and court decisions restricting hunting rights.[11] More generally, the CPNT argued in favour of the protection of national rural traditions and the rural poor against Parisian and European technocratic diktat and Green 'fundamentalism'. The CPNT also set out to challenge the Greens directly, with some of the more violent party activists forcing the cancellation of several Green meetings. The potential success of the CPNT forced the other lists on the right to address hunting issues in order to avoid losing votes. Leading Socialists revelled in the opportunity provided to weaken the Greens and blamed the Green Environment Minister for being inefficiently sensitive to the concerns of the rural population.

The Commission's forced resignation and fraud were largely ignored in the partisan debate, probably because the reasons behind the resignation were considered less significant in France than in some of the other member states. At the European level, only the European Liberal Party group – which had just one French MEP – focused upon the Commission's recent downfall. The RPFIE, FN and FN–MN all complained about the problems associated with Commission technocracy, but apart from an initial debate among the parties in the immediate aftermath of the crisis, the issue of Commission reform was not the subject of partisan debate. Likewise, enlargement failed to spark much interest in the campaign despite the potential impact upon French farmers. Moreover, EMU and the single currency did not excite much debate, except insofar as they denoted the move towards a federal Europe. With the French economy performing relatively well and unemployment starting to drop, it was difficult to target EMU on economic grounds. None the less, the extreme right and left and Move Europe lists argued against EMU as an element contributing to economic liberalisation at the expense of French workers.

Differences on European integration corresponded largely to deeper ideological differences concerning the appropriate role of the nation-state and the degree of acceptable decentralisation, and the economy. The more pro-European lists tended to be the most in favour of greater decentralisation. Opponents to further integration also tended to be hostile towards economic liberalisation, the most virulent being the

RPFIE and the extreme right and left lists. In addition to this division, the pro-European Socialists, Left Radicals and Greens divided, with the more liberal right claiming to defend a 'Social Europe' with more Keynesian style infrastructural projects and classic French *volontarisme*. The Socialist platform called for the negotiation of a 'Social Treaty' which would include convergence criteria of precise goals for the European countries to meet. This reflected the priority of protecting national social rights from the impact of competition from other EU countries with lower social expenditures and greater labour market flexibility. It also reflected the French preoccupation with a persistently high rate of unemployment and the preference of the Socialist-led government for interventionist job creation strategies. The list's platform included the establishment of a 35-hour week throughout Europe. Under the leadership of Philippe Séguin, the RPR–DL list presented a balancing act of images: its leader's reputation was Euro-sceptic and less liberal, while the DL leader was pragmatically pro-European and liberal. Séguin's presence prevented many neo-Gaullist Euro-sceptics from moving to the Pasqua camp. With Séguin's resignation and the appointment of Sarkozy, the image of lists became decidedly less populist and more liberal.

The transnational party group manifestos played a minimal role in the French campaign. They were not distributed, and neither were they normally referred to during the French campaign. Members of four French parties played a leading role defining the agenda of and drafting the election manifesto of their EP party group. The European Socialist Party manifesto was written under the authority of the former French Socialist minister Henri Nallet and the British foreign minister Robin Cook. This involved a great deal of cutting and pasting in an effort to find a version acceptable to the French, Germans and British. Even though the French formed a small proportion of the Socialist seats, the assumption was that a common platform agreed upon by the ideological poles of the party group (the British and the French) would be acceptable to the rest. The manifesto, the first ever for the Socialist Party Group, was signed by the national party leaders in Milan on 1 March. This represented their minimal common campaign positions. It addressed several French preoccupations, including a commitment to establish a European Employment Pact and an economic growth strategy, and to protect social rights. However, the differences between the parties, eleven of which were now in government, were considerable. On budget and CAP reform they de-aligned according to national interest. Also, it was clear that some European socialist parties were more in favour of integration and 'federalism' than others, with the French falling towards the less integrationist side of the spectrum. The Socialists' proposal for a more Keynesian-oriented European Employment Pact was rejected yet again by the June Cologne European Council in favour of one which effectively continued the previous policies, emphasising flexibility and adaptability. Moreover, the agreement of the Blair–Schröder pact for a 'New Centre' and a 'Modernised Europe' publicly excluded the Jospin Government which could not officially embrace such a market-oriented vision. These two developments took place only a few days prior to the June elections, allowing right-wing opponents to claim that French Socialists were isolated from the more 'progressive' left-wing parties in Europe, the policies of which more closely approximated those of the mainstream French right. However, it is unlikely that the French Socialists would lose any votes for appearing too left-wing!

The French had a dominant presence in only three of the minor lists: the UFE party group, with the RPR MEPs constituting half of its 34 MEPs; the federalist European Radical Alliance, dominated by Left Radical MEPs elected on Tapie's defunct list and for the most part destined not to be re-elected; and the Independents of a Europe of Nations party group, dominated by the MEPs from de Villiers's list, now participating in the RPFIE. French MEPs dominated the preparation of the three groups' platforms, although none participated in any significant transnational campaign. It had been mooted in RPR circles that its MEPs should join the EPP group, the second largest, in order to increase their influence in the European Parliament. The RPR had previously resisted joining the EPP as many leading members disliked the Christian Democratic label and would have difficulty belonging to a party group which, according to its charter, was a federalist organisation. The decision by the British Conservatives to join the EPP, however, had undermined the logic behind this refusal. The organisation of the RPFIE list and the resulting decline of the Euro-sceptic presence in the RPR list, together with the co-operation with DL, further encouraged a reconsideration of EPP membership.

UDF MEPs participated in the EPP, but with only small groups of MEPs in this the second largest European party group – dominated by Germans and the Italians – they did not weigh very heavily in the organisation of its transnational campaign. The French Greens joined with their European counterparts to announce on 25 February a common campaign and a minimum common platform. However, the national differences among the Green parties were simply too great and, in practice, they maintained a separate national campaign and a distinct platform.

The results

The most noteworthy result, particularly in terms of the democratic legitimacy of the European Union, was the abysmally low voter turnout. The rate of abstention reached 53 per cent (52.28 per cent in metropolitan France and 78 per cent overseas), 6.5 per cent higher than in 1994 and 1.7 per cent higher than 1989 and the second highest for any election ever held in France.[12] The EP's increased publicity was not an effective investment.[13] The rate of abstention was predictably highest among those expressing no party preference (82 per cent).[14] Abstentionism affected all the parties, although unequally. It was highest among supporters for the extreme right (58 per cent) due principally to disillusion with the fighting between the FN and FN–MN. Abstention was also surprisingly high among Green and Socialist supporters (52 per cent and 49 per cent) and above that for the divided RPR. It was lowest among Communist sympathisers (23 per cent), in large part due to a very active mobilisation of the vote during the campaign. Abstention was much lower among those that voted 'Yes' in the Maastricht referendum (42 per cent) than those who voted 'No' (56 per cent), which helped the pro-European parties. A great surprise was the very high rate of abstention among farmers, at 68 per cent the highest of any major socio-professional grouping. In previous EP elections, farmers were by far the least likely to abstain (31 per cent in 1989 and 34 per cent in 1994). This high rate can perhaps be explained by the general disillusion with the CAP and hostile resignation in agricultural communities. Another surprise was the relatively low rate of abstention (47 per cent) among the youngest voters (18–24 year olds). This arguably represents a victory for the EP office in Paris which targeted these voters in its campaign to encourage participation. However, it is also clear that the youngest voters were disproportionately drawn to the Green list and, principally, its provocative leader. Also unusual was the relatively high rate of abstention among retired people (52 per cent), normally the most likely to vote. This adversely affected the Extreme Right.

A second manifestation of voter disaffection was the high percentage of blank and spoiled votes (5.9 per cent or 1.1 million in total) which exceeded all previous elections in France. The number of such votes had increased from 2.8 per cent in the 1989 EP elections, to 5.3 per cent in 1994. See Table 10.1 for the results.

Table 10.1 Results of the French EP Election, 13 June 1999

List	List Leader	Votes	% of votes	Seats
Total Right		6 206 078	35.16	
RPFIE Rally for France, Independence from Europe	Pasqua	2 304 106	13.05	13
RPR-DL Rally for the Republic – Democratic Liberale	Sarkozy	2 263 415	12.82	12
UDF Union of French Democrats	Bayrou	1 638 557	9.28	9
Total Left		6 785 168	38.44	
PS – MRG – MDC Socialists – Left Radicals – Citizens Movement	Holland	3 873 728	21.95	22
Greens	Cohn-Bendit	1 715 284	9.72	9
Move France (PCF + others)	Hue	1 196 156	6.78	6
Extreme Right		1 583 860	8.97	
National Front	Le Pen	1 005 113	5.69	5
FN-MN National Front-National Movement	Mégret	578 747	3.28	–
Extreme Left Workers' Struggle + Revolutionary Communist League	Laguiller	914 657	5.18	5
Other*		2 161 797	12.25	
Hunting, Fishing, Nature and Traditions (CPNT)	Saint-Josse	1 195 500	6.77	6

*Apart from the CPNT none of the other lists won more than 2% of the vote.
Eligible to vote: 40 144 816 (including French expatriates)
Number of voters: 18 766 582
Valid votes cast: 17 651 560
Abstentions: 53.25% (52.28% in metropolitan France and 78.02% overseas).
Source: *Le Monde*, 16 June 1999.

The elections were considered a relatively impressive victory for the 'Plural Left' Government, contrary to the tendency for governing parties to fare badly in European elections and the results in Germany and the UK. The seat total increased slightly from 35 to 37, even though the parties of the government lost votes from 1994: 38.5 per cent compared to the 40.9 per cent in 1994, with votes sliding to the extreme left.[15] In spite of divisions and party infighting, the right's seat total also increased from 31 to 34, even though its vote dropped (to 35 per cent versus the 37.9 per cent in 1994 for the joint UDF–RPR and de Villiers lists). These contradictory results were due to the division of the FN which lost seats. The elections saw a further fragmentation of the vote with the top two lists winning only 35.1 per cent versus 40 per cent in 1994, and the top five winning 66.8 per cent as opposed to 74.9 per cent in 1994. More than 49 per cent of voters chose to support lists other than those dominated by the four historically largest parties (PS, PCF, UDF, RPR). However, it is important to note that the fragmentation of the vote did not help the ten smallest lists, which received a smaller percentage of the vote in 1999 compared to 1994 (5.5 per cent versus 6.4 per cent).

The majority of the counted votes (53.7 per cent) was for lists with a moderately positive or very positive position in favour of limited further integration (PS, Greens, RPR–DL and UDF),[16] a substantial drop from the 1994 elections (57.4 per cent). However, these figures are somewhat misleading as several 'pro-European' lists included more hostile elements (notably the UDF–RPR lists in 1989 and 1994 and the PS list in 1999). MDC deputies, for example, voted against the ratification of the Amsterdam Treaty. Regional voting, in terms of support for pro-European lists, conformed to previous elections and the Maastricht Referendum. In metropolitan France, the highest support for these lists was in Brittany (62.4 per cent) and Alsace (59.2 per cent).[17] The lowest support for these lists was in Picardie (43.9 per cent), where the CPNT performed well, and Provence-Alpes-Côtes d'Azur (45.7 per cent), where support for the RPFIE and the extreme right was high.

François Hollande and the government presented the vote for the Socialist-led list as a great success: well above Michel Rocard's score in 1994 (14.5 per cent), nearly at the level of Jospin's first ballot vote in the 1995 Presidential elections and well above the Socialists' nearest rival. This was also the first EP election with a Socialist-led list in first place, a position previously secured by joint RPR–UDF lists. However, it is problematic to make such comparisons. Rocard did poorly in 1994 because of rival lists run by the MDC and Bernard Tapie (Left Radicals), which attracted many Socialist votes.

The Green victory was more impressive, although again not unqualifiedly so. Public support for the Greens had actually declined somewhat in the polls during the unofficial campaign but they profited from the dioxin scandal in Belgium and the Netherlands in the weeks prior to the elections. Their vote was below that of 1989 (10.6 per cent and eight seats) but was well above the combined Green vote in the 1994 elections (4.95 per cent) and the preceding Presidential and National Assembly elections. The good score led many Greens to demand an additional ministerial portfolio and a change in Government policy on illegal immigrants and nuclear energy. More surprising was the relatively poor showing of the Move Europe list. Despite having opened itself to non-Communists, investing a great deal of money and modifying its message, the list won 6.8 per cent of the vote, slightly less than the vote for the PCF in 1994 (6.8 per cent) and considerably less than Hue's first ballot vote in the 1995 Presidential elections (8.5 per cent). On the extreme left, the joint list LO–LRC exceeded the 5 per cent threshold and won seats for the first time in the EP. However, the 5.2 per cent – which barely exceeded that won by list leader Laguiller in the first ballot of the 1995 Presidential elections – disappointed many party activists.

Many expected that the RPR–DL list would drop below the disastrous RPR 1997 National Assembly result (15.5 per cent). However, the drop exceeded the worst predictions. The vote was a major defeat for Sarkozy, whose chances of maintaining the party leadership were seriously weakened. Given the relative success of the UDF, it was argued that Alain Madelin and his DL made a strategic mistake in siding with the RPR. Most were equally surprised that the RPFIE beat the RPR, winning 0.38 per cent more of the vote and one more seat. Pasqua's victory boosted his decision to declare a new political party, the RPF (Rally for France), which was the most significant medium-term impact of the elections upon the national party system. The claim of a great victory of the UDF is somewhat legitimate. Its share of the vote was considerably lower than in the 1997 National Assembly elections, but this was due to the departure of two of the UDF component parties led by Madelin and de Villiers. The poor showing of the two extreme right lists was somewhat to be expected. Mégret's list performed particularly badly and, not having reached the 5 per cent threshold, faced the prospect of bankruptcy without large contributions from its members. Le Pen had won 10.5 per cent of the vote in 1994, with 11 seats, while the combined extreme right vote had dropped by 1.5 points due in large part to the high rate of abstention. The CPNT scored its greatest victory to date in any nation-wide election, performing particularly well in the rural south-west. This was a considerable increase from its vote in 1989 (4.1 per cent) and 1994 (3.9 per cent). This success encouraged the CPNT to transform itself into a larger 'rural' movement with the aim of extending its power base in rural areas in the municipal elections of 2001.

The elections were also construed as a victory for women. The number of French female MEPs jumped from 26 (out of 87) to 35, still less than half, but a larger percentage than for any French assembly. The left-wing lists were very close to achieving parity and the mainstream right-wing lists had more women elected. Women outnumbered male MEPs only on the extreme left (four out of five).

The results in France had a limited impact on the European party groups. The success of the Socialist-led list and the poor showing of their counterparts in Britain and Germany would tend to increase French influence within the European Socialist Party (with 22 out of 180 group seats – including the MDC and Left Radical MEPs – as opposed to 16 out of 214). However, one should doubt claims that the poor showing of the left in Britain and Germany suggests a rejection of the policy recommendations of the Blair and Schröder governments and the likelihood of greater ideological influence of the French Socialists at the European level. None the less, as a manifestation of their increased weight in the party group, Michel Rocard was subsequently appointed to lead the EP's Employment and Social Affairs Committee, which examined issues of particular political and ideological importance to the French Socialists. This was France's only committee presidency.

The French EP elections had a much more significant impact upon the European Radical Alliance which, because of the departure of its French MEPs, ceased to fulfil the criteria of a European party group. The two remaining left-wing Radical MEPs joined the Socialist party group. The French were also in a position to play a more important role in the Green party group in which they now had the largest number of members. However, this would not necessarily change the focus of the group. The French RPFIE MEPs also maintained their dominance within the marginal Euro-sceptic party group, now labelled the Union for a Europe of Nations. William Abitbol, the chief adviser to Charles Pasqua, became vice-president of the Committee on Economic and Monetary Affairs.

Alone, the UDF MEPs would be a very small group in the EPP. The subsequent decision by the Union for Europe MEPs to join the EPP would give the French increased influence within the party group (21 out of 224 seats) although they would remain the fifth largest national delegation.[18] The presence of the UFE MEPs in the EPP would also work to dilute further the 'federalist' line of the party group, already undermined by the presence of the British Tories. An RPR MEP (Roger Karoutchi) led the French delegation to the EPP for the first time, and his party succeeded in assuming two of France's four committee vice-presidencies: a rather impressive presence given that the RPR only won four seats![19] With the departure of the RPR MEPs, the UFE party group collapsed as it was unable to fulfil the criteria as a party group. Nicole Fontaine, a UDF MEP and former EP vice-president, profited from her pro-European credentials and the EPP victory, and was positioned as the leading candidate for EP president.

Conclusions

The election results will probably not have much of an impact on the Jospin Government's European policy. The reticence to accept substantial further integration is shared by all the political parties with the possible exception of the UDF, which is always contained within more reluctant government coalitions. Likewise, despite the widely shared interest in the creation of a common European defence, none of the French parties recommends the surrender of the national veto in this area. The slightly enhanced position of the Greens within government will probably not have any major impact upon the government's European policy.

On the right, the relative success of the 'sovereignists' and the creation of a new party does not necessarily mean that future centre right governments will be more reluctant to accept further integration. Hostile RPR and UDF ministers have been previously contained within centre-right governments. The major difference is that a party dedicated to blocking further integration might be one – and perhaps the largest – of the three main blocs in a future centre-right government. This might require this government to assume a somewhat more hostile approach to further integration, in rhetoric if not in practice. However, opposition to previous developments does not indicate that Pasqua and other RPF politicians would ever advocate any backtracking on the *acquis commuautaire*. Unlike the extreme right, the RPFIE has never advocated the renegotiation of the Maastricht Treaty.

One can speculate about the impact of the election results upon future developments in French politics. The establishment of the RPF has transformed the party system and might further undermine support for both the RPR and the FN. The departure of many of the more Eurosceptic and populist members of the RPR, and the decision by the RPR to join the EPP group, suggest the possibility of an eventual unification of the RPR, DL and the UDF. However, even if ideological differences are less pronounced, the ambitions and personal differences of the party leaders might be too great to overcome division. Given the two ballot majoritarian electoral system, the presence of a second neo-Gaullist party does not necessarily weaken the chances of the centre right in the next elections.[20]

The EP elections saw the erosion of the hard core of President Chirac's electorate, with only 28 per cent of those voting for Chirac in the first ballot of the 1995 presidential elections remaining with the RPR.[21] Even though the President remained popular in the opinion polls, the erosion of his vote did not bode well for his re-election bid in 2002. However, a great deal can change in French politics in three years! The high rate of abstention temporarily sent alarm bells ringing in the EP's French office. Had abstention clearly disadvantaged the largest parties, the possibility of a dramatic reconsideration of election strategy would be greater. However, abstention affected the Greens and the Extreme Right even more. Parties will seek to get their vote out, but given the already high level of campaign activity it is difficult to see what other tactics they can employ. The claim that the personalisation of politics and the participation of party leaders will normally increase public participation suggests that the adoption of the new rules on the accumulation of mandates – thus blocking the candidacy of the more famous politicians – would increase the rate of abstention even further.

Except for the very small possibility of significant steps to further integration which might motivate voter participation, it appears unlikely that abstention will decline considerably in the next elections. There is an intractable divide between interest in the EU and European integration, on the one hand, and lack of interest in the EP elections, on the other. The remedies are not particularly clear and no one appears to be looking very hard. Some have suggested the impact of proportional representation and the alienation of the population from national lists as part of the explanation for the consistently high rate of abstention. However, attempts in 1998 to introduce a bill to create regional lists for the EP elections met with determined opposition from the smaller parties in the government.[22] Given the rising rate of abstention in countries with regional based lists, and rising abstention rates for all French elections, it is unlikely that such tampering will address a more fundamental malaise which undermines the democratic legitimacy of the governing institutions of both the EU and France.

Appendix 1: Lists and leaders

Lutte Ouvrière–Ligue Communiste Révolutionnaire (Workers' Struggle–Revolutionary Communist League, LO–LRC): Arlette Laguiller
Bouge l'Europe (BE, or Move Europe, PCF, etc): Robert Hue
Les Verts (L'Ecologie, Les Verts: Daniel Cohn-Bendit et Dominique Voynet) (Greens): Daniel Cohn-Bendit

Construisons Notre Europe (Let's Build Our Europe, Socialists, MDC, Parti radical de gauche (PRG): François Hollande
Union pour l'Europe, l'opposition unie avec le RPR et DL (Union for Europe, RPR–DL): Nicolas Sarkozy
Avec Europe, Prenons une France d'advance (Union for French Democracy, or UDF) (With Europe Let's Move France Forward): François Bayrou
Rassemblement pour la France et l'Indépendance de l'Europe (RPFIE) (Demain la France/Mouvement pour la France) (Rally for France and Independence from Europe): Charles Pasqua
Front National (LE PEN – Front National avec Jean-Marie Le Pen, pour une France libre changeons d'Europe) (National Front, or FN): Jean Marie Le Pen
Européens d'accord, Français d'abord, Mégret l'avenir (Europeans OK, French First, Mégret the Future Mouvement National, or MN): Burno Mégret
Chasse, Pêche, Nature, Traditions (CPNT) (Hunting, Fishing, Nature, Traditions): Jean Saint-Josse

The smaller lists

Mouvement Ecologiste Indépendant–Ecologisme: le choix de la vie (Independent Environmentalists–Environmentalism: the choice of life): Antoine Waechter
Combat pour l'emploi (Combat for Employment – The Union for a Four Day Week): Pierre Larrouturou
Vive le Fédéralisme (Long Live Federalism): Jean-Philippe Allenbach
mi ou, mi mwen (Martinican List): Joseph Jos
Moins d'impôts maintenant! Liste indépendante des partis soutenue par les contribuables français (Less Taxes Now!): Nicholas Miguet
Liste de la Ligue nationaliste (Nationalist League List): Guy Guerrin
Politique de vie pour l'Europe Collectif Associations Citoyennes (A Life Policy for Europe, collective citizen associations): Christian Cottene
Mouvement Vivant Energie–France (Living Energy List–France): Gérard Maudrux
Liste du Parti Humaniste (Humanist Party List): Marie-Laurence Chanut-Sapin
Liste du parti de la loi naturelle (Natural Law Party List): Benoît Frappé

Notes

1 As an indication of this fragmentation, the vote for the top two lists had dropped from slightly less than 60% in 1979, to around 50% in 1989 and 40% in 1994.
2 Other possible motives for Séguin's resignation include his opposition to Chirac's support of French participation in the NATO bombing of Serbia and Kosovo, in addition to the danger of a successful Pasqua list.
3 Although the UDF imposed a specific list of European policy demands as its condition for a joint list, it is likely that its decision to maintain a separate list was motivated as much, if not more, by political considerations.

4 Prior to the 1 March poster ban, Pasqua's supporters put up 10 000 4 m by 3 m posters of himself proclaiming: 'Left; Right: Let us march together against Euroland'.
5 Max Gallo, the well-known author, leading member of the left-wing MDC and a former Socialist government spokesperson, announced his support for Pasqua–de Villiers list on 8 June. Didier Motchane, another leading MDC intellectual, also announced his support for the RPFIE list.
6 The UDF and RPR–DL presented a superficial respect of parity with, respectively, 45 and 48 female candidates. However, on these lists, there were proportionately fewer women in the higher places (although more than previously). While it refused to respect the principle of parity, the RPFIE none the less placed several female candidates in higher places. Only the extreme right and CPNT lists made little effort to increase the presence of female candidates substantially (especially in electable places). Only the extreme left list, LO–LRC, was led by a woman, the indefatigable election veteran, Arlette Laguiller.
7 The rate of abstention in 1979 was 39% and in 1984 nearly 43%.
8 For a good summary of the research on abstentionism applied to the EP elections see Pascal Perrineau, 'L'abstention du 13 juin démontre l'ampleur du malaise démocratique', *Le Monde*, 1 July 1999.
9 During all of May, the EP ran a trailer in French cinemas encouraging the young to vote. A brochure about Europe and women was distributed by the Ministry of Women's Rights. Posters advertising the elections were placed in 72 train stations, in the Paris metro and airports in the DOM. During Ascension and Pentecost weekends, brochures about the elections with a prize winning game were distributed at motorway toll booths.
10 The UDF called for the creation of a European president elected initially by the EP and national parliaments and later by universal suffrage and a strengthened EP. The Socialist Minister for European Affairs, Pierre Moscovici, argued, in a personal capacity, in favour of a functionalist approach with a reform of the institutions to improve their efficiency, through the increased use of QMV, a new weighting of national votes and the election of an EU president for 2 or 3 years. Moscovici also argued that it was necessary to build up the legitimacy of the EU prior to giving it a Constitution. He sought a Charter of Civic and Social Rights which would regroup the existing rights in Europe and define new ones: the right to health care, housing, education and minimum revenue (*Le Point*, 6 March 1999, no. 1381).
11 Specifically, the CPNT focused upon the May decision of the French Council of State to ban the night hunt of the water wood pigeon to conform with European directives (this followed bans on the night hunting of other species); the rejection by the European Human Rights Court of the Verdeille Law permitting hunting on private property of less than 20 hectares; the restrictions placed on hunting turtle doves; and the decision to shorten the hunting season in some departments.
12 Abstention rates were highest in the department of South Corsica (69.09%) and lowest in the rural Corrèze department (43.47%), that of the President and where Hollande, the head of the Socialist-led list, had his National Assembly seat.

13 Given lack of space, the reasons for this high rate of abstention are not explored here.
14 All the following information is drawn from a Sofrès poll, cited in *Le Monde*, 15 June 1999.
15 In 1994, the votes for left-wing parties were Rocard (Socialists), 14.49%; Tapie (Radical Energy), 12.03; Wurtz (PCF), 6.88; Isler-Béguin (Greens), 2.94; Chevènement (Other Policy), 2.54; Lalonde (The True Environmentalists), 2.01.
16 This does not include the Move Europe list which presented a less hostile line on integration than the PCF position in 1994.
17 Support for these pro-European lists was even higher in the four overseas regions/departments, reaching 71.22% in Martinique and 74.22% in Réunion.
18 DL MEPs were previously members of the EPP.
19 Marie-Thérèse Hermange, Vice-President, Committee on Employment and Social Affairs; and Hughes Martin, Vice-President, Committee on Fisheries. Joseph Daul, a member of the RPR–DL list, and leading member of the French agricultural union (FNSEA), also became a Vice-President on the Committee on Agriculture and Rural Development. The other candidates elected on the UFE (RPR–DL) list belonged to the DL or smaller right-wing parties.
20 See Jean-Claude Casanova, interviewed in *Libération*, 12 August 1999.
21 Sofrès, *Le Monde*, 15 June 1999. In fact 21% voted for Pasqua, 17 for Bayrou, 10 for the extreme right and 6 for the CPNT.
22 On 1 July 1998, the Government chose to reject a bill to create regional lists for the EP elections. This would have made the French system similar to that in place in Germany and the new system introduced in the UK. The Greens and Communists feared that such a system would dramatically diminish their seats because the threshold to elect MEPs would invariably be higher in regional elections.

References

Benoit, B. (1997) *Social-Nationalism: An Anatomy of French Euroscepticism* (Aldershot: Ashgate).

Gaffney, John (1996) 'France', in Juliet Lodge (ed.), *The 1994 Elections to the European Parliament* (London: Pinter).

Guyomarch, Alain, et al. (1998) *France in the European Union* (London: Macmillan), esp. Chapter 3, 'Parties and Public Opinion', 73–103.

Information provided by the Paris information office of the European Parliament (*Bureau d'Information pour la France du Parlement européen*).

Robert, Dominique (1999) EP official responsible for the organisation of the EP elections campaign, response to written questions to the Paris information office of the European Parliament, 25 August.

Other French party abbreviations used

DL	Liberal Democracy
FD	Democratic Force
FN	National Front
MDC	Citizens' Movement
PCF	Communist Party of France
PS	Socialist Party
RPF	Rally for France
RPR	Rally for the Republic
UDF	Union for French Democracy
UFE	Union for Europe

11
Ireland

Edward Moxon-Browne

Introduction

It has become customary to regard EP elections as 'second-order' elections in the sense that their function is to provide an indication of the current popularity of political parties and politicians between general elections.[1] In Ireland EP elections are taken seriously by the political parties since they are normally fought on domestic issues and the results are therefore perceived as important political barometers from which accurate assessments of the public mood can be ascertained. Voters, on the other hand, appear to take an increasingly relaxed attitude towards EP elections. First, turnout is generally lower than for national elections. For example, while the national election in 1992 attracted a 68 per cent turnout, the 1994 EP election persuaded only 44 per cent of the electorate to go to the polls. In fact, in Ireland, EP elections have generally been held simultaneously with another test of public opinion: local elections in 1979; a referendum in 1984; and a general election in 1989. The apathy that attends EP elections in Ireland is partly a reflection of the fact that many European issues are still distant from the minds of voters or too abstruse to be neatly expressed in sound bites or slogans. But the apathy is also due to the fact that the three main parties agree over the principle of EU membership and differ only on tactics. On major issues of national interest, there is little disagreement: agriculture; regional policy; neutrality; and institutional reform. As one newspaper headline in 1994 put it: 'Euro-candidates struggle to bury their similarities'. To the extent that Irish voters do participate in EP elections, participation has more to do with local personalities, local issues and intra-party rivalries than with the broad issues of European integration around which the EP election is reputedly focused.

The function of EP elections in Ireland appears to be to provide a political stage on which both the parties and individual politicians can compete for approbation. The European language of the campaign has to be seen, therefore, as a convenient alibi behind which the real issues of personality and political preference are played out. The voters and the candidates collude in this rhetorical charade. Speeches on monetary union, neutrality, beef exports and structural funds are made and listened to, with tacit acknowledgement on both sides that the words are merely the currency of political exchange but that the real issues on which people vote, or abstain, lie elsewhere and are seldom articulated. This is not to say that the parties do not express divergent views on EU issues in these campaigns. Party manifestos do put forward a variety of divergent, and even conflicting, views but these differences do not, in themselves, sway the vote.

Another widely accepted view is that EP elections allow voters to 'punish' the government without actually changing its political complexion. Looking back at the elections of 1979, 1984 and 1989, the percentage of the vote received by the government of the day was substantially lower than that received in the previous (in 1989, simultaneous) general election. If we accept this argument, and we shall shortly suggest why we may not, its corollary is that small parties usually do better than in the analogous general election since voters are prepared to 'take a risk' with their European ballot paper in a way that is not the case with their general election choice. EP election results may not be an accurate indication of general election performance. In 1989, when the EP and national elections were held on the same day, nearly half the voters voted for different parties in the two elections.

It may be time, however, to question the conventional wisdom that EP elections are mainly mid-term tests of the government's popularity. Opinion polls before the EP election in 1999, and indeed the results themselves, showed that the ruling Fianna Fail party had not suffered at all in terms of popularity despite serious scandals relating to the probity of leading politicians. The view that voters pass judgement on governments in EP elections is actually very dubious. Although Fianna Fail's poor showing in the first EP election in 1979 was seen, at the time, as a rebuke to the government, with hindsight Fianna Fail's performance in that election was broadly in line with its performance in the subsequent three EP elections. There has been a general tendency for Fianna Fail to do worse in EP elections than in national elections irrespective of whether it has been in government or not. Fianna Fail's relatively good showing in the EP election is not evidence that the party has thrown off its European albatross. Just as it was an over-simplification to interpret Fianna Fail's EP performance to date as a judgement on its role in government, so it would be a mistake to regard its currently promising performance in the EP election as an endorsement of the current government's policies or probity.

The actors in the campaign

For the purposes of the EP elections in 1999, the Republic of Ireland was divided into four constituencies: Connacht-Ulster (three seats) with 11 candidates standing; Dublin (four seats) with 13 candidates; Leinster (four seats) with eight candidates; and Munster (four seats) with ten candidates. The principal political parties contending for these seats were Fianna Fail, Fine Gael, the Labour Party, the Green Party, and Sinn Fein. There were also a number of independents, two of whom were eventually elected. Before considering the range of independent candidates, it may be worth depicting briefly the policies each of these parties was putting before the electorate.

Fianna Fail indicated that the twin challenges of enlargement and the adoption of the euro would require careful management if the major opportunities they offered were to be satisfactorily exploited. Emphasis was placed on balanced regional development in Ireland; the modernisation of the road and rail network to European standards by 2005; the reform of the CAP in a 'manner that which protects the integrity of the family farm unit'; and improvements in the Rural Environment Protection Scheme (REPS). In foreign policy, Fianna Fail pledged itself to negotiate Irish membership in the Partnership for Peace (PfP) 'as a logical extension of our existing policy, and not as a departure from it'. Such a decision would not have any implications for Irish neutrality, it was argued, and there was no reason why Ireland should become 'more neutral than all the other neutral countries of Europe'. Fine Gael reiterated its 1996 EU Presidency objectives – peace; safe streets; secure jobs; and sound money – as the campaign slogan for 1999. Arguing that only the EU was strong enough to deliver these objectives, Fine Gael argued that its EPP membership guaranteed it a real say in how the future of the European Union is shaped. The Labour Party advocated economic and social cohesion, a fair share in the fruits of prosperity for all citizens, with job creation at the top of the agenda. Labour favoured the single currency, since it cuts interest rates, and stimulates economic growth. Labour was also committed to a 'citizens' Europe', to a clean environment and to cutting greenhouse gas emissions as well as improving the safety and quality of the nation's food.

The Green Party laid considerable emphasis on its EU-wide connections and on the legislative impact that the Group had already made in the EP. Refuting their allegedly anti-EU stance, the Green Party argued that it was simply 'critical of the direction in which the EU is moving'. They were against the 'militarisation' of Europe, and against an EU that was too 'economically-driven' and not sufficiently 'people/social issue driven'. Ireland should refuse to participate in any of the so-called Petersberg tasks and should not join the NATO-led PfP. Neither of these groupings required a UN mandate and would in effect (as shown in Kosovo) serve to usurp the UN's role in peacekeeping and crisis management. Besides a broad raft of measures to extend their achievements on environmental matters further, the Greens also proposed an ambitious programme to advance the position of women in society: legally-binding measures to close the gender gap in salaries so that it became zero by 2010; more extensive 'mainstreaming' to all policy fields and decision-making processes at EU level and in member states; and a quota system in all decision-making bodies, starting with a 40 per cent ratio of women in the European Commission. Sinn Fein was the only party standing in all constituencies in the Republic of Ireland as well as in Northern Ireland and used this to advocate 'an all-Ireland political and economic identity within Europe'. The party was looking for institutions in the EU that were transparent, open and accountable and dedicated to advancing the causes of democracy and social justice. Sinn Fein, like the Greens, opposed 'militarisation' of the EU and also specifically opposed Irish membership of the PfP. The CAP should be reformed to keep the 'greatest numbers of farmers on the land' and there should be a total ban on genetically-modified foods.

In addition to these main party platforms, the voters were also offered an exotic array of independent candidates, among them a priest in Connacht-Ulster who promised to 'replace the bureaucracy with practical, workable structures'; a lecturer in Limerick who was standing to 'restore honour, integrity and probity to the government of our country'; a 27 year old from County Roscommon who advocated the legalisation of all prohibited substances, and the cancellation of all Third World debt; and a Natural Law Party representative whose radical agenda included a 'disease-free' Europe, a 50 per cent reduction in European crime, the abolition of value added tax, a ban on genetically-modified foods, and last, but not least, 'the creation of peace in Yugoslavia through the deployment of 7000 experts in Transcendental Meditation and Yogic

Flying who would radiate peace throughout the region'! There were two other independents whose platforms deserve more serious consideration because they were elected: Pat Cox, a sitting MEP in the Munster constituency, and Dana Rosemary Scallon, a newcomer in Connacht-Ulster. Pat Cox was, of course, experienced in Irish politics and a leading figure in the EP, having been elected leader of the ELDR. His appeal to the electorate was of someone knowledgeable of the EU, and determined to take it seriously. In his own words, he had 'the energy, capacity and experience to contribute substantially' to the shaping of major issues at the level of the EP 'in a manner consistent with Ireland's emerging interests and needs'. The appeal of the other independent to be elected, Dana Rosemary Scallon, was totally different. She made a virtue of the fact that she was 'willing to learn' and, like the ordinary people she aspired to represent, she was suspicious of the 'controlling consensus'. Describing her campaign as 'driven by a passion for people and a respect for life at all ages', she dismissed economic development as an end in itself. Although she understood the importance of regional development in this, the poorest part of Ireland, she vowed to 'fight the thinking that starts and finishes with economic development, disallowing hopes and dreams, characteristics and traits, traditions and behaviours that don't fit into the economic model. The priority has to be to keep European thinking at a human scale.' This populist message resonated well in a part of Ireland where 'traditional values' still matter and where the rapid secularisation of Irish society is still resented and resisted.

The EP election of 11 June was held simultaneously with both local government elections and a referendum on the subject of local government. The bundling together of two types of election, plus a referendum, was intended, and was widely perceived, as a way of enticing the electorate to vote. It would, however, be difficult to say definitively which of the three polls was envisaged as being sufficiently exciting to 'pull up' the other two: the turnout for all three tests of public opinion was remarkably low. Local government elections, as well as EP elections, have been conspicuously unsuccessful at attracting substantial voter attention. The referendum on local government was held in order to insert into the Constitution a statutory basis for local government, which in Ireland has long been seen as the poor relation to a highly centralised government in Dublin. The most important provision being mooted for Constitutional protection was the requirement that local government elections be held at least once every five years in order to avoid the deferrals of elections that had occurred, sometimes for lengthy periods, under governments of both major parties. The proposed

Constitutional provisions (Article 28A), designed to give statutory recognition to local authorities, were not controversial and were supported by all political parties. However, although the referendum itself was not the subject of much public debate, still less of any interest, it served to draw attention away from the EP election towards local issues and personalities. In the media, much more attention was given to the local government elections than to the EP elections; and there was, inevitably perhaps, considerable confusion as to which election was being addressed when politicians and parties made pronouncements on current issues. By contrast with local government issues, European issues seemed remote, more diffuse, less immediate, and more arcane. Often there were deliberate, as well as unintended, attempts to juxtapose local and European themes so that the relevance of the former to the latter might reinforce the perceived legitimacy of the European discourse.

As in previous EP elections in Ireland, there were few serious issues relating to European integration that featured prominently in the campaign. The Government's announcement that it was intending to join the Partnership for Peace was probably the single most provocative issue to emerge during the campaign, but most of the excitement centred on Prime Minister Ahern's reversal of his promise, made while in opposition, to hold a referendum on the topic, rather than on any deep-seated concern that PfP membership might undermine Ireland's neutrality.

The election results

The main political parties found various reasons to be satisfied with the outcome of the EP elections. The party that made the most progress, and saw the EP election as a step towards greater acceptance in the political system, was Sinn Fein. In the local elections, the party made a good impression in the inner cities where voters who felt excluded by the wealth of the 'Celtic Tiger' turned to a party that seemed to be in touch with their concerns. More surprisingly, perhaps, this progress was matched by a creditable performance in the EP elections despite the fact that Sinn Fein had little hope of getting an MEP elected. The success of Sinn Fein was, perhaps, all the more surprising because several pundits had predicted that the potentially gruesome but largely futile search for bodies of the 'disappeared' (victims of the Irish Republican Army, or IRA, during the Troubles) would rebound badly against Sinn Fein's electoral chances. If the electoral success of Sinn Fein, leading to an increased commitment to the politics of the ballot box, can be seen as the 'silver lining' of elections marred by apathy and spoiled votes, there was little for the other major parties to be especially cheerful about. Fianna Fail held on to six of the seven seats it had secured in the previous EP election but only with difficulty and despite an increase in its share of the vote; and it was clear that much of the cynicism that characterised voter behaviour in these elections stemmed from disillusionment with the 'government by tribunal' ethos that seemed to be gnawing away at the body politic. By contrast, Fine Gael could feel pleased at their EP election performance in view of their increased share of the vote over the 1994 level. The party performed well in Dublin, not its traditional stronghold, and in Leinster it challenged the support of the Greens. The Labour Party fared badly in the EP election with deflated electoral support in Munster, Leinster and Connacht-Ulster. In Dublin, the last seat went to Proinsias de Rossa but, generally speaking, the merger between Democratic Left and Labour failed to produce the intended electoral bonus. Finally, the Greens managed to retain both of the seats that they had won in 1994.

The emphasis on personalities, and the apathy displayed towards run of the mill politicians, was probably epitomised and exemplified by the success of Dana (a former Eurovision song contest winner who had returned from living in the Deep South of the USA) who stood in this election for the western constituency of Connacht-Ulster. This constituency is the largest in geographical area and the smallest in terms of population and has been characterised by maverick candidates in national and EP elections in the past. Dana's roots in Donegal, an area that had supported her strongly in the presidential election in 1997, helped to cement her performance in a constituency that attracted a more colourful array of candidates than anywhere else in the state: a priest; a development lobbyist; a supporter of legalising cannabis; and a candidate who described himself as a 'recycler'. This assorted group of candidates reflected a constituency that has tended to feel left behind by the Celtic Tiger, victimised by Dublin, and proud of its attachment to traditional family and religious values. Against this background, the apparently guileless but acute political performer, Dana, was bound to succeed: few predicted how well. During the campaign, Dana had made a virtue of her relatively low level of knowledge of the EU by claiming that she 'would learn from the people' and the way to succeed was to have a 'heart for the people'. This kind of message, which in a more sophisticated political system might have attracted derision, found fertile ground in a political system still reeling from the disclosures of corruption in high places. Dana's obvious distance from the orthodox political style stood her in good stead. Her ignorance of the wheeling and dealing mentality and her aloofness from it found a resonance in the rural west that was ever eager to 'cock a snoot' at the smug complacency of the urban east.

Aftermath

Although turnout across the EU for these elections was the lowest ever, turnout in Ireland rose from 44 per cent in 1994 to 51 per cent. This was due most probably to the decision to hold the local government and EP elections simultaneously. The general 'swing to the right' noticed in the elections was endorsed, but not exaggerated, in the Irish result. Fine Gael held on to its four seats and its representation in the EPP was reinforced by the addition of Dana Rosemary Scallon whose alleged attachment to traditional 'family values' is probably closer to the ideological Christian Democratic roots of the EPP than the more secular Fine Gael contingent. Fianna Fail, despite increasing its share of the vote, returned only six MEPs instead of seven. These now participate in the Union for the Europe of Nations (UEN) group which will, together with the ELDR, form a natural right-wing alliance with the EPP. In fact, these three groups can muster 314 votes between them and thus can collectively express the 'will of Parliament' in crucial votes. The ELDR, led by Irish independent MEP, Pat Cox, is now in a strategically key position in the context of an EP that has swung noticeably to the right, albeit in a European Union led by governments that are purportedly left of centre. The remaining three Irish MEPS, two Greens and a Labour member, belong to a left that is now in a position of unprecedented weakness, epitomised most obviously by the decline of the Socialist group following a poor showing by Labour in Britain, but reflecting also the improbability of all leftish groups in the EP uniting on any single issue (see Table 11.1).

Table 11.1 Euro-election results in Ireland

Party	Euro-Party	Seats
Fianna Fail	UEN	6
Fine Gael	EPP	4
Greens	Greens	2
Labour	PSE	1
Independent	ELDR	1
Independent	EPP	1

Although overall there was a high turnover of MEPs in this election (54 per cent had not been members of the previous Parliament), Irish turnover was much lower at 20 per cent. Avril Doyle, Dana Rosemary Scallon and Proinsias de Rossa are the only new additions to the Irish team. The relatively greater experience of the Irish MEPs is expected to put them at an advantage in the new EP, and they are expected to scoop some of the more important rapporteurships in the Committees. Already, besides having a group leader in the person of Pat Cox (ELDR) and a likely President for the Parliament in 2002, an Irish MEP (Mary Bannotti) topped the poll in the election of Quaestors. In a number of ways, therefore, the Irish MEPs, although few in number, are likely to punch well above their weight in the new Parliament.

Within the newly consolidated committee structure, the Irish MEPs will continue their previous policy of concentrating their relatively scarce resources onto a number of committees considered especially pertinent to Irish interests. Thus, while several committees have no Irish representation at all, the Foreign Affairs committee has four covering the whole spectrum of political opinion in Ireland from Green pacifism to a former Fianna Fail foreign minister's orthodoxy on matters concerning defence, security and neutrality. One committee, that concerning trade, industry, energy and research, has an Irish vice-chair (Nuala Ahern) plus one other Irish member. Social Affairs has three Irish MEPs, as do the committees on the environment, public health and consumers, and the regional policy committee. Other committees have one or two Irish MEPs, often reflecting personal or constituency interests, such as Pat the Cope Gallagher on the fisheries committee and representing an Atlantic coastal region, or wheelchair-bound Brian Crowley on the Equality Committee. Within the committee system, where the seminal legislative work is carried out, Irish MEPs can be expected to play a role out of all proportion to their sparse numbers.

Conclusion

The election was marked by even more apathy than had existed in 1994 despite the fact that the turnout was slightly higher. There were no issues of substance that came to the fore during the election, despite the fact that there were significant differences between the parties on at least one issue: Irish membership of PfP. As we have seen, the Fianna Fail party, and Fine Gael, supported Irish membership, while the Greens and Sinn Fein did not. Doorstep canvassers reported that this issue did not come up in discussions and it may be that the voters either did not care about the PfP or, more likely, realised that MEPs were unlikely to make much of a difference one way or the other. There was, in fact, more concern, even within the ranks of Fianna Fail, about the government's decision to go ahead with Irish membership in the PfP without holding the referendum that had been promised by the party when it was in opposition. Newspaper columnists speculated long and hard in the wake of the election about the lack of interest in voting, and the lack of concern for political issues. Explanations ranged from those who said that exposures of corruption within high political circles had alienated voters from the political process generally – 'a plague on all your houses' – to those who suggested that the Celtic Tiger syndrome meant that most people were highly content, absorbed with making money, and unable to discern much difference between the parties on issues that mattered to them. In other words, a sort of 'feel good' factor was militating against political debate or controversy and making party politics seem irrelevant. Still other pundits pointed to the fact that turnout in these elections was low all over Europe and simply reflected a postmodern society where parties were becoming obsolete, and decisions were being taken by lobbyists, media personalities, 'focus groups' and committees of 'experts'.

The newly-elected EP is the most powerful in its history, and can draw on the talents of prominent politicians who have been elected to it. Irish MEPs are likely to be in the forefront of the EP's desire to seek greater powers *vis-à-vis* the other institutions and, indeed, to seek a more transparent and accountable policy-making system. In Ireland, EP elections have become an integral part of the domestic political process. Governments are judged; personalities and local issues loom large; and European issues are mediated through the lenses of domestic political perceptions. If some allowance is made for the idiosyncratic features of European elections, they provide a valuable insight into the workings of the political system.

Note

1 The author wishes to acknowledge the research assistance provided by Laura Bolger, Centre for European Studies, University of Limerick.

12
Italy

Philip Daniels

The fifth elections to the EP were held in Italy on 13 June 1999. The elections took place against a backdrop of an uncertain domestic political situation. The centre-left government of Massimo D'Alema, which had taken office in October 1998, was inherently precarious, based as it was on a broad, multi-party parliamentary majority including several MPs who had been elected with the opposition centre-right coalition in the 1996 national elections. The tenuous hold of the government reflected the fragmentation of the party system and the fluid nature of political alignments. Since the early 1990s, the party landscape has undergone a profound transformation which has seen the disappearance of a number of traditional parties and the emergence of a large number of new political formations. This turbulence in the party system was clearly evident in the 1999 EP elections in which 26 parties and movements presented lists, many contesting European elections for the first time. The proportional electoral system, with its relatively low threshold for representation, encourages the proliferation of party lists in European elections.[1] In contrast to national parliamentary elections, in which the shift towards a largely majoritarian system has encouraged the formation of electoral alliances and greater bipolar competition, in European elections the parties typically stand alone.

The principal novelty in the 1999 Euro-elections was the entry of a new party, the Democrats for the Olive Tree (the *Asinello*), led by the former prime minister Romano Prodi. With the support of leading city mayors and national figures such as the former investigating magistrate Antonio Di Pietro, the new political party (formed in February 1999) represented an attempt by Prodi to strengthen his position on the centre-left of the political spectrum. The collapse of the Prodi government in October 1998, when Communist Refoundation (RC) withdrew its parliamentary support, had provoked a crisis in the governing Olive Tree alliance and exposed Prodi's political vulnerability in the absence of his own secure political base. The new D'Alema government, supported by a more centrist-oriented parliamentary majority, was viewed by Prodi as a betrayal of the Olive Tree alliance. The formation of the Democrats and Prodi's decision to contest the EP elections intensified conflicts within the centre-left coalition. In particular, relations between Prime Minister D'Alema and Prodi worsened significantly despite the Democrats' continued support for the government. D'Alema feared that the challenge from the Democrats would fragment the centre-left coalition and undermine his governmental majority. The conflict between D'Alema and Prodi was, however, attenuated by the Italian government's support for Prodi's nomination as the new President of the European Commission. Prodi's success in securing the position in March 1999 effectively removed him from the domestic political scene in the longer term but in the run-up to the European elections it served to give him and the Democrats welcome publicity and national media exposure. In particular, in the context of the European elections, Prodi's appointment established a visible link between the Democrats and the European political arena.

The elections also saw the entry of another novel formation, the Emma Bonino List, which was in essence the Radical Party in a new guise. Bonino, a European Commissioner in the Santer Commission, gave the party list a strong European identity. In addition, there had been a popular campaign in support of Bonino's efforts to become a candidate for the office of President of the Italian Republic in spring 1999. Although unsuccessful in her bid to become Italy's first female president, she had nevertheless enjoyed many weeks of largely favourable media exposure.

The campaign

As in earlier European elections, national political issues, which were of more immediate concern to voters, overshadowed European themes. Although the parties paid lip-service to European issues in their election programmes and campaign literature, they tended to focus on domestic themes which differentiate the parties much more than their largely consensual pro-European positions. For the most part, parties only alluded to the European dimension of the vote when it served to reinforce their positions on divisive issues in the domestic political arena. European themes were not, however, entirely absent from the campaign.

In the run-up to the European elections, three of the issues at the centre of national political debate had clear European dimensions. The first of these issues was the appointment of Romano Prodi as the new EU Commission President and the controversy this provoked regarding his long-term future in national politics and, more immediately, the role that he should play in the Euro-election campaign. Second, the Government's continued efforts to introduce pension and welfare reforms were made more urgent by the fiscal constraints imposed by Italy's membership of the single currency. Third, Italy's support for NATO's intervention in the war in Kosovo divided (and threatened to bring down) the D'Alema government. The importance of the election for the domestic political arena was clearly evident in the way many of the parties campaigned. For example, Silvio Berlusconi, the leader of the opposition Forza Italia, attempted to turn the election into a referendum on the D'Alema government, contending that a combined vote of below 40 per cent for the parties in the governing coalition would indicate that it had lost its legitimacy and should resign from office. In addition, the most intense competition was between proximate parties (e.g., the National Alliance, or AN, against Forza Italia and the Democrats against the Democrats of the Left, or DS), indicating the over-riding importance of the vote in terms of domestic political competition.

As in previous EP elections, many prominent politicians stood as candidates for election in more than one constituency. This represents an attempt by the parties to attract support to their lists by fielding their better-known candidates. In addition, many of the party lists included the usual sprinkling of 'flower in the buttonhole' candidates from the worlds of sport, entertainment, culture and the arts: the actress Gina Lollobrigida, for example, stood unsuccessfully as a candidate for the Democrats; and three retired footballers, including World Cup winner Paolo Rossi (who stood for the National Alliance), failed to get elected.

The turnout of 70.8 per cent was the lowest in Italy since the introduction of direct elections to the EP in 1979 and continued the trend of decline in electoral participation in European elections: in 1979 the turnout was 84.9 per cent, in 1984 83.4 per cent, in 1989 81.5 per cent and in 1994 74.8 per cent. In comparison with the 1996 national elections, there was a decline in the turnout of more than 6.5 million voters. The rise in abstention might have been worse had the administrative elections not been held on the same day. In addition, the entry of the Bonino List and the Democrats mobilised voters who had abstained in the 1996 national elections. The decline in electoral participation in Italy is not exceptional, however, and turnout in European elections remains markedly higher than in all other EU states where voting is non-mandatory (only Belgium and Luxembourg recorded higher turnouts in 1999). Nevertheless, the rise in rates of abstention is consistent with a general long-term trend in Italy recorded in both national parliamentary and local administrative elections. The significant decline in turnout in 1999 is attributable, in part, to national specific factors: the impasse in Italy's political transition is likely to have contributed to a sense of voter disillusionment; the uninspiring campaign failed to mobilise voters; and the proliferation of parties, both old and new, and the shifting party alliances are likely to have left many voters bewildered and disenchanted. In addition to the national factors, the generalised decline in turnout across the EU in the 1999 EP elections indicates that other influences were at work. These include a growing disenchantment with the project of European integration; the EP's continued lack of visibility and political weight despite cumulative increases in its powers introduced by successive treaty revisions since 1987; and a negative public reaction to the corruption scandals involving the EU Commission in the months leading up to the elections.

The results

Nineteen of the 26 lists contesting the elections won seats in the EP. This compares with the thirteen lists which secured representation in the 1994 EP elections. The 1999 elections brought a significant turnover in the Italian delegation in the European Parliament: of the 87 MEPs elected in 1994, only 22 (25.3 per cent) were re-elected in 1999. The most significant features of the results were Forza Italia's emergence as the largest single party; the electoral breakthrough of the Democrats; and the success of the Emma Bonino List (see Table 12.1). These formations were the three principal winners in the Euro-elections, and significantly each of them is a non-traditional party. Forza Italia was the only major party to see an increase in its vote compared to the 1996 national legislative elections, rising from 20.6 per cent to 25.2 per cent in the 1999 EP elections. The party achieved the highest vote in four out of the five multi-member constituencies, recording its best performance of 29.6 per cent in the North-West constituency. Berlusconi came top of the poll in each of the five constituencies and achieved a record of just under 3 million preference votes. In contrast to most of the other parties, Forza Italia devoted large resources to the Euro-election campaign; in particular, the party spent heavily on television election spots, significantly outspending all other parties.[2] The party's strong performance, coupled

Table 12.1 EP elections results, 1994 and 1999, and the Chamber of Deputies, 1996

Party lists	European 1999 %	European 1999 Seats won	National 1996 %	European 1994 %	European 1994 Seats won
Left Democrats	17.3	15	21.1	19.1	16
PdCI	2.0	2	–	–	–
Democrats	7.7}		–	–	–
SVP	0.5}	7	–	0.6	1
Union Valdotaine	0.1}		0.1	0.4	–
PPI	4.2	4	6.8	10.0	8
U.D.Eur.	1.6	1	–	–	–
CDU	2.2	2	5.8*	–	–
RI-DINI	1.1	1	4.3	–	–
PRI–Lib–ELDR	0.5	1	–	0.9†	1
SDI	2.2	2	–	1.8‡	2
Greens	1.8	2	2.5	3.2	3
RC	4.3	4	8.6	6.1	5
Forza Italia	25.2	22	20.6	30.6	27
AN–	10.3	9	15.7	12.5	11
Segni Pact			–	3.3	3
CCD	2.6	2	5.8§	–	–
MS–Tricolore	1.6	1	0.9	–	–
Socialist Party	0.1	–	–	–	–
Northern League	4.5	4	10.1	6.6	6
Emma Bonino List	8.5	7	1.9¶	2.1	2
Pensioners' Party	0.7	1	–	–	–
Others	1.0	–	1.6	2.8	2
Total	100.0	87		100.0	87

Note: See p. xxx for a full list of party abbreviations.
* Joint list with CCD in 1996.
† PRI and PLI (Liberal Party of Italy) stood separately in 1994.
‡ In 1994 PSI (Socialist Party of Italy) Democratic Alliance.
§ Joint list with CDU in 1996.
¶ Pannella–Sgarbi List in 1996.
Source: Author's elaboration of Ministry of the Interior data.

with the electoral decline of the National Alliance, strengthened Berlusconi's claims for leadership of the centre-right coalition. The party's success also vindicated Berlusconi's strategy of moving the party towards the political centre and acting as a responsible and moderate opposition; this was evident, for example, in Forza Italia's support for the government during the war in Kosovo and in its vote for Carlo Azeglio Ciampi as the new President of the Republic.

For the National Alliance, which stood on a joint list with the Segni Pact, the European election results were a major setback. The joint list polled 10.3 per cent compared to the 15.7 per vote for the National Alliance in the 1996 legislative elections; the party's vote was down by more than 2.5 million votes compared to 1996. The National Alliance lost ground in each of the five multi-member constituencies but the greatest losses were in the party's traditional areas of strength in the south. The party's decline indicates that the attempts by Gianfranco Fini, the party's leader, to project a new image and political identity as a modern conservative party has so far failed to convince. In addition, the Euro-election results effectively ended Fini's efforts to challenge Berlusconi as the leader of the right-wing pole. In the aftermath of the elections Fini offered his resignation as leader but he was persuaded to stay on. The Northern League also suffered a significant electoral reverse with a decline from 10.1 per cent of the vote in the 1996 legislative elections to 4.5 per cent in the June 1999 European elections. In the North-West constituency the party's vote fell from 21.8 per cent in 1996 to 10.5 per cent in 1999, while in the North-East constituency the party declined from 18.7 per cent to 6.9 per cent over the same period. The party's poor performance was attributed in part to the support it gave to Serbia in the Kosovo conflict.

The elections were also disappointing for many of the parties in the governing centre-left coalition. The vote for the Left Democrats fell from 21.1 per cent in the 1996 legislative elections to 17.3 per cent in the European elections. The party declined across the territory with the heaviest losses of support in the south and the islands and in the party's traditional areas of strength in central Italy. In common with the pattern of EP elections in other member states, incumbency in office during a difficult period of government is likely to have eroded support for the Left Democrats. In addition, post-electoral analyses of the flow of the vote indicated that the party had lost votes to Prodi's Democrats.[3]

The Democrats made a significant electoral impact, winning 7.7 per cent of the vote and picking up support from parties across the centre-left of the political spectrum. The Emma Bonino List also achieved a notable success, emerging as Italy's fourth largest party with 8.5 per cent of the vote. This represented an impressive advance on the 1.9 per cent polled by the Radicals in the 1996 legislative elections. Bonino had enjoyed months of media coverage and, in contrast to most of the parties contesting the election, she had brought a distinctively European appeal to her campaign. The party campaigned vigorously, making much of Bonino's success as a European Commissioner, and spending heavily on television electoral spots (only Forza Italia spent more). Bonino attracted support from across the political spectrum, but primarily from the centre-left, and also appealed to voters detached from the traditional parties (just over 10 per cent of the vote for the Bonino List came from individuals who had abstained in the 1996 national elections).[4]

The EP elections represented a further setback for the direct descendants of the Christian Democrats. The Italian Popular Party (PPI) and United Christian Democrats (CDU) polled 4.2 per cent and 2.2 per cent respectively, down from the 10 per cent they won as a single party (the PPI) in the 1994 European elections. Of the other major parties, Communist Refoundation saw its vote halved from 8.6 per cent in the 1996 national elections to 4.3 per cent in 1999. Two policies appear to have cost the party votes: first, the return to opposition in October 1998, a move which brought down the Prodi government and led to a party split and the creation of a new breakaway party, the Party of Italian Communists (which polled 2 per cent of the vote in the 1999 Euro-elections); second, the party's opposition to NATO's intervention in Kosovo. The Greens also declined from 2.5 per cent in the 1996 national elections to 1.8 per cent in 1999; this went against the trend in support for Green parties in most other EU member states in the 1999 Euro-elections. As in the cases of the electoral setbacks for the Northern League and Communist Refoundation, the Italian Greens' opposition to NATO intervention in Kosovo appears to have damaged the party electorally.

The post-election analyses focused almost exclusively on the importance of the results in terms of the domestic political arena and their impact on inter-party competition. Prime Minister D'Alema claimed that the combined vote for the coalition parties (41.2 per cent) was a satisfactory performance. He could point to the fact that a decline in the vote for governing parties is a common pattern in Euro-elections across the EU, particularly in those cases where a government is in the middle of its term of office or beyond. Although the powers of the European Parliament have been enhanced in recent years, European elections remain essentially 'second-order' elections: voters are not choosing a national government and they are more willing to abstain, to switch votes and to vote against incumbent parties. Despite D'Alema's upbeat interpretation of the results, there were some worrying signs for his government. The two largest parties in the coalition, his own Left Democrats and the PPI, both declined significantly; at the same time, the Democrats' electoral success rebalanced the centre-left alliance and increased the new party's bargaining power in the coalition. Perhaps most worryingly of all for

D'Alema, the EP elections highlighted the fragmented nature of his parliamentary majority; of the twelve separate lists from the centre-left coalition which contested the Euro-elections, nine polled a vote of 2.2 per cent or less, while the remaining three lists won 4.2 per cent (PPI), 7.7 per cent (the Democrats) and 17.3 per cent (the Democrats of the Left).

The EP election results confirmed that the Italian party system remains highly fragmented, a tendency exacerbated by the proportional representation system used for the Euro-elections. Fragmentation was further reflected by other features of the results: only five parties got more than 5 per cent of the vote and three got little more than 4 per cent; of the remaining eighteen lists, the highest recorded vote is 2.6 per cent. The fragmentation of the party system provoked new calls for electoral reform for national elections. It is difficult to see how a fundamental restructuring and simplification of the Italian party system could be achieved, however, if the proportional system used for EP elections is not also reformed.

The European dimension to the elections featured less in the post-electoral analyses but is also important. The nineteen lists winning EP seats were dispersed among each of the EP's seven party groups. As Table 12.2 shows, membership of the party groups in the newly-elected EP is often anomalous in terms of alliances in the domestic political arena. For example, Forza Italia, the principal party of opposition in national politics, sits in the same group as a number of the small parties which form part of the centre-left governing coalition. Prodi's Democrats have joined the ELDR, traditionally a centre-right party group, which is at odds with the party's position in the centre-left governing coalition in national politics. These anomalies reflect in part the fluid nature of party political identities in Italy and the competition among the parties to occupy distinctive political space. Membership of the EP's party groups may help a party to project an image and a political identity in the domestic political arena. In the late 1980s, for example, the Italian Communist Party's 'Euroleft' strategy made much of its links with European socialist parties in an attempt to enhance its own domestic political legitimacy. Eventually the party, with its new name and identity as the Democratic Party of the Left, was admitted to full membership of the Socialist group. More recently, Forza Italia has joined the newly-formed EPP (Christian Democrats) and the European Democrats, thus locating itself in the mainstream of European centre-right parties. This is important for Forza Italia's identity and is consistent with the party's domestic political strategy of occupying the centre-right political space which was once the domain of the Christian Democrats. Aware of this political

Table 12.2 Membership of EP party groups

EP party group	Italian parties/lists	Seats
Group of the EPP (Christian Democrats) and European Democrats	Forza Italia	22
	PPI	4
	CDU	2
	CCD	2
	SVP	1
	Pensioners' Party	1
	U.D.Eur.	1
	RI–Dini	1
Union for a Europe of Nations Group	AN–Segni Pact	9
Confederal Group of the European United Left/ Nordic Green Left	RC	4
	PdCI	2
Group of the Party of European Socialists	DS	15
	SDI	2
Group of the European Liberal, Democrat and Reform Party	Democrats	6
	LN (Northern League)*	1
	PRI–Lib–ELDR	1
Group of Greens/European Free Alliance	Greens	2
Non-attached	Bonino List	7
	LN (Northern League)	3
	MS-Tricolore	1
		87

* One northern League MEP did not join the party's other three MEPs in the Group of the Non-attached.
Source: Based on information from the EP and the Ministry of the Interior.

challenge, the descendants of the Christian Democrats, and in particular the PPI, tried for many years to block Forza Italia's membership of the EPP. Having failed, these small parties now find themselves reduced to a minor role within the EP party group and Forza Italia has become the principal party of the EPP (Christian Democrats) and the European Democrats in Italy. Forza Italia's centrality in the 1999 EP contrasts with its experience when it first entered the EP in 1994. Then in a governing coalition with the neo-fascists at national level and Eurosceptic in outlook, Forza Italia was unable to find allies in the EP and was forced to create its own party group, Forza Europa. The experience of the National Alliance, the other main party of the right in Italy also illustrates the potential importance of EP party group membership for a party's domestic image. Following the 1999 elections, the National Alliance's principal ally in the EP, the French Gaullists, joined the Group of the EPP (Christian Democrats) and European Democrats; the National Alliance sits in the Union for a Europe of Nations Group, a choice which leaves it outside the mainstream European right and at odds with its efforts to project an image of itself as a modern conservative party.

Conclusions

In terms of both the conduct of the campaign and the importance attached to the results, the 1999 Euro-elections in Italy were essentially about domestic politics. This is now a familiar pattern in Euro-elections with domestic political issues and inter-party competition in the national arena taking priority over specifically European themes. While the major Italian parties allude to European themes, the cross-party consensus on key aspects of European integration gives them little scope for distinctive European appeals to the electorate. The Euro-elections are viewed by the contestants principally as a barometer of their support in their national political setting. This tendency in Euro-elections is reinforced by the use of proportional representation which encourages the proliferation of political formations and offers no incentives for the parties to form electoral alliances. While domestic political concerns and competition were central to the Euro-elections, the European dimension to the contests was not entirely absent. Party divisions over the war in Kosovo and the debate about welfare and pension reform were issues which straddled the domestic/European divide. In addition, two of the principal winners in the elections, the Bonino List and Prodi's Democrats, had clear European dimensions to their appeals.

In the longer term, the most important feature of the Euro-election results is the potential impact which the EP's nascent party system will have on the highly fluid Italian domestic political alignments. At present, Italian parties' alliances in the EP are often inconsistent with their alliances in the national setting. In their attempts to construct clear political identities, many of the leading Italian parties recognise the value of ties with established parties and groups in the EP. In those cases where more than one Italian party sits in the same EP group, this may encourage the parties to merge into a single formation. Such a development would be desirable in terms of simplifying the Italian party system with a reduction in the number of parties and a more clear-cut centre-left/centre-right political competition.

Notes

1 For EP elections Italy is divided into five multi-member constituencies with the 87 seats allocated on the basis of population: the North-West returns 23 MEPs, the North-East 16, the Centre 17, the South 21, and the Islands 10. The parties present lists of candidates in each constituency (up to a maximum of the number of seats available) and seats are allocated to the parties on the basis of proportional representation. Voters may also indicate preferences for a candidate or candidates on the list; the number of preference votes available to the voter varies according to the number of MEPs elected in each constituency.
2 See *La Repubblica*, 15 June 1999.
3 See *La Stampa*, 15 June 1999.
4 See *Corriere della Sera*, 15 June 1999.

List of Italian parties/party abbreviations used

AN	National Alliance
CCD	Christian Democratic Centre
CDU	United Christian Democrats
DS	Left Democrats
Forza Italia	Forza Italia
I Democratici	Democrats
Lega Nord	Northern League
Lista Bonino	Emma Bonino List
MS-Tricolore	Social Movement-Tricolour
Patto Segni	Segni Pact
Partito Socialista	Socialist Party
Pensionati	Pensioners' Party
PdCI	Party of Italian Communists
PDS	Democratic Party of the Left
PLI	Italian Liberal Party
PPI	Italian Popular Party
PRI-Lib-ELDR	Republicans-Liberals-ELDR
PSI	Italian Socialist Party
RC	Communist Refoundation
RI-Dini	Italian Renewal–Dini List
SDI	Italian Democratic Socialists
SVP	South Tyrol Peoples' Party
U.D. Eur.	Union of Democrats for Europe
Union Valdotaine	Valdotaine Union
Verdi	Greens

13
The Netherlands

Henk van der Kolk

Introduction

The European elections in the Netherlands, held on Thursday, 10 June, were contested by a large number of parties. The largest party is the Social Democratic Partij van de Arbeid (PvdA). With 45 out of 150 seats in the Dutch Parliament it has been the biggest party in the present 'purple' coalition led by PvdA party leader and Prime Minister, Wim Kok, since 1994. The European list of the PvdA was headed by the well-known Max van den Berg. In the 1970s van den Berg was an alderman in Groningen and (in)famous for his left-wing opinions. He was PvdA chairman until 1986 when he became chairman of the NOVIB, an important Dutch organisation involved in foreign aid. He failed in 1998 to become Minister of Foreign Aid in the second purple cabinet. The PvdA is part of the EP Socialist group.

The Liberal Volkspartij voor Vrijheid en Democratie (VVD) with 38 seats in the Dutch Parliament is the second largest party of the second purple coalition. Frits Bolkestein, nominated as Euro-commissioner in June 1999, was party leader of this liberal party, although he stepped aside soon after the national elections of 1998. The head of the European VVD list was Jan Kees Wiebenga. An MEP since 1994, he was a member of the Dutch Second Chamber between 1982 and 1994, and of the First Chamber between 1977 and 1982. The VVD is part of the EP's Liberal and Democratic Reformists.

Democraten 66 (D66) is the smallest of the three parties in the second purple coalition. By refusing to form a coalition with both Christen Democratisch Appèl (the Christian Democrats, or CDA) and VVD in 1994, D66 was able to form a cabinet without a Christian party for the first time since the introduction of universal suffrage in 1917. Despite the popularity of the first purple coalition, D66 was unable to attract the voters again: it lost 10 of its 24 seats in 1998, almost falling back to the number of seats it had won after the the national elections of 1989. The D66 list for the European elections was led by Lousewies van der Laan. The D66 selection committee put van der Laan third on the party list, but D66's strong democratic practices allowed party members to put her at the top. Having done a stage (internship) with Dutch Euro-Commissioner Andriessen and worked for the Dutch CDA Eurocommissioner van den Broek, she was familiar with Brussels. Like the VVD, D66 sit with the EP's Liberal and Democratic Reformists.

The main opposition party in the Netherlands, the CDA, was founded in 1977 after a merger of the large Catholic party and two considerably smaller Protestant parties. Until the 1960s these three religious political parties had been dominant, winning about half the seats in Parliament. Owing to secularisation in the 1960s and early 1970s, the CDA lost many seats, but retained its dominant position in the 1970s and 1980s because of intense conflicts between the PvdA and VVD. In 1994, however, the CDA lost 20 of its 54 seats and was forced into opposition by D66. In the Second Chamber, the CDA has 29 of the 150 seats. The CDA party group in the EP is a member of the EPP. Hanja Maij-Weggen led the CDA's Euro-election list. She became an MEP in 1979, was Minister of Transport between 1989 and 1994, and returned to the EP shortly thereafter. She chairs the EP 'intergroup' on animal welfare.

Minor parties represented in Parliament

After the merger of communists, pacifists and some radical Christians in 1990, Groen Links (Green Left) became a small but significant party with 11 seats in Parliament. Joost Lagendijk was one of the driving forces behind the merger of parties and was co-founder of the Green Left publication *De Helling*. He became an MEP in September 1998. Green Left is not the only leftist party in the Dutch Second Chamber because in 1994 a new socialist group entered the Second Chamber. The Socialistische Partij (SP) was founded in the 1970s as the successor to a Marxist-Leninist group. It presents itself mainly as an 'action party' focusing on specific problems, and won five seats in the Second Chamber in 1998. The EP list was headed by Erik Meijer. After a long career in different left-wing parties in the Netherlands he joined the SP in 1996. He was member of a submunicipal council in Rotterdam. The SP had yet to win EP seats. Three different orthodox Christian parties together have eight seats in Parliament. Attempts to merge into one Orthodox Christian party failed, but the Reformatorische Politieke Federatie (RPF), the

Staatkundig Gereformeerde Partij (SGP) and the Gereformeerd Politiek Verbond (GPV) ran a combined list for the European elections led by Johannes Blokland (GPV). Blokland was chairman of the GPV between 1984 and 1994. He became an MEP in 1995.

Parties not represented in the Dutch Parliament

Some minor political parties, not represented in the Dutch Second Chamber or the EP, participated in the EP elections. First of these was the Conservatieve Democraten (CD), characterised by its negative attitude towards immigration. In 1998 the CD (which then stood for 'Centre Democrats') lost its three seats in the Dutch Parliament. The European list was led by Chiel Koning, CD Treasurer and former council member. The second minor party was the Europees Verkiezers Platform Nederland (EVPN), which was organised around J. Janssen van Raay. Elected to the EP in 1979, he left the CDA after being accused of using EP letters in favour of a business partner. The CDA tried to expel Jansen van Raay, but he left the party before the procedure was completed whilst keeping his seat in the EP. The main point of the Europese Partij, founded in 1998 by Roel Nieuwenkamp, was the organisation of European-wide elections. The party strongly favours Europe and wants the EU to become a genuine (federal) state. Finally, only one person stood independent of a party organisation: Luc Sala, a physicist involved in many things related to computers and software.

The players in a domestic context

Of the political parties represented in the Second Chamber, D66 most strongly favours further extension of EU powers. At the other end of the spectrum is the SP, which opposes further integration. The orthodox parties share the SP position, although their background and rhetoric clearly differ. The VVD is said to be against further integration, although it strongly supports integrating the European market. The CDA, traditionally a party representing many people working in the agricultural sector, is in favour of the EU and does share this general opinion with the PvdA, although these parties differ widely on specific issues. Green Left is not completely against the EU but opposed EMU and the Amsterdam Treaty because it feared an uncontrolled growth in the power of EU bureaucracy. This party stresses further democratisation of Europe more than the others. With respect to EU policies, Dutch political parties clearly divide along the traditional left–right dimension with the PvdA stressing the importance of employment policies and a Europe-wide minimum wage, the VVD focusing upon entrepreneurs, a free market and a small European bureaucracy, and D66 and the CDA somewhere in the middle.

In 1994 a coalition of PvdA, VVD and D66 kept the CDA out of government after a very long period of Christian Democratic domination. After the national elections of 1998 this coalition was able to continue its work after winning a majority of 97 seats out of 150 in the Second Chamber. Maybe partly due to the fact that the pro-European D66 and reluctant European VVD are both members of the present coalition, there are no serious party rifts with respect to the EU between the major political parties. Within the parties, party representatives within the EP are generally more in favour of furthering European integration than national party representatives, but this does not cause major tension within political parties. Therefore, most discussions in the Dutch Second Chamber with respect to the EU are about the best ways to promote Dutch interests within the EU and about filling the democratic gap at the European level by extending and improving parliamentary control at the national level. Two parliamentary discussions with respect to this seem to be important. First, on 29 March 1999, the Second Chamber asked the Dutch government to review the procedures used to inform the Second Chamber about European issues. In short, the Second Chamber asked for more, more standardised and quicker information about draft European legislation. On 25 May 1999 the government said it would respond to these demands, and stated that most information is easily available. More important, therefore, were some proposals on MEPs' activities in the Dutch Parliament. On 28 May 1999, the committee on internal procedures of the Dutch Parliament presented a list of possible ways to tighten links with the MEPs in order to enhance the role of the Second Chamber and the involvement of MEPs in respect of Dutch policy-making. Some of the proposals required constitutional revision, but most could be accommodated by existing rules and procedures. The proposals ranged from the right of MEPs to advise parliamentary committees to a limited involvement of MEPs in parliamentary debates. The proposals were debated on 8 June 1999. The opponents of these proposals (VVD, SP and SGP) were reluctant to admit MEPs to parliamentary debates because they could not discern a genuine difference between MEPs and representatives of other groups in Dutch society. Proponents stressed the importance of the democratic gap and expressed the hope that involving MEPs in national debates would improve control of the EU. The proposals were accepted by a majority in Parliament.

The campaign: mobilising the vote

Voting is not compulsory and turnout for the EP elections was very low, relative to other Dutch elections. It confirmed a clear downward trend since the first EP elections in 1979. In 1979, 58.1 per cent voted. This fell in 1984 to 50.9 and in 1989 to 47.5 per cent. In 1994, turnout was only 35.7 per cent and many EP candidates expressed their anxiety about an even lower turnout in 1999. Predictions by polling bureaux varied significantly. Gallup, for example, expected a turnout of 58 per cent. The Dutch bureau *Interview*, weighing the opinion data, on 2 June predicted a 28 per cent turnout.

Several means were used to mobilise the vote. The Ministry of the Interior and Kingdom Relations, in collaboration with the EP information office, organised a campaign to increase turnout in the EP election. This was part of larger campaign aimed at increasing turnout in all Dutch (national) elections in 1998 and 1999. The general idea behind the campaign was represented by a logo depicting a voting pencil in a knot. For the European elections this knotted pencil was surrounded by European stars. The campaign also included small advertisements with pictures of party representatives urging the public to vote for one of them, and a television spot showing a young woman and her grandfather talking about the EP elections (Grandfather states that everyone should vote; the granddaughter tells her grandfather to get lost, using a Dutch proverb telling him to go and climb a tree, after which the grandfather climbs a tree and shouts that everyone should vote). A less visible and much more criticised effort to increase turnout was the sponsoring by the EP information office of television programmes featuring MEPs. The EP information office provided 100,000 Dutch guilders (about 45000 euros) for a television version of 'Trivial Pursuit'. Some MEPs were not aware of the fact that the programme they participated in was sponsored by the EP and were furious the moment this was revealed.

In the Netherlands, political parties are not subsidised for their electoral campaigns. State funding of political parties is limited to educational and scientific activities. Political parties tried to mobilise their voters using different means, including television spots, Internet pages and traditional party meetings. Because the number of party members has decreased in the past decades, parties face difficulties in funding their campaigns and have to find alternative sources. At least one party, the PvdA, used EP money earmarked to inform the public about the EP for its political campaign, something explicitly forbidden by EP regulations. This was made public in the months preceding the EP elections. The PvdA representative Max van den Berg explained that nothing illegal was done, but that the party, nevertheless, would act differently in the future.

The issues

The European elections were almost completely overshadowed by two major national events. The first event was the end of the second 'purple' coalition. On 19 May 1999, D66 could not accept a VVD representative in the First Chamber voting against a proposal to change the constitution in order to introduce a *corrective referendum* (a popular referendum to vote down a bill accepted by Parliament). The D66 ministers decided to resign, after which the whole cabinet handed in its resignation. All coalition parties, however, again expressed their support for the promises made in the coalition programme, and because D66 was promised a 'consultative referendum' – which did not require a change of the constitution – on Monday, 7 June, Prime Minister Wim Kok was able to ask the Queen to withdraw the resignation. Because EP elections are generally seen as second-order elections, the aforementioned coalition crisis was welcomed by some EP candidates. Retiring PvdA MEP Piet Dankert said the crises would 'politicise the campaign' and 'boost turnout levels'. Retiring D66 MEP Jan-Willem Bertens expected the campaign to have a more 'national' character. Maij-Weggen (CDA) hoped her party would benefit from the crisis within the purple coalition.

In the meantime the Second Chamber had to discuss a parliamentary report about the so-called 'Bijlmer disaster'. On 4 October 1992, an Israeli El Al aeroplane crashed into a block of flats, killing 43 people. Soon after this incident, rumours started to spread. The plane was said to be loaded with biochemical products used for chemical warfare. Some people claimed to have seen 'men in white suits' not belonging to either the police or the rescue team searching the place, and the flight recorder of the aeroplane disappeared. Some people connected with the disaster (policemen, firemen, inhabitants of the flats) complained about severe fatigue. A commission from the Second Chamber, including all relevant parties, studied the disaster and the way it was handled by the government. After its hearings, the commission criticised many ministers, including the minister of health (D66), the minister of transportation (VVD) and even the Prime Minister (PvdA). The ministers warned that an acceptance of all conclusions made in the report by a majority in Parliament would endanger the continuation of the purple coalition. Despite criticism, on Thursday, 3 June, a majority ignored the most important conclusions, thereby saving the coalition. The fact that the three coalition parties were unwilling to accept the conclusions of the parliamentary report may have contributed to support for the opposition.

During the campaign, most major parties used their 1998 national election platforms in conjunction with the European programmes of the European party groups. D66, however also adopted an extension of the ELDR manifesto. Green Left, the SP and the Orthodox Christian parties wrote a new manifesto for the EP elections. The manifestos were easily available via the Internet. Moreover, many newspapers provided a systematic comparison of promises. A comparison of programmes reveals that the main political parties differed little over the role of the EU: they favoured greater democracy and transparency, and enlargement to Central and Eastern Europe providing they could meet the conditions of democracy and the rule of law. Most parties wanted to see EU agreement on asylum policy and crime. Despite this, the campaign was not about these general European issues. Even the national events mentioned were not very important because they offered no clear choice between political parties. Nevertheless, some issues clearly related to the EU entered the media agenda in the months preceding the European elections and may have influenced the general climate of opinion.

A first clear European issue discussed in the media and in Parliament was fraud by MEPs. Only a few months before the EP. elections, in March 1999, a book was published about the functioning of the EP. Parts of this book were published in Dutch newspapers and were cited and discussed in the media. The main point of his publication was that most MEPs profited from double pensions and extensive expense accounts. This was of course preceded by some foreign and Dutch documentaries showing the so-called SiSo (Sign in Sod off) system: signing in on Friday and then leaving without attending the meeting. To this the book added an overall analysis of attendance between September 1994 and July 1998, revealing many representatives of all Dutch political parties signing in and leaving without actually attending the meeting. Other aspects discussed in the book were the pensions of MEPs, paid to a large extent by the EP, and the extensive declaration of bills. These issues were not only discussed in the media, but also led to questions in Parliament. As a result of these discussions in the media the EP party groups of PvdA, VVD, Green Left, D66 and the Orthodox Christian Parties voluntarily adopted a stricter regulation of expense accounts. The CDA opposed this. Maij-Weggen, heading the CDA list during the campaign, decided not to co-operate, but after being criticised for this, she changed her opinion.

The second EU event of some import was the resignation of the Santer Commission. 'Whistle-blower' Paul van Buitenen is Dutch, so his suspension at the end of 1998 attracted a lot of attention in Dutch newspapers. Most newspapers defended van Buitenen. All Dutch political parties criticised the Commission, but the motion aimed at forcing the Commission to resign in January 1999 was not supported by the CDA and PvdA. The CDA was not willing to sack honest members of the Commission, including its own Commissioner, van den Broek. A motion against Cresson (Socialist) was supported by all but the PvdA. The motion to examine the Commission's behaviour was supported by many Dutch parties, although many representatives decided not to vote. Despite the fact that Dutch political parties acted differently and despite the fact that this was mentioned a few times during the campaign, it did not play a very important role, partly because newcomer van den Berg (PvdA) explained he would vote against Commissioner Cresson.

After the resignation of Santer Commission, Commission President-designate Prodi asked for several candidates to be nominated by the member governments so he could have a wide choice from which to pick the ones most suited for the new Commission. The Dutch government decided to nominate only one person and before the European elections were held, former VVD leader Frits Bolkestein was mentioned as the only serious candidate. The announced nomination of Bolkestein was not welcomed by the leaders of the other EP lists. The VVD, as explained before, was not a proponent of the deepening of Europe; it was said that Bolkestein considered Europe 'finished', although he more or less denied ever having held such an opinion. Moreover, Bolkestein was a rather controversial party leader, who was not liked by many people from other parties. Van den Berg (PvdA), Maij-Weggen (CDA), Lagendijk (Green Left) and van der Laan (D66) all questioned the nomination of Bolkestein in the electoral debate on 6 June 1999.

The result

Turnout the Euro-elections was as dramatically low as predicted by some of the pessimists: 29.9 per cent. This low turnout was generally ascribed to the absence of a choice between policies at the European level and the absence of national issues. Partly due to the extremely low turnout, a big difference exists between the outcome of Dutch national elections and the results of the European elections. This discrepancy favours the CDA and, to a much lesser extent in absolute percentages, the orthodox Christian parties. The voters of these religious parties are probably more attached to their political party, and are somewhat more traditional, resulting in a relatively higher propensity to vote. The low turnout also prevented a clear translation of the electoral outcome in terms of domestic politics.

The Euro-elections did not cause a major shift in the division of seats within the EP (see Table 13.1). Compared to 1994 and 1998, D66 (5.9 percent), CDA (3.9 percent) and the PvdA (2.8 percent) all fared badly. The fall in support for D66 was in line with the national outcome in 1998. D66 was criticised for its invisibility in the first purple coalition and is a very 'modern' party with a low level of ties with its members and voters. The loss of the CDA is in line with a national trend caused by the weakening of ties with voters due to a combination of secularisation and weak party leadership. Its loss in seats, compared to just prior to the elections, is less obvious, because Janssen van Raay was no longer a CDA representative in 1998 (see above). The gains of Green Left and the SP can both be attributed to their clear role in opposition in the national parliament, strong party leaders and their relatively critical attitude towards Europe.

Table 13.1 Outcome of national and EP elections in the Netherlands, 1994–99

	National elections 1994 (%)	European elections 1994 (%)	Seats EP (1998)	National elections 1998 (%)	European elections 1999 (%)	Seats EP (1999)
CDA	22.2	30.8	9	18.4	26.9	9
PvdA	24.0	22.9	7	29.0	20.1	6
VVD	20.0	17.9	6	24.7	19.7	6
D66	15.5	11.7	4	9.0	5.8	2
RPF/GPV/SGP	4.8	7.8	2	5.1	8.7	3
Green Left	3.5	3.7	1	7.3	11.9	4
CD	2.5	1.0	–	0.6	0.5	0
Luc Sala					0.3	0
EVPN			2		0.4	0
SP	1.3	1.3	–	3.5	5.0	1
Europese Partij					0.7	0
Other	6.2	2.9	–	2.4		
Total	100	100	31	100	100	31
Turnout	78.7	35.7		73.3	29.9	

Conclusion

Where no real choices are offered, no real choices are made. The whole EP campaign was determined by some big political parties agreeing on most issues related to the EU and a collective fear among politicians that turnout would be low. Despite all attempts to increase turnout by means of advertisements and television spots, turnout dipped below 30 per cent. The main cause of this low turnout is probably a combination of *parteiverdrossenheit* (a low level of party attachment) and the absence of a real choice between different European policies because the EP is not an executive power. However, even though one might have expected some compensatory turnout to result from mobilisation on national issues because EP elections are second-order elections, this did not materialise. But despite the crisis in the purple coalition in the weeks preceding the EP elections, and the publication of the report about the Bijlmer disaster, a clear national issue was absent, partly due to the fact that the present coalition includes both a party from the right (VVD) and a party from the Left (PvdA), which are both willing to continue their collaboration.

References

Writing this, I used party material published on the Internet, material published by the EP and articles from many Dutch newspapers, including *NRC Handelsblad*, *De Volkskrant* and *Tubantia* (many still available on the homepages of these newspapers). The book by Joep Dohmen (1999), *Europese Idealisten, Een chronique scandaleuse van het Europees Parlement* (Sun Kritak), offered insight into the background of scandals in both EP and EU Commission. National Party platforms are all available in A. Lipschits (1998) *Verkiezingsprogramma's, SDU uitgevers* (Den Haag). Information about the most important Dutch political parties can be found on the Internet (for links, see http://www.ub.rug.nl/dnpp). Information about the background of the low levels of turnout were communicated by Kees Aarts, working on a study about turnout levels in the Netherlands. Survey data about opinions of members of national parliament and the EP collected in the context of the European representation study (A. Schmitt and B. Thomassen, editors, *Political Representation and Legitimacy in the European Union*, Oxford University Press, 1999) were used to confirm data from other sources.

List of Dutch party abbreviations used

CD   Conservatieve Democraten
CDA   Christen Democratisch Appèl
D66   Democraten 66
EVPN   Europees Verkiezers Platform Nederland

GPV   Gereformeerd Politiek Verbond
PvdA Partij van de Arbeid
RPF   Reformatorische Politieke Federatie
SGP   Staatkundig Gereformeerde Partij
VVD   Volkspartij voor Vrijjheid en Democratie

14
Portugal

José Magone

Introduction

The Portuguese party system has remained very stable since the founding elections of 25 April 1975. Since then most of the general and secondary elections have produced a four-party system, in which two major parties, the Socialist Party (Partido Socialista, or PS) and the Social Democratic Party (PSD), have alternated in power. Two smaller parties, the Communist Party and the People's Party, were able to gain representation throughout this period, although their electoral strength declined from election to election (Aguiar, 1985; Magone, 1999). No other parties have been successful in EP elections. The 1999 Euro-elections confirmed this basic pattern of the Portuguese party system. A short overview of the internal atmosphere of each party shows that the Socialist Government was able to take the initiative and win an important electoral contest before the general election in October 1999. Strategies for the general election were at the forefront of the EP election campaign, and even led to the revival of internal factionalism in the liberal PSD and the conservative Partido Popular (PP). The EP elections were certainly a victory for the left-wing parties, the Socialist Party and Communist Party which preserved continuity and coherence throughout the campaign period.

European Parliament elections in Portugal are defined in electoral law 14/1987 of 29 April. Twenty-five MEPs are elected in a single constituency by the d'Hondt system of proportional representation. After thirteen years of EU membership, the Portuguese political elite has become more assured and assertive concerning the EU – something observed for the first time in these EP elections – and is better integrated in EU decision-making processes. Portugal has benefited from the EU economically via the structural funds. Rural infrastructures have been modernised. While regional disparities have grown over the past fifteen years, national GDP per capita relative to the EU average has risen from 55.1 per cent in 1986 in terms of purchasing power parity (PPP) to 68.4 per cent in 1995 (Pires, 1998, 214). Compared to other west European countries, unemployment remained relatively low (4–7 per cent during the 1990s, and now around 4.3 per cent), the second lowest in the EU after Luxembourg, excluding the informal or 'black economy' (Eurostat, April 1999). Average unemployment in the EU is 9.6 per cent, with the eurozone having an average of 10.4 per cent. Portugal's participation in EMU enhanced the political class's self-confidence.

The desire to play an active part in building the EU has been an important element of policy-making since 1986. The return of Mário Soares to the political arena was the most remarkable single issue of the 1999 EP elections in Portugal and a clear sign of willingness to play an important role in the EU institutions. After his term as president of the Republic ended at the end of 1995, Mário Soares retired from politics and dedicated most of his time to the establishment of the Mário Soares foundation. In spite of his age, he is still able to mobilise the population for the SP. His re-entry into the political fray led to a general crisis among the opposition parties, who almost withdrew when they heard the news. Their sentiments were best summarised by the comment that having Mário Soares in the EP elections was like having Amalia Rodrigues a *fado* singer in a song contest. According to well-informed weekly newspaper *Expresso*, Mário Soares went to the Prime Minister and asked him to lead the EP list. This was unusual and says a lot about his continuing power in the Socialist Party (*Expresso*, 15 May 1999, 3). This seems to be confirmed by an interview with Prime Minister Antonio Guterres (*Expresso, Revista*, 22 May 1999, 69–70). Reference is made to the Soares political family because his son, João Soares, is the mayor of Lisbon and also enjoys special status in the Socialist Party. An analogy with the Papandreou political family before the death of Andreas Papandreou can be made as noted by PSD head of list, José Pacheco Pereira, and PSD leader, José Manuel Barroso. Internal factionalism is a major problem for the PSD which is still dominated by regional barons. These so-called *baronatos* need a charismatic strong leader to give them direction. Such leaders included Anibal Cavaco Silva (1985–95) and Francisco Sá Carneiro (1974–80). Consequently, leadership with authoritarian traits is common. Since Anibal Cavaco Silva returned to academic life, the party has already had three more leaders: Fernando Nogueira (1995–6), Marcelo Rebelo de Sousa (1996–8), and now José Manuel Durao Barroso. The stakes were upped when the SP announced at its party conference held in February that Mário Soares would be the next EP president if it won the elections. This was seen as a first salvo in the general election. SP Leader and Prime Minister Antonio Guterres was able to unite the whole party once again around the highly controversial and difficult figure of Mário Soares. This was certainly a major success for Guterres because Mário Soares had opposed the SP line in the campaign on the referendum on regionalisation held on 8 November 1998. Guterres's conciliatory political instinct increased the SP's chances of winning outright in the EP elections. The other parties clearly had difficulties in finding leading candidates for their lists (see Table 14.1).

The PSD was riven by internal factionalism. Party leader Marcelo Rebelo de Sousa was challenged in a television interview on 14 February 1999 by the former Minister of Foreign Affairs, José Durao Barroso, who opposed the confrontation strategy chosen by the party leader. He advocated a more sophisticated approach to exclude any possible alliance with the conservative People's Party (Centro Democratico Social/Partido Popular, or CDS/PP). This led finally at a party conference in Coimbra in early May to the replacement of Marcelo Rebelo de Sousa by José Durao Barroso as the new leader. This was an extraordinary change of policy because five weeks earlier Marcelo Rebelo de Sousa had been reconfirmed as leader at a party conference in Porto. One of the first moves of new leader Durao Barroso was to nominate as head of the list the well-known PSD MP, José Pacheco Pereira, after the previous head of the list Leonor Beleza had resigned over the change in leadership (*Expresso, Revista*, 8 May 1999, 98–108). This internal factionalism was clear at the Coimbra

Table 14.1 The main political parties

	Socialist Party (PS)	Social Democratic Party (PSD)	Unitary Democratic Coalition (CDU)	People's Party (PP)
Leader	Antonio Guterres	José Durao Barroso (1999)	Carlos Carvalhas	Paulo Portas
Head of List	Mário Soares	José Pacheco Pereira	Ilda Figueiredo	Paulo Portas
Euro-Group	European Socialist Party (PSE)	European People's Party (EPP/PPE)	Union of European Left (EUL-NGL)	Union for Europe (UFE)
Membership	70 000 (1992)	183 630 (1996)	100 000 (1996)	23 856 (1992)

Sources: Frain (1997), 94, 98; Cunha (1999), 35; *Expresso*, 4 July 1992, 20–1.

conference when about 40 per cent of the delegates still opposed the election of Durao Barroso as new PSD president. Others simply changed sides, and there was a strong majority for Durao Barroso. Internal division and the insecurity of the leadership certainly became one of the major factors leading to EP election losses in June (*Expresso, Revista*, 8 May 1999, 98–108). With the rise of Durao, the conservative PP abandoned its strategy of forging an alliance with the second largest party. The project of former PSD leader, Rebelo de Sousa, to form an preelectoral alliance called the Democratic Alternative (Alternativa Democrática, or AD) with the PP and so break the Socialist majority was no longer advocated by Durao Barroso. After the PP losses in the local elections of December 1997, the new leader, Paulo Portas, moderated the highly radical Euro-sceptic approach of former leader Manuel Monteiro (Magone and Stock, 1998, 511). The PP was robbed of a strategy they had pursued since 1998, just one month before the elections. This was a major factor in the PP's electoral decline. Internal factionalism still affects its performance as a credible alternative to the other major parties.

The Communist Party united around its Secretary-General, Carlos Carvalhas. It continued its steady decline in membership and electoral support. It continues to follow a Marxist-Leninist ideology and identity and is at present one of the largest orthodox Communist parties in the EU, along with the French Communist Party. Its candidates have been in coalition with the tiny Green Party (Partido 'Os Verdes', or PEV) since 1987. Seven tiny parties took part in the EP elections. The most original one was the new Bloc of the Left (Bloco de Esquerda, or BE) founded in March by a group of these parties, including the Trotskyist Socialist Revolutionary Party (Partido Socialista Revolucionario, or PSR) under the leadership of Francisco Louca, the Democratic People's Union (Uniao Democratica Popular, or UDP) and Politica XXI. The main candidate was the brother of Miguel Portas, the PP/CDS leader. The Maoist Revolutionary Movement for the Reconstruction of the Party of the Proletariat/Communist Party of the Portuguese Workers (Movimento Revolucionario do Partido do Proletariado/Partido Comunista dos Trabalhadores Portugueses, or MRPP/PCTP) had existed since the 1970s, but failed to get re-elected in any elections (*Expresso*, 22 May 1999, 8). Another Trotskyist party losing support from election to election was the Workers' Party for Socialist Unity (Partido Operário para a Unidade Socialista, or POUS). The People's Monarchic Party (Partido Monarquico Popular, or PPM), the Movement of the Party of the Earth (Partido Movimento da Terra, or MPT), the pro-pensioners' Party of National

Solidarity (Partido de Solidariedade Nacional, or PSN) and the regionalist Democratic Party of the Atlantic (Partido Democrático do Atlantico, or PDA), based in the Azores, contested the elections without any real chance of winning EP seats.

The campaign

Most parties contested the EP elections nationally and on predominantly national issues. The lack of domestication of European politics in EU member states is the major obstacle to mobilising the vote (Fenner, 1981, 35–41; Hix and Lord, 1997, 15). According to the Eurobarometer's autumn 1998 survey, only 35 per cent of the Portuguese thought the European Parliament was doing a good job (see Table 14.2).

The Portuguese also had the lowest knowledge about the EU, even though support for membership was still the highest in the EU (Eurobarometer, 49, 1998, 3). Although Portuguese parties took part in the joint events organised by the European People's Party (EPP) in Madrid and by the European Socialist Party (PSE), this did not translate into a European campaign (*Expresso*, 22 May 1999, 4). Moreover, party

Table 14.2 What is your opinion on the performance of the European Parliament?

Country	Good	Bad	Don't know
Ireland	57	18	25
Luxembourg	54	25	21
Italy	41	28	31
Belgium	39	34	27
Denmark	39	48	13
Austria	38	29	33
France	37	38	25
Netherlands	37	39	24
Finland	37	52	11
Greece	36	51	13
Spain	35	30	35
Germany	35	32	33
Portugal	*35*	*37*	*28*
UK	34	33	33
Sweden	23	38	39
EU 15 average	37	34	29

Source: Eurobarometer, 50 (October–November 1998).

strategies were designed as a dry run for the legislative elections in October 1999. The EP elections were seen as of secondary importance to legislative and local elections (*Expresso*, 5 June 1999, 7), as had been confirmed by turnout in elections since 1987. The most interesting feature of the Portuguese case, mirrored in Spain, was the growing interest of former national leaders such as Mário Soares and Felipe Gonzalez in EU-level politics.

The campaign started officially one month before the day of the elections on 13 June 1999. The EP tried to mobilise the youth vote by raising public awareness through a simultaneous poster campaign in Lisbon, Brussels and Berlin. The presidential approach of Mário Soares to the EP elections made it very difficult for the opposition to draw attention to other problems. As in most European countries, the campaign was dominated by the Kosovo war and the Portuguese contribution towards NATO bombing of Yugoslavia, which was opposed by both Mário Soares and José Pacheco Pereira, and seconded by the anti-war campaign of the Communist, Ilda Figueiredo. Among the four major parties only Paulo Portas, the head of the list of People's Party, supported NATO action (*Expresso*, 15 May 1999, 4). During the campaign, Lisbon's mayor, João Soares, went to Belgrade to show solidarity with the Serbian people against NATO, even while declaring himself opposed to Milosevic. This was seen as an affront to the Government (*Expresso*, 15 May 1999, 6). In spite of this individual action, Prime Minister Guterres saw this as part of democracy, where people of the same party may have different opinions in relation to the Kosovo war. (*Expresso, Revista*, 22 May 1999, 69). Shortly before the elections, the Government announced that it would send 290 soldiers to the peace-keeping forces in Kosovo (*Expresso*, 5 June 1999, 1).

Another campaign issue was the move towards European taxation raised by Mário Soares in the Paris meeting of the European Socialist Party. Most of the parties attacked Mário Soares on this score. In the end, Prime Minister Guterres intervened to deny that European taxation was on the agenda. This issue was raised in the television debate with the four main candidates and rejected by Ilda Figueiredo, Pacheco Pereira and Paulo Portas. In the same debate the creation of a European Army was considered as positive step by both main parties. Ilda Figueiredo rejected this, but Paulo Portas felt that a Euro-Atlantic alliance should remain as close to the USA as possible. Mário Soares also broached the question of a European constitution and the establishment of a federal Europe. This was strongly rejected by Paulo Portas, a supporter of a Europe à la carte, and advocacy of national interests above any consideration of moves towards new forms of integration (*Expresso*, 15 May 1999, 4). There was little difference between the manifestos of the two main parties, but both the Communist and the People's Party tended to have more salient programmes opposing the EU's current development. The Socialist Party was well integrated into the structures of the European Socialist Party, and most of its manifesto had a European dimension. The PS slogan was 'A team in the heart of Europe' (*Una equipa no coracao da Europa*). It offered a confidence pact for the future of Europe, borrowing from the stable and successful Guterres government whose achievements it catalogued. It noted, in particular, the Government's success in getting a good financial outcome for Portugal in the Agenda 2000 negotiations. European citizenship, the fight against unemployment and enhancing the role of the European Union in playing a role in the international arena were the other main topics in the manifesto. Moreover, the party programme clearly wanted to place Portugal at the centre of EU decision-making: the choice of Mário Soares and the strategy of co-ordinating efforts between Portuguese socialist MEPs and the Guterres government have to be seen in this context (PS, 1999a).

The Programme of the PSD had to be written during the campaign. The late election of José Manuel Durao Barroso to the leadership of the PSD in early May 1999 and the divisions inside the party led to a difficult start for the head of the list, José Pacheco Pereira. He had to find a substitute for the early choice of Leonor Beleza when she declined to head the list under Manuel Durao Barroso's leadership. Carlos Pimenta, the most prominent social democratic MEP, also decided against being part of the group, which reduced Durao Barroso's options (*Expresso*, 22 May 1999, 9). The programme of the social democrats is quite thorough. It really asked the voters to return to the idealism of the EU's founding fathers. Solidarity with Central and Eastern Europe was portrayed as essential to European unity. The manifesto also stressed the need to strengthen the concept of European citizenship and to defend Portuguese interests. The programme was directed against socialists in Portugal and other EU member states. This anti-partisan attitude was evident throughout the manifesto, whose main objective was to break the socialist dominance of Europe (PSD, 1999; *Povo Livre*, 26 May 1999, 3–8; *Povo Livre*, 12 May 1999, 18). José Pacheco Pereira was not able to enthuse the population with this programme (*Expresso*, 22 May 1999, 6). Resignation to the inevitability of the outcome was mirrored in one of several similar slogans: 'Soares is already elected. Now it is necessary to find somebody to work' (*Expresso*, 5 June 1999, 2; *Povo Livre*, 26 May 1999, 3–8).

During the campaign Paulo Portas reminded the population that Pacheco Pereira could not be trusted because he had been part of the revolutionary radical MRPP during the Revolution of Carnations and he had just published a biography of Alvaro Cunhal. In this way, Paulo Portas tried to win over some conservative and right-wing voters from the PSD. The PP programme did not change very much from that of its predecessor, Manuel Monteiro. A Europe of Nations with maximum flexibility to defend the sovereignty of the country remained its central plank (*Expresso*, 15 May 1999, 8). The Communist coalition led by Ilda Figueiredo issued a programmatic declaration opposing movement towards a federal Europe, and supporting a strengthening of national sovereignty, social Europe, European citizenship and the role of national parliaments in European integration. Throughout the campaign, the CDU opposed the Kosovo war (CDU, 1999; *Avante*, 1 April 1999, 5–6). The smaller parties had to campaign with very limited means. They had some air time on television, but protested because it was significantly less than that allocated to the four larger parties (*Expresso*, 11 June 1999, 7). The Bloc of the Left led by Miguel Portas stressed the need to draft a manifesto of citizens and peoples in order to augment democracy at the supranational level and to give more opportunity to national citizens to actively shape European policies. The MRPP/PCTP campaigned against the Balkan War and Portuguese secession from NATO, which it wanted to see dissolved without a new alliance being put in its place. This position was also supported by the POUS. The MPT advocated strengthening the EU budget, particularly for agriculture, and taxing capital movements. The PDA wanted reform of the EP election law, while the PPM sought health sector reform by arguing for traditional medicine and reinforcing action against alcoholism and drug consumption. Last but not least, the PSN remained faithful to its origins by seeking to be the voice of the pensioners, of migrants, and small shopkeepers (*Expresso*, 15 May 1999, 4).

On the eve of the elections, the president of the National Commission of Elections (Comissao Nacional de Eleicoes, or CNE), the main supervising body of electoral processes in Portugal, intervened on television. Armando Pinto Basto suggested that abstainers could not be seen as good citizens: only those who voted deserved such a label. This was opposed by the political class as being a paternalistic attitude (*Expresso*, 19 June 1999, 6).

The results

From the outset the main enemy of the largest parties has been abstention, which has been the most popular option in the past 12 years of EP elections in Portugal. In the end the abstention rate was lower than in 1994. Nevertheless, 60 per cent did not vote, showing that these elections are seen as second-order elections. Abstention has risen throughout the 1980s and 1990s in all Portuguese elections. The improvement in turnout was interpreted as a success, however, when the Portuguese levels were compared to those in other member states where abstention had increased while falling in Portugal (see Table 14.3).

Opinion polls published by the *Expresso* one week before the elections predicted that the PS would get 44.8 per cent of vote, the PSD only 25.3 per cent, the PP 4.8 per cent, the CDU 5.2 per cent and the Bloc of the Left 2 per cent. The actual result was less spectacular. In spite of the overwhelming PS victory, the other opposition parties were able to withstand the dominance of Mário. Although all three main opposition parties lost votes, they did not lose as many as predicted, which each party interpreted as a victory (see Table 14.4).

José Pacheco Pereira was very pleased that the PSD secured 30 per cent of the vote and kept its nine MEPs. The CDU won more than 10 per cent of the vote and lost one MP while becoming the third largest political group in the country, while the Paulo Portas could argue that the PP's predicted demise was off the mark as its share of the vote had risen since the local elections of December 1997, when it got 5.6 per cent of the vote. A more thorough analysis undertaken by *Expresso* shows that the only winner of the elections was the PS, which gained not only in terms of the absolute number of votes won since 1994, but also in terms of the share of the vote (see Table 14.5). Territorially, the PS became the only party with balanced support across the mainland and the Azores. Only in Madeira did the PS win less than 30 per cent of the vote, but even there its share rose to 29 per cent from 25 per cent in 1994. The main reason for this is that Madeira is still dominated by the charismatic president of the regional government, Joao Jardim, who uses tough language against

Table 14.3 Electoral turnout (and abstention)

Year	1999	1994	1989	1987
Turnout (abstention) in percentage	40(60)	36.6(64.4)	48.9(51.1)	72.4(27.6)

Table 14.4 The results

	1999		1994		1989		1987	
	Votes(%)	Seats	Votes(%)	Seats	Votes(%)	Seats	Votes(%)	Seats
PS	43.05	12	34.8	10	28.52	8	26.9	8
PSD	31.10	9	34.4	9	32.70	9	37.4	9
CDU	10.32	2	12.5	3	14.40	4	11.5	3
CDS/PP	8.12	2	11.2	3	14.13	3	15.4	4
B.E.	1.79	–	–	–	–	–	–	–
PSR	–	–	0.59	–	0.76	–	0.5	–
UDP	–	–	0.63	–	1.08	–	0.9	–
Politica XXI	–	–	0.4	–	–	–	–	–
MRPP	0.88	–	0.78	–	0.64	–	0.3	–
POUS	0.16	–	–	–	0.27	–	0.3	–
MPT	0.39	–	0.41	–	–	–	–	–
PSN	0.25	–	0.37	–	–	–	–	–
PDA	0.15	–	0.23	–	–	–	–	–
PPM	0.44	–	0.27	–	2.03	–	2.08	–
Total	–	25	–	25	–	24	–	24

Source: Ministério da Administra cao Interna (1987, 1989, 1994); http://www.europeias.dgsi.pt

the PS. The alliance between Antonio Guterres and Mário Soares also contributed to this balanced result. Indeed, Mário Soares was able to mobilise the voters in the north, while Guterres was successful in the south (*Expresso, Revista*, 19 June 1999, 76–9). During the Cavaco Silva years the PSD was regarded as the only nation-wide party but, after he quit politics, the party lost about 20 per cent of the vote. The presence of a charismatic figure to unite the political barons and give a sense of purpose is still important. José Durao Barroso has a long way to go in asserting himself inside the party and as a credible alternative to the Prime Minister, Antonio Guterres. The PSD's strongholds are in the north and in Madeira. In the south the party is quite weak, apart from the Algarve (*Express. Revista*, 19 June 1999, 80).

The CDU retains its stronghold among farm labourers in the southern region of Alentejo. The desertification of this region is leading to a steady decline in this loyal electorate. The CDU is traditionally very weak in the north, due to the fact that the population is more conservative and opposed the CDU's dominance during the revolutionary process (*Expresso, Revista*, 19 June 1999, 82) In spite of change of strategy, the PP remains in the doldrums. Its best results of over 10 per cent were achieved in the centre of Portugal (in Aveiro, Viseu and Guarda) and in

Table 14.5 Voters' mobility between 1994 and 1999

	PS '94	PSD '94	CDU '94	PP '94	Other, blank and nil voting	Abstention '94	New voters '99	Total '99
PS 99	1036	99	31	8	–	296	20	1490
PSD 99	–	932	–	26	–	105	13	1076
CDU 99	–	–	302	–	–	54	1	357
PP 99	–	–	–	281	–	–	2	283
Other, blank and nil votes	12	3	6	2	215	14	3	255
Abstention	–	–	–	59	–	4969	94	5122
Total 94	1048	1034	339	376	215	5438	133	8571

Source: Expresso, 19 June 1999; Revista, 85.

Table 14.6 Continuity of Portuguese MEPs

	PS	PSD	CDU	CDS/PP	Total	%
Elected before 1999	6	3	1	–	10	40
Elected for the first time	6	6	1	2	15	60
Total	12	9	2	2	25	100

some regions of the north. It lost heavily in urban centres and most parts of the north. The party disappeared completely from the south of the country, but was able to retain a presence in the islands of Madeira and the Azores. The smaller parties became almost 'virtual' parties in that they were not able to make any impact on the four-party format of Portuguese politics. The total vote share of these seven 'dwarfs', as one of the leaders of the MRPP called it, was 4.09 per cent.

Some 40 per cent of MEPs were re-elected, while 60 per cent are new to the EP (see Table 14.6). Continuity is highest among the Socialists and lowest among the social democrats, who were certainly affected by the fact that many candidates gave up after the election of the new party leader, Manuel Durao Barroso, and internal squabbling in the PP. The percentage of female MEPs rose from 8 to 20 per cent, but remained well below the EP average.

Most MEPs have plenty of political experience in Portuguese politics. Most were MPs in the Portuguese Assembly of the Republic. The PS group comprises several state secretaries from the present Government. The

PSD group includes a former minister of Agriculture, Arlindo Cunha. Among the most experienced MEPs is Luis Marinho. The infighting within the PSD led to the self-exclusion of Carlos Pimenta, a very experienced MEP from the PSD list. The group also has many academic MEPs, the most prominent scholar being Maria Carrilho, a prominent promoter of political science in Portugal before becoming an MP in 1995.

Conclusions

The electoral losses of the European Socialist Party in most member states, particularly Germany and the UK, did adversely affect Mário Soares's prospects of becoming the next EP president. He did try to suggest that securing his election was a matter of national importance, and that therefore all Portuguese MEPs should vote for him. This approach backfired and led to an intensive discussion about this misinterpretation of the national interest (*Expresso*, 3 July 1999, 9; *O Publico*, 20 July 1999, 2–3). Mário Soares's position was difficult to assess. In the media, he was described as a big loser of Portuguese elections because, as a former president of the Republic, he lost his supra-partisan authority by standing for the Socialist Party. In comparison to his 70 per cent share of the vote in the presidential elections, the 43 per cent secured in the EP election was seen as a defeat for Soares and as a victory for Antonio Guterres in his preparation for the next legislative elections on 10 October 1999 (*Expresso*, 19 June 1999, 3). For the political class, in general, a seat in the European Parliament is highly coveted even though Portuguese MEPs are paid less than most of their counterparts apart from the Finns and Spanish (the Austrians being the most highly paid).

The EP is, however, for the people very distant. Portugal, along with the UK, records the lowest level of knowledge about the EU. The low turnout confirmed that the electorate preferred to use the long weekend (10–13 June) to go to the sunny beaches of the Portuguese coast (10 June is a national holiday). In 1999, it fell on a Thursday. In such cases the Portuguese take Friday off and 'make a bridge' (*fazer ponte*) to the weekend, so having a four-day holiday. As elsewhere, the EP elections seemed of marginal significance.

References

Aguiar, J. (1985) 'The hidden fluidity in a ultra-stable system', in E. Sousa Ferreira and W. C. Opello Jr (eds), *Conflict and Change in Portugal, 1974–1985* (Lisbon: Editorial Teorema), 101–26.

Barroso, J. M. (1999) *È Tempo de Mudar* (Lisbon: PSD).
Coligacao Democratica Unitaria (CDU) (1999) 'Mais CDU', *Declaracao Programática do PCP* (Lisbon: CDU).
Cuhna, C. (1999) 'The Portuguese Communist Party', in T. C. Bruneau (ed.), *Political Parties and Democracy in Portugal* (Boulder, Colorado: Westview Press), 23–54.
Fenner, C. (1981) 'Die Grenzen einer Europäisierung der Parteien: Europa kann man nicht wählen', *Politische Vierteljahresschrift*, 22, Heft 1: 26–44.
Frain, M. (1997) 'The Right in Portugal: The PSD and the CDS/PP', in T. C. Brunneau (ed.), *Political Parties and Democracy in Portugal* (Boulder, Colorado: Westview Press), 112–37.
Hix, S. and Lord, C. (1997) *Political Parties in the European Union* (London: Macmillan).
Magone, J. M. (1996) 'Portugal', in Juliet Lodge (ed.), *The 1994 Elections to the European Parliament* (London: Pinter), 147–56.
Magone, J. M. (1997) *European Portugal. The Difficult Road to Sustainable Democracy* (London: Macmillan).
Magone, J. M. (1999) 'Party System Installation and Consolidation', in D. Broughton and D. Mark (eds), *Changing Party Systems in Western Europe* (London: Pinter), 232–54.
Magone, J. M. and M.-J. Stock (1998) 'Portugal', in *Political Data Yearbook*, Special issue of *European Journal for Political Research*, 34, 507–11.
Ministério da Administracao Interna (1987) *Eleicao para o Parlamento Europeu 1987* (Lisbona: STAPE).
Ministério da Administracao Interna (1989) *Eleicao para o Parlamento Europeu 1989* (Lisbon: STAPE).
Ministério da Administracao Interna (1994) *Eleicao para o Parlamento Europeu 1994* (Lisbon: STAPE).
Partido Social Democrata (1999) *Eleicoes para o Parlamento Europeu. Programa Eleitoral 1999*. (Lisbon: PSD).
Partido Socialista (1999) Mário Soares, *Portugal no Coracão da Europa* (Lisbon: PS).
Partido Socialista (1999a) *Uma Equipa no Coracao da Europa* (Lisbon: PS).
Pires, L. M. (1998) *A Politica Regional Europeia e Portugal* (Lisbon: Fundacao Calouste Gulbenkian).
Sites with final 1999 EP election results:
 HYPERLINK http://www.stape.pt; http://www.stape.pt
 HYPERLINK http://www.europeias.telepac.pt; http://www.europeias.telepac.pt

List of Portuguese party abbreviations used

AD Alternativa Democrática (Democratic Alternative)
BE Bloco de Esquerda (Bloc of the Left)
CDS Centro Democratico Social
CDU Unitary Democratic Coalition
MPT Partido Movimento da Terra (Movement of the Party of the Earth)

MRPP/PCTP	Movimento Revolucionario do Partido do Proletariado/ Partido Comunista dos Trabalhadores Portugueses (Revolutionary Movement for the Reconstruction of the Party of the Proletariat/Communist Party of the Portuguese Workers
PDA	Partido Democrático do Atlantico (Democratic Party of the Atlantic)
PEV	Partido 'Os Verdes' (Green Party)
POUS	Partido Operário para a Unidade Socialista (Workers' Party for Socialist Unity)
PP	Partido Popular (People's Party)
PPM	Partido Monarquico Popular (People's Monarchic Party)
PS	Partido Socialista (Socialist Party)
PSD	Social Democratic Party
PSN	Partido de Solidariedade Nacional (National Solidarity Party)
PSR	Partido Socialista Revolucionario (Socialist Revolutionary Party)
UDP	Uniao Democratica Popular (Democratic People's Union)

15
Spain

John Gibbons

Spain's transition to democracy was helped by political and economic aid emanating from inside the EU (Gillespie, Rodrigo and Story 1995, 33). Democratic Spain repaid the EU by a steadfast loyalty to the goals of European integration. Since joining on 1 January 1986, it has been in many respects, a 'model' member. Major Community initiatives, including the SEA, the TEU, EMU and the Amsterdam Treaty, were welcomed and adopted in Spain without serious political reservations. Even a change of government in 1996 from left to right – from the PSOE (*Partido Socialista Obrero Español*/Socialist Workers' Party of Spain) to the PP (*Partido Popular*/People's Party) – barely caused a ripple in Spain's relations with the EU. The new government settled down to the task of making sure Spain qualified for EMU membership. It did. But there was some political disquiet – especially within IU (*Izquierda Unida*/United Left) – about the high social costs involved in the process of economic convergence, among other matters.

In view of Spain's relatively quiescent history of EU membership, one of the most intriguing questions to be answered in this chapter is why the 1999 elections for Spain's 64 seats in the EP attracted a myriad of Spanish, minority nationalist, regional and single issues parties and electoral blocs? The simple answer to this question is that for all participants, this election served purposes other than the obvious task of returning 64 MEPs to Strasbourg. In the case of the PP and its leader, Prime Minister José Maria Aznar, a straight victory in the European elections was not enough. Preferably, that victory would be large enough to provide unambiguous evidence to the PP and the public of the success of his prime ministership (*ABC*, 14 June 1999, 13). While maintaining PP's presence in the EP was important, much more significant was the task of winning the general elections due in 2000, preferably with enough MPs to form a majority government, rather than the minority government Aznar currently headed with the support of minority nationalists. In short, he hoped that the Euro-election results would provide clear evidence that a second term in office was within PP's reach. Aznar did not anticipate maintaining the 10.46 per cent lead over the PSOE that it had achieved in 1994, but opinion polls predicted a 5 per cent margin, and he declared that he would be satisfied if PP achieved this.

The PSOE was working to a rather different agenda. With its candidate for prime minister (José Borrell) forced to resign a few months before the European elections due to an awkward entanglement in one of Spain's many corruption '*casos*', the outlook for the general election in 2000 was not great for the PSOE in June 1999. Under the circumstances, the Socialists were looking for any ray of hope on the rather gloomy electoral horizon. In particular, they sought to achieve some improvement on the dire results of the 1994 European and 1996 general elections to serve as evidence that the PSOE had stepped back from the edge of the abyss and was, once more, on its way towards government, or at least as soon as it appointed a replacement candidate for Prime Minister, a potentially explosive task wisely postponed until after the European elections.

IU, under the leadership of Julio Anguita, took the moral high ground in these elections by loudly condemning the NATO campaign in Kosovo (and EU member states' contribution to it) as more evidence of what it considered to be American imperialism in Europe. It had attracted 13.4 per cent of the vote in 1994 (including many disillusioned Socialists) and envisaged that it would also absorb the anticipated widespread opposition in Spain to NATO's war in Kosovo (the IU was born out of such opposition in the mid-1980s). Ultimately, IU aimed to expand its presence on the left of the political spectrum, as the PSOE moved to the centre ground.

CiU (*Convergencia i Unió*/Convergence and Union), the Catalan nationalist coalition of parties led by Jordi Pujol, emerged from the 1994 elections with 4.6 per cent of the vote. Success in the 1996 general election enabled it to bargain strongly for various concessions to Catalonia, in exchange for supporting Aznar's minority government (Gibbons, 1999, 75–8). Its main priority therefore in those elections was to ensure that the electorate did not punish it for its intimacy with the PP government over the previous three years. More specifically, CiU sights were trained, beyond the Euro elections, on the all-important regional elections to the government of Catalonia – the *Generalitat* – scheduled for October 1999.

Basque nationalists also had a lot at stake in the 1999 European elections, and not just their influence in the European Parliament. The PNV (*Partido Nacionalista Vasco*/Basque Nationalist Party) was the biggest party in the nationalist bloc of self-styled 'nations without states', which contested the elections on a joint ticket: the *Coalición Nacionalista/Europea de los Pueblos* (Nationalist Coalition/People's Europe). Another Basque party contesting the elections was *Euskal Herritarrok* (EH, or Basque Nation) led by Arnaldo Ortegi. It was the electoral bloc for ETA's political wing HB (*Herri Batasuna*/Basque Homeland). EH was formed in September 1998 and had made a big impact on the Basque regional elections that autumn. Subsequently, it created, with PNV and EA (*Eusko Alkartasuna*/Basque Solidarity), the first all-nationalist Basque regional government. Paramount for the Basque nationalist parties in these elections was winning public approval for developments to date in the peace process there and credit for their various political strategies following the ETA ceasefire of September 1998.

Contesting the European elections alone rather than in an electoral formation with other nationalist parties was the BNG (*Bloque Nacionalista Galego*/Galician Nationalist Bloc) led by Xosé Manuel Beiras. BNG had commanded growing success during the late 1980s and early 1990s in local and regional elections of the 'historic nationality' of Galicia. In 1996, it won a seat in the Spanish *Cortes* (Parliament). A seat in the EP would further enhance the legitimacy of the cause of Galician nationalists.

Among other contestants of note were right-of-centre regional parties of a non-nationalist persuasion which tended to group under the banner of *Coalición Europea* (European Coalition): for example, PA (*Partido Andalucista*/Andalusian Party) and CC (*Coalición Canaria*/Canary Islands Coalition). The Greens were in a left-of-centre grouping of *Los Verdes/Isquierda de los Pueblos* (The Greens/People's Left) led by trade unionist Antonio Gutiérrez. Among the many other competing parties and groups in these elections was an all-women's coalition, whose all-women list of candidates was a unique feature of the European elections.

The players in a domestic context

Whereas the 1994 EP election coincided with regional elections in Andalusia, in 1999 the municipal, regional (in 13 of the 17 regions) and EP elections took place on the same day. These circumstances added momentum to the events of the elections which in Spain were on Sunday, 13 June. In both the 1994 and 1999 EP elections the anticipation of an imminent general election helped to concentrate the minds of voters, who were frequently reminded by media commentators and some politicians that the results would be treated as a measure of the government's popularity. However, there was probably more at stake in 1994. At the time, the PSOE government was hanging by a thread, in the face of a myriad of cases of sleaze and scandal, whereas in 1999 the PP was touched by comparatively little controversy. In Catalonia, where the socialists were likely to make a strong challenge for the CiU-dominated regional government in the elections of November 1999, there was an added piquancy to these elections.

The most overt discussions on EU policies before and during the elections undoubtedly took place between the IU coalition and the other main Spanish and nationalist and regional parties and indeed, to some extent within IU itself. The IU offered the nearest Spain had to Euro-scepticism in a major political party. Among other matters, it expressed reservations about the social effects of the rush to further economic integration, the levels of democratisation in the EU and the role of EU states in the Kosovo crisis. Javier Solana, the Spanish Secretary-General of NATO, was picked out for particular opprobrium by Julio Anguita, but the aggressive rhetoric of Anguita caused tensions within the IU.

The identities of who would be formally nominated by the government as Spain's EU Commissioners were not confirmed until after the elections, but the two nominees of the main parties were well known. They were Pedro Solbes of the PSOE and Loyola de Palacio of the PP (the latter was also the PP's chief candidate for the EP elections). As the appointments had yet to be confirmed, their suitability for the role of Commissioner became a source of election controversy between the main parties. While the PSOE alluded to de Palacio's failings as Spain's agriculture minister, some in the PP questioned whether Solbes was suitable to serve in a new Commission committed to 'cleaning up' politics in Brussels, given his association with corrupt PSOE administrations of the past. The possibility that the newly empowered EP might even reject Spain's nominees to the Commission was also heatedly debated. Oddly, nothing was said during the campaign by any party about the outgoing Commissioners or the allegations of mismanagement of some EU projects specifically levelled earlier in 1999 at one of them: Manuel Marín.

Mobilising the vote

The designation of candidates to head the various party lists reveals something about the campaigns waged by those parties. PP, as we have seen, chose Loyola de Palacio, the ex-minister of agriculture (a highly Europeanised portfolio), to symbolise the importance of EP representation to it and to emphasise the tough EU negotiating skills it was wielding in Europe. The fact that she was also widely expected to be the PP nominee to the Commission did not detract from her EP campaign. Palacio's campaign launch was draped in the Spanish and EU flags to symbolise PP's primary determination to defend Spanish interests in the EU firmly. In this respect, PP portrayed itself as more nationalistic (i.e., for Spain) than its main opponent, the PSOE.

The PSOE, which had long been used to regarding the PP as a lightweight on the European stage, also designated a woman, Rosa Diáz, to head its list. She was a Basque with a lot of political experience, not just within that region but also in its conduct of relations with the EU. She was also mentioned in some PSOE and media circles as a possible PSOE candidate for the next general election, and heading the PSOE list for the EP would certainly raise her profile nationally in Spain. (Interestingly, the head of the PP list in the 1994 EP elections, Abel Matutes, had gone on to become the PP Foreign Minister, while Fernando Morán, the former PSOE head of list, was the PSOE candidate in the election of the mayor of Madrid.) IU's candidate, Alonso Puerta, was an outgoing MEP seeking re-election to the EP. He emphasised his experience, continuity and commitment to European parliamentary representation, as did some of the minority nationalist party candidates.

PP's desire not to seem complacent in what was such an important test of support for the party nationally was evident from the seriousness with which Prime Minister Aznar took the EP elections. After all, the PP was able to declare following the 1994 campaign that it had for the first time won more votes nationally than the PSOE, following which landmark event it went on to win the general elections of 1996. In 1999, it needed to prove that much of the support gained then had consolidated behind the PP during its period in government and that the party was still well ahead of the Socialists in the run-up to the general elections of 2000. Despite a buoyant economy, a ceasefire in the Basque country and other positive indicators, opinion polls did not confirm a fail-safe lead for the PP over the PSOE, its nearest rival. Hence Aznar's vigour in these elections and the frequent warnings to supporters to be vigilant against a false sense of security about the outcome.

PSOE were in some difficulty during this campaign. The hasty resignation of its candidate for prime minister, José Borrell (alluded to above), undermined the party's effort to stage a comeback. However, of some comfort to the Socialists was the knowledge that its debilitation increased pressure on the PP to produce a good result. During the campaign, the former PSOE prime minister, Felipe González, played a high profile role. In so doing he attracted resentful accusations, especially from the PP, that he neither led nor represented anybody and that he should therefore desist from such high profile involvement. But González was in his element, playing a quasi-leadership role in the campaign by addressing around a dozen meetings and protesting vigorously to the Central Electoral Board, which regulates elections in Spain, when television coverage of his speeches was less than he considered to be appropriate.

Familiar refrains from elections of the past appeared in PSOE election material, PP's fascist antecedents and a legacy of intolerance being common themes. PP, for its part, cited its record in government, claiming that it had recovered Spain's reputation in Europe. It was now, according to the PP, a country best known for its skill in job creation, rather than for the corruption presided over by the PSOE governments (Heywood, 1995, 726). The PP dismissed PSOE attempts to undermine the integrity of its candidate, Loyola de Palacio, by its frequent references to the EU flax production subsidies scandal (*caso de lino*), which occurred in the Ministry of Agriculture, Food and Fisheries while she was in charge (*La Vanguardia*, 4 June 1999, 20–2).

PP sought also to provide evidence during the campaign of its growing stature in the EU. Helpful in this respect were events such as the trip by Aznar to Dublin to help launch the Fine Gael campaign there. More controversial was the attendance of EPP leader Wilfred Martens at a PP campaign meeting in Bilbao. Martens's visit also provoked an outcry from the PNV – itself a founder member of the EPP – for its insensitivity to PNV's own Euro-election campaign in the same region. *Unió Democratica de Catalunya* (part of the Catalan party coalition, CiU), which was also a member of the EPP bloc, warned Martens to stay clear of the Euro-election campaign in Catalonia.

Minority nationalist parties focused some attention during the campaign on the impact of the electoral system in use for the EP elections. (The system involved just a single constituency for Spain, returning 64 members, by a closed list system of proportional representation.) They campaigned for what they considered would be a fairer electoral system based on smaller regional constituencies. Ironically though, under the current system some regionally concentrated parties stood to gain from this electoral system, embracing as it did the diaspora of support from workers who had migrated from those regions across Spain.

The IU criticised the highly personalised tone of the campaign being waged by the two major parties and their failure, as IU saw it, to address the major issues of the EU. But IU was itself sucked into inter-party squabbles. In particular, it was the subject of the PSOE's strategy since the last election to characterise it as an ally of the PP. The PSOE resented deeply the defection of frustrated Socialists to the IU at the last general election and sought to claw back that lost support by sniping vigorously at IU. (Despite such heated clashes during the European election campaign, the irony was that IU and the PSOE would find themselves engaged in coalition negotiations for the running of some municipal governments after 13 June.) The CiU projected the image that it was the only party bloc seriously engaged in a European campaign. For evidence of its claim, it pointed to the Socialists' failure to present a full list of candidates (64 plus three supplements) to the voters in Catalonia, offering only the names of the 13 Catalan candidates and the head of the list (Rosa Diáz). CiU, in contrast, offered voters a full list, corresponding, it would boast, to the spirit of the EP elections. In the Basque country, the campaign was dominated by the peace process and the ETA ceasefire. EH, because of its ETA links, reaped early rewards from the campaign in the form of favourable opinion polls, at the expense of its moderate nationalist rival, the PNV.

The issues

Party programmes dutifully outlined the aspirations, policies and performances of their authors in the various EU arenas, but were not in themselves a major source of conflict or debate during the campaign. The PP programme outlined its achievements in government and in the EU since its elevation to government in 1996 (*El Pais*, 12 June 1999, 24). It took full credit for the rapid improvement in the Spanish economy and public finances which enabled Spain to qualify comfortably for participation in the single currency. The skill of its negotiators in winning concessions at EU Summits and Council of Ministers meetings was also alluded to (e.g., the budgeting deal agreed at the Berlin Summit of March 1999). Loyola de Palacio was singled out for praise in negotiating agriculture subsidies for Spain when she served as agriculture minister. The PSOE programme consisted simply of the 21 points agreed with others in the European Socialist group. It was a highly succinct and fully European manifesto. Its theme emphasised the highest aspirations of European integration, such as European citizenship and greater social cohesion in Europe. The PSOE programme praised Spain's role (under the leadership of previous PSOE governments) in the construction of the EU. By implication it was also diminishing the significance of PP's brief and comparatively less consequential EU role thus far.

The IU programme offered the most trenchant critique of EU political and economic institutions. The failure to incorporate a Constitution of Europe into the Amsterdam Treaty was one of its main themes, as was the issue of the imbalance of power between the Commission, the Council of Ministers and the European Parliament: to redress that imbalance, IU advocated greater powers for the EP. Among other matters highlighted by IU were its opposition to the high concentration of power in the European Central Bank (*Cambio 16*, 11 June 1999, 38–9). The programme of the nationalist coalition led by the PNV advocated a federal European Union in which the Basque country, for example, would be able to play a full and equal part with other nations. It also promoted a second regional EP chamber and was favourable to further enlargement of the EU to embrace the peoples of Central and Eastern Europe. The Catalan nationalists in CiU had a similar programme to that outlined, echoing the common principles contained in what was called the Declaration of Barcelona, signed by PNV, CiU and BNG. Concerns specific to Catalonia were also raised, notably the call for full recognition of the *Generalitat* as Catalonia's legitimate negotiator at all EU levels (instead of via the Spanish government). The call for Catalonia to be given the right to challenge the Spanish government before the European Court of Justice (and in the process bypassing the Constitutional Court of Spain) created an interesting question regarding judicial review for the other Spanish parties. CiU also campaigned to have Catalan adopted as an official language of the EU, making the reasonable point that since it was spoken by a higher percentage of the population in Europe than Danish or Swedish, its status should at least be equivalent to those languages. The BNG emphasised Spain's multi-national character and the existence there of the 'nations without states', such as Galicia. Though it shared common ground with CiU and the nationalist bloc led by the PNV, BNG also pointed to the comparative disadvantages experienced by Galicia compared to other Spanish regions. This was evident in EU agricultural subsidies to Spain, 30 per cent of which went to Andalusia while only 1.2 per cent went to Galicia. Specific weaknesses in the EU dairy subsidies regime affecting Galicia, a major dairy products producer, were also highlighted in BNG's programme.

The dominant 1999 EP election campaign issues in Spain were considerably more narrowly focused than those highlighted in the party programmes. This was particularly true in the cases of the two main parties which rarely made reference to their EU policy statements. With their sights set firmly on the general elections of 2000, the most controversial issues in their campaigns were the (past and present) reputations of the PSOE and PP governments. PP repeatedly cast doubt on the PSOE's fitness for high office (in Europe or elsewhere) due to the many scandals which dogged its leadership. Allegations against Pedro Solbes, its nominee for the European Commission, and José Borrell, PSOE's ex-candidate for prime minister, in the months preceding the elections allowed PP to claim during this election campaign that the PSOE was still riddled with corruption. The PSOE reacted with vehement accusations against Loyola de Palacio concerning the *caso de lino* scandal (three senior de Palacio appointees had resigned over allegations of involvement of improperly benefiting themselves or their families from EU flax subsidies). The PP and de Palacio felt it necessary to provide a detailed rebuttal of the PSOE's allegations and insisted, in turn, that it was the (Socialist) regional government of Castilla–La Mancha and not the ministry which was responsible for mismanaging flax subsidies. PP also instructed a parliamentary committee to investigate the matter after the European elections.

The *caso de lino* issue also forced Loyola de Palacio, against her instincts, to respond to the PSOE allegations in a face-to-face television debate with her chief opponent: the PSOE candidate, Rosa Diáz. They both accused each other of endangering, by their words and actions, Spain's highly important agricultural subsidies, but de Palacio's tough rebuttal of the accusations levelled against her was generally regarded to have reinforced her position at the head of the PP list and also to have strengthened her candidacy for the European Commission post. Wider substantive questions about the evolving role of the Spanish regions in implementing EU policies which emerged from the *caso de lino* scandal were marginalised by the attention paid in the campaign to personality traits and political point scoring.

The IU successfully drew attention to the issue of the EU and Spain's role in the Kosovar War but it disastrously miscalculated the effect its campaign would have on public opinion. Spain had a long history of opposition to NATO going back to the 1950s when Franco permitted NATO bases to be located on Spanish soil. In the spring of 1999, the PP government had been pursuing a rather secretive policy on the Kosovo conflict and was most reluctant to disclose the nature and amount of Spain's contribution to the Allied cause. IU attempted to galvanise that body of public opinion which was opposed to NATO behind its election campaign. During a campaign of widespread condemnation of the bombing campaign in Kosovo, Javier Solana, the NATO chief at the time (and EU representative-designate for foreign and security policy), was referred to by IU as a 'war criminal'. This tactic seriously backfired and repelled many Socialists who had previously voted for IU. Despite a midcampaign opinion poll evidence that this policy was counterproductive for IU, Julio Anguita continued to pursue it vigorously, convinced of its ultimate soundness.

The peace process in the Basque country was also an issue which was close to the surface. A number of parties hoped to gain electorally from the possibility of a negotiated settlement. With the ceasefire still in place, the PP and the nationalist parties in the Basque country attempted to win as much credit as possible for it. Press revelations during the campaign that senior Spanish government officials had been engaged in dialogue with ETA were timed to give the PP credit for handling the Basque question in a pragmatic way. Electoral success for EH would also strengthen the hand of the 'peace faction' in ETA but might be won at the expense of PNV's moderate nationalists (*The Economist*, 19 June 1999, 44).

The Blair–Schröder Third Way (New Centre) Declaration of June 1999 prompted some debate during the campaign about the ideological leanings of the PP and the PSOE. *ABC*, the conservative Spanish daily newspaper, pointed to the many similarities between it and the Aznar–Blair Declaration of April 1999. PP too was keen to identify itself as a party of the centre ground occupied by the European Christian Democrats in EPP. Aznar emphasised the need for more compassionate government and tended to tone down the emphasis on the free market. PP promoted these ideas as 'progressive' and 'modern' and complementary to those of Blair and Schröder. When Rosa Diáz acknowledged that she identified more with the socialism of Lionel Jospin than Gerhard Schröder, the PSOE was caricatured by PP as living in the socialist past together with their French Socialist party counterpart.

The results

Overall turnout in Spain was 63.0 per cent, compared to 59.1 per cent in 1994, though this hid important low (55.5 per cent in Catalonia) and high (75.4 per cent in Castilla–La Mancha) pockets (*La Vanguardia*, 14 June 1999, 49). The high overall turnout can be explained mainly by the intensity of the campaigns surrounding the municipal and regional elections held on the same day as the EP elections. Both of the major parties claimed some comfort from the results. Although PP lost a seat, it still had a clear lead nationally, though it fell short of its own declared target of 5 per cent more votes than PSOE (see Table 15.1). Notwithstanding a clear victory and evidence of public approval for Aznar's government, the PP leadership was not in the mood for wild celebration (*El Mundo*, 14 June 1999, 3). The EP elections did not give it the evidence that it craved for an easy passage into a second term of office after the 2000 general election. The PSOE found sufficient evidence in these elections to suggest that it was beginning to recuperate following the many corruption allegations and scandals which had rocked the Socialist government leading up to its defeat in 1996: it won two extra seats and a reduction in the gap that separated it from the PP. The PSOE also made a major recovery in Catalonia and interpreted it as a good omen for the regional elections there later in the year. Given the disarray in its leadership, the PSOE could gain considerable comfort from these results. The PSOE's gain had been entirely at the expense of the IU, whose vote collapsed in the EP elections (it shed around 1.5 million of the vote harvested in the 1994 elections). In the process IU lost five seats in the EP. Blame for this state of affairs was directed by most observers at the anti-NATO, anti-Solana strategy adopted by Julio Anguita (the 'last victim of Kosovo', as one media report dubbed him). PSOE could also take some of the credit for its strategy of branding IU and Anguita as unlikely bedfellows of the PP and Aznar (or, as Gonzalez crudely put it to a private PSOE meeting during the campaign, subsequently leaked to the press, they were 'two parts of the same turd': a choice of words that, unsurprisingly, provoked huge indignation in PP and IU circles).

The CiU won the same number of seats (three) as in 1994 on a slightly smaller vote. However, it could take comfort from the fact that its close association with the PP government did not appear to have damaged it significantly. Of the other free-standing nationalist parties, EH and BNG performed most impressively. BNG broke through to win its first seat at the European level and EH regained the seat held by HB between 1989 and 1994. Both results were landmarks in the evolution of Galician and Basque nationalist politics. The nationalists (*Coalición Nacionalista*) and regional (*Coalicion Europea*) blocs also did well winning two seats each. The nationalist seats were taken by the two Basque parties, PNV and EA. The two regional seats went to Andalusia (PA) and the Canary Islands (CC). Though comparison with the previous European elections is complicated by variations in the make-up of the coalitions, the achievement of the nationalist bloc was somewhat marred by the fact that its percentage vote did suffer a small decline. The all-women's list did not return any MEPs but, none the less Spain had 22 women MEPs in the new

196 Spain

Table 15.1 Results of the 1999 EP Elections in Spain

Parties	1999			1994			
	%	Seats	Groups	%	Seats		
Partido Popular (Conservatives)	39.74	27	EPP/ED	40.12	28	28	EPP
Partido Socialista Obrero Español (Social Democrats)	35.32	24	PES	30.79	22	22	PES
Izquierda Unida (Left Party)	5.77	4	EUL/NGL	13.44	9	9	EUL/NGL
CiU (Convèrgencia I Unió): (Catalan Parties)	4.43	3		4.66	3		
UDC (Unió Democrática de Catalunya) – CDC (Convèrg. Democrática de Catalunya)			(1) EPP/ED (2) ELDR			1 2	EPP ELDR
Coalición Europea (Regional Parties)	3.20	2					
Coalición Nacionalista Europa de los Pueblos (Regional Parties)	2.90	2	Green/EFA	2.82	2	1 1	EPP ERA
Bloque Nacionalista Gallego (Regional party: Galicia)	1.65	1	Ind	0.75			
Euskal Herritarrok (Regional party: Basque)	1.45	1		0.97			
Los Verdes (Greens)	1.42	0					
Others	4.12			6.45			
Total	100.00	64		100.00	64	64	

Electorate: 33 841 211   Valid votes: 20 807 269
Turnout: 21 334 125   Abstensions: 12 507 086
Turnout: 1999, 63.04 %; 1994, 59.1%   Spoiled Votes: 526 249
Sources: Ministry of Interior (Junta Electoral) 1 July 1999; European Parliament (1999).

Parliament (i.e., 34 per cent of all Spanish MEPs), marginally more than in 1994.

Conclusion

The results in Spain placed the slightly smaller PP contingent in a strong position in the much enlarged EPP/ED group. As the only prime minister from a major EU country with a party in the group, Aznar was in a good position to play a more assertive leadership role in the group, particularly in challenging the Socialists claims over the political centre ground. Aznar quickly convened a meeting of the heads of Europe's Christian Democratic parties in Marbella in July to celebrate the EPP's triumph in the European elections generally and to plot its future strategy in the forthcoming parliament. The PSOE, in contrast to PP, went away from the elections as a marginally stronger presence in a weaker Socialist group in the EP. However, its main preoccupation was with matters closer to home, especially the resolution of its leadership crisis in time for the general election in 2000. IU also faced a leadership crisis and a crisis of confidence generally after the disastrous Euro-election results presided over by Julio Anguita. The IU's election results had much more impact domestically than they could ever have in the European Parliament where its four MEPs sat with the EUL–NGL group.

The new MEPs for the smaller parties spread themselves across the range of party groups in the EP. The Greens/EFA group was expanded by the presence of two Basques (from the PNV and EA), plus an Andalusian (PA) and Galician (BNG) MEP. One of the three CiU MEPs (from the UDC) joined the EPP group, while the other two (from the CDC) joined the ELDR, where they were also joined by the Canary Islands MEP (from the CC). The Radical Basque nationalist MEP from EH retained independent status. In conclusion, in a country of frequent election contests, political parties digested the EP election results and quickly turned their attention to Catalan October regional elections which, for very different reasons, were also of great importance to Spain.

References

ABC (1999) 14 June 1999, 13.
Cambio 16 (1999) 11 June, 38–9.
The Economist (1999) 19 June, 44.
El Mundo (1999) 14 June, 3.
El Pais (1999) 12 June, 24.
European Parliament (1999) *EP Elections: June 1999* (Brussels: DG for Information and Public Relations).
European Voice (1999) 24–30 June, 6.
Gibbons, J. (1999) *Spanish Politics Today* (Manchester: Manchester University Press).
Gillespie, R., F. Rodrigo and J. Story (1995) *Democratic Spain: Reshaping External Relations in a Changing World* (London: Routledge).
Heywood, P. (1995) 'Sleaze in Spain', *Parliamentary Affairs*, 48 (4).
La Vanguardia (1999) 4 June, 20–2; 14 June, 49.

List of Spanish party abbreviations used

BNG   Bloque Nacionalista Gallego (Galician Nationalist Bloc)
CC   Coalición Canaria (Canary Islands Coalition)
CDC   Convergencia Democrática de Catalunya
CiU   Convergencia i Unió (Convergence and Union)

EA	Eusko Alkartasuna (Basque Solidarity)
EH	Euskal Herritarrok (Basque Nation)
HB	Herri Batasuna (Basque Homeland)
IU	Izquierda Unida (United Left)
PA	Partido Andalucista (Andalucian Party)
PNV	Partido Nacionalista Vasco (Basque Nationalist Party)
PP	Partido Popular (People's Party)
PSOE	Partido Socialista Obrero Español (Socialist Workers' Party of Spain)
UDC	Unió Democrática de Catalunya

16
Sweden

Karl Magnus Johansson

Introduction

For the first time ever, Sweden was part of EU-wide elections in June 1999. The Swedish EP election of September 1995 had, in several respects, been a special Swedish event, with the purpose of electing directly 22 Swedish MEPs (Gilljam and Holmberg, 1998). At that time, 41.6 per cent of the electorate turned out to vote. This time, turnout dropped to 38.8 per cent. This is much lower than the 81.4 per cent turnout in the 1998 general election, and was the lowest turnout in a Swedish national election for over 40 years. Against this background, it seems that the Swedes, at least in elections to the distant and little appreciated EP, are 'reluctant voters'. By and large, we may also characterise the Swedes as 'reluctant Europeans' (cf. Jerneck, 1993; Johansson, 1999b).

European elections have traditionally been referred to as second-order elections (see, for example, Reif, 1984). In such elections voters exhibit different behaviour from general elections, which generally are considered more important by the electorate. In second-order elections, which provide a referendum and test of popularity for governments, fewer voters turn out to vote and those who do tend to experiment and vote for smaller parties, often of the opposition, and thereby vote in protest against the ruling party or parties (Johansson, 1996, 214–17). In what ways does Sweden fit into this picture? This chapter surveys and analyses the 1999 European election, focusing on the peculiarities of Sweden. However, the election is 'contextualised' by setting it in a domestic political context, with a view to political institutions and issue cleavages. Both the domestic and the European settings more broadly will be taken into consideration to illuminate the origins of the issues on the political agenda. Was the campaign in Sweden run on a domestic agenda and on the 'wrong' (that is, national rather than truly European) issues?

The primary material on which the analysis is based consists of party publications and media reports, especially press articles and then mainly in the Stockholm daily newspapers of *Dagens Nyheter* and *Svenska Dagbladet*.[1] I have also had access to some material from a study by Observer Media Intelligence, Stockholm, of the role of the media in the election. Likewise, I have benefited from the exit poll conducted by Swedish Television and from statistics provided by the National Tax Board (Riksskatteverket), which is the Swedish administrative agency with responsibility for electoral matters.

The players in a domestic context

As for the main players, it has been commented that 'the media won the election' (*Dagens Industri*, 14 June 1999). There is more than a grain of truth in this, but the impact of the media was not all healthy given the nature of media reporting on the EP and the EU more broadly. This is examined in more detail below. Suffice it to note here that the media, rather than the judiciary, is referred to as the 'third state power' in Sweden, with the Parliament, the Riksdag, and the government as respectively the first the second state powers.

Sweden is a system of parliamentary and party government. The political parties, primarily the Social Democratic Party (SAP), have traditionally held the key to the evolution of the Swedish democracy and the welfare state. However, the consensual party system of the past has become increasingly conflictual and fragmented, and the historically almost hegemonic position of SAP is challenged. Sweden has mostly been governed by one-party social democrat governments, but nowadays they need the support of other parties to muster majorities in the Riksdag. In the 1994–98 Parliament there was an informal coalition between the SAP and the Centre Party (C). After the September 1998 general election an alliance, of convenience rather than conviction, was instead built with the Left Party (V) and the Green Party (Mp). All three alliance parties, of the past or the present, act as guardians of the Swedish policy of neutrality and non-alignment, and the two new allies campaigned against EU membership in the November 1994 referendum and are still, at least in their rhetoric, calling for Sweden to leave the EU. Overall, the attitudes of the public and also of the political parties still reflect the division over membership from the time of the referendum. Since the two alliance partners, especially Mp, hardened their anti-EU profile in connection with the European election, the social democrats' alliance with them has become increasingly fragile. In terms of domestic politics, the stability of the Swedish government decreased even further as a result of the alliance parties' behaviour during the campaign. Moreover, it has become even more difficult for the governing party to come out in favour of the single currency. The consequences for domestic politics will be discussed further below. All four non-socialist parties remain in favour of EU membership, even though the Christian Democratic Party (Kd) especially is divided over most, if not all, matters related to the EU. C is also divided, whereas the Liberal Party (Fp) and the Moderate Party (M) are the most pro-EU parties. However, some objections have been raised against EMU within the latter party, which is a Liberal-Conservative (and also a highly disciplined) party.

Significantly, the parties' responses to Europe reflect a growing tension within them (cf. Johansson, 1999a; Johansson and Raunio, 1999). European integration has provoked an intra-party activity that it is not always easy for party managers to contain. In short, the EU has given rise to an increasing factionalisation of Swedish political parties. Except for Kd, SAP is probably the party most split over EU matters.[2] This is the most important reason for the undecisive position of the governing party over participation in the third stage of EMU (the single currency). Both SAP and Kd adopted a 'wait-and-see' position. This intra-party discord is also reflected in the EP. The two Christian Democrats elected in June 1999 embraced strikingly different policies, and the SAP delegation remains divided (but less so than during the previous parliamentary term). One member of the delegation even said that the split then was 'a disaster'. One of the most high-profile SAP Euro-sceptics, Maj Britt Theorin, is still in the team, providing a familiar face for all those social democratic voters opposed to the EU.

In a deeper sense, however, party leaders have been able to maintain organisational coherence. It may even be argued that European elections serve as a convenient outlet for tensions to be released within parties (Johansson and Raunio, 1999). One indication of this is the fact that candidates have run their own campaigns, albeit within certain limits. Sometimes they actually seemed ignorant of party positions. It is important to emphasise, however, that those candidates employed party labels. Thus, parties remain principal actors.

Unlike Denmark, no single EU-issue party has emerged in Sweden, even though there have been some attempts to form such a party for European elections, notably by people of the cross-political 'No to EU' movement. Since this movement has a leftist orientation it has been difficult for non-socialist Euro-sceptics to associate themselves with it. The failed attempts to establish such a party can be explained by the nature of cross-cutting cleavages. These include the European dimension and the socio-economic dimension, in addition to a cleavage such as the rural–urban. The left–right scale remains the most important in Sweden, guiding party and electoral behaviour. In particular, the two largest parties and main antagonists, SAP and M, seek to structure competition on the left–right dimension, thereby uniting their own troops, and to evade debates about the further development of European integration. The problem, for pro-EU parties and governments, is that all such debates are more or less sensitive in Euro-sceptic Sweden. Eurobarometer data shows that the Swedes are the most reluctant of all EU citizens to further European integration. Euro-sceptical public opinion as well as inter-party rivalry and intra-party discord, strain the government and result in weak political leadership, at least in regard to EU affairs more broadly.

The campaign

This section sets out to show how the parties contested the European election in the Sweden, how the vote was mobilised, and what the main issues were. Media coverage was more extensive than in 1995, at least in the leading daily Stockholm newspapers. They contained a series of reports practically every day during the three weeks before polling day. However, quantity is one thing, quality another. Perhaps neither the media nor the political class had been able to explain what this election was really about, in the light of the extension of EP powers since the entry into force of the Amsterdam Treaty and the new institutional balance since the resignation *en masse* of the Commission, or perhaps the conduct of European election campaigns, and of the media and political parties, is primarily (if not inevitably) national rather than transnational?

Mobilising the vote

Voting is voluntary. This provides a challenge for political parties in terms of the mobilisation of the vote. It is important to note that the parties represented in the Riksdag are themselves in control of election funding, since parties and elections are state-funded in Sweden. Some candidates have raised money from alternative sources for their personal campaigns, but there is public money for expenses such as advertising. Non-partisan advertising was also rare in this election. In fact, the EP office in Sweden was not allowed to advertise on the public service televised channels on the grounds that the EP is not a Swedish administrative agency. Accordingly, the National Tax Board could advertise and they did so in a provocative manner. Appealing to emotion and a sense of democracy and duty, attention was drawn to the dark totalitarian history of Europe, showing the face of Hitler. The moral was clear: voting, however voluntary, is a democratic right that should be exercised.

Politicians were more cautious when trying to mobilise the vote. By and large, there was little engagement of national political leaders, and ministers were conspicuously absent from the campaign. One obvious exception was the then minister for development and deputy foreign minister, Pierre Schori, who was the SAP's top candidate. As the governing party was more or less doomed to do badly, it seemed that individual ministers deliberately stayed away. The prime minister himself, Göran Persson, said that this election 'lived a life of its own' and that it would have no consequences whatsoever for domestic politics, regardless of the outcome. From his perspective, this argument made sense, considering the second-order nature of European elections.

The parties, notably the governing SAP, apparently wanted a brief campaign. The party leaders' debate in the Riksdag was held as late as 2 June.[3] Overall, and with the anti-EU parties as notable exceptions, the parties ran campaigns with the involvement of candidates rather than party leaders. Unlike 1995, when C, Kd and SAP offered mixed lists, containing the names of both supporters and opponents of the EU, parties presented single lists. Nevertheless, voters could not fail to see how different were the attitudes of candidates on the lists of C and Kd; perhaps the SAP's list was slightly more in agreement, whereas the other parties presented candidates who were largely unanimous over EU matters.

Given the brief campaign, it is perhaps not that surprising that a large proportion of the electorate, shortly before polling day, were reportedly ignorant of the date for the election or even that there would be an election to the EP at all.[4] The campaign did not really help to give voters an idea of the role and powers of the EP and the work of MEPs. Media attention was still on their salaries and generous allowances. The focus should have been on what MEPs really do, and how effective they are in their broad range of activities which goes beyond asking questions to commissioners and speaking in the plenary. All in all, it was a myopic campaign, with little analysis of the long-term development of the EU in general and the EP's evolution in particular, against the background of the entry into force of the Amsterdam Treaty. Although the number of legislative procedures has been reduced, it remains difficult to explain the relationship between the institutions and the 'nature of the beast': that is, the kind of political system the EU is. Unlike Sweden, the EU is not a parliamentary system of government, which implies that voters cannot see a government being formed on the basis of the composition of the EP they are being asked to vote for.

Given that journalists, noticeably in televised interviews with leading party candidates, often appeared to be ignorant of the powers and role of the EP, it is perhaps even more serious that government ministers did not even attempt to explain them, whatever their knowledge. We are dealing with a huge communication problem, with implications for the state of democracy. The relationship between EP and national institutions is also a communication problem insofar as the members have to spend most of their working time in Brussels or Strasbourg and often experience difficulties when trying to communicate back home, in this case to Sweden, the nature of transnational politics and of their European existence within a Parliament and party groups with multi-national memberships (cf. Petersson *et al.*, 1999, ch. 4). In that setting, members have to compromise to reach all-EU deals affecting 375 million citizens rather than just the Swedes, a population of some 9 million.

In European election campaigns, parties and the media reveal how nation-bound they are as institutions. Their prestige and rationale are as Swedish institutions and they are responsible for, and report to, domestic electorates. However, at a time of increasing globalisation and Europeanisation, there is not really such a thing as a distinctly domestic agenda, fully controlled by the national political class and media. Instead, ideas and issues travel across national borders, and their routes and destinations are often unpredictable.

The issues

National issues were not predominant. Instead, the campaign was primarily run on European issues. Although the governing party in particular engaged in issue avoidance, trying to steer away from sensitive issues, it was a victim of the fact that more or less unexpected issues on the EU's political agenda were placed on the one within Sweden. Unfortunately for the government, issues of a European defence identity were brought up in connection with the Cologne European Council in early June 1999. Conversely, this was very timely for those parties acting as guardians of the policy of neutrality and non-alignment. At a time of NATO bombings over Yugoslavia, it was claimed that Sweden was about to enter a 'war alliance'. Whatever the exaggerations in their rhetoric, it might be argued that the arguments of the anti-EU parties, including V and Mp, served to 'Europeanise' the campaign and to challenge the party government to clarify its own positions and argument.

There was also a clear European dimension in the questions of consumer protection, and specifically the topical issue of hormones in beef. This was the single issue that took up most of the time during the final debate, two days before the election, to which the two front-runners of each party had been invited. Somewhat surprisingly, EU enlargement remained a low-profile issue throughout the campaign, though this might be explained by the broad consensus among the parties on enlargement. Neither was there much dispute with the governing party over the single currency. Like the SAP, others argued that this was not an issue for the EP, but an issue to be decided by the Riksdag or in a referendum.

As in 1995, the social democrats prioritised jobs and the fight against unemployment. The emphasis on questions related to the labour market is perhaps the major evidence of the attempt by SAP and M to structure the campaign along the left–right scale. This is a divide between a traditional active labour market policy versus one characterised by deregulation and liberalisation. To some extent, therefore, ideology was important. At the same time, however, it was not always easy for voters to distinguish the main contenders from each other. Interested in the content rather than the form of politics, voters often seemed to embrace different issues from 'their' representatives. In Swedish television's exit poll, reported as early as the live broadcast in the evening of the election, the sample of voters prioritised the following: peace in Europe, democracy in the EU, employment, social welfare, the economy, the environment, national independence, equality between men and women, EU defence, EMU, conditions for enterprise, agriculture, refugees/immigration, and, finally, EU enlargement.

It is important to emphasise, however, that those in favour of the EU and Swedish membership were over-represented among those who turned out to vote. A clear indication of this is the response in regard to the single currency, with 46 per cent in favour of Swedish participation, 36 per cent against, and 18 per cent undecided. More often than not, a majority has been against in different opinion polls.

The passages on making the euro a success were difficult for SAP in the context of the drafting of the PES manifesto, a common programme with 21 commitments for the twenty-first century. As a PES vice-president, the Swedish deputy prime minister and former foreign minister, Lena Hjelm-Wallén, negotiated the text on behalf of SAP. She and the prime minister himself were present when the manifesto was adopted by the PES congress in Milan in early March 1999.[5] The manifesto had no significant impact on the campaign, but it is interesting to note that one of the social democrat candidates even said that the PES manifesto was better than the Swedish platform. Social democrats never lost a chance to point out how instrumental they had been, acting through the PES, in the realisation of the separate title on employment in the Amsterdam Treaty (cf. Johansson, 1999c). Now the goal was a pact for employment, with a view to the Cologne summit in June.

The M leadership, including the then party leader, Carl Bildt, was present when EPP held its congress in Brussels in February 1999. This time, the 'moderates' voted in favour of the action programme, but there was a clear dissent between the more Liberal-Conservative parties, such as M, and traditional continental Christian democrats. Neither was the EPP platform much heard of in the campaign in Sweden. The same applies to the platform of the ELDR, which held its congress in Berlin. However, the Fp leader actually claimed, in the party leaders' debate in the Riksdag on 2 June that the platform of the Fp corresponded '98 per cent' with the working aims of 'our European party group'.

It is noteworthy that the political parties to a very limited extent engaged leading personalities of the European parties or their member parties in the campaign. However, the names of both Swedish top candidates and chairmen, or vice-chairmen (as Hans-Gert Pöttering then was of the EPP group) of the EP party groups figured in joint newspaper articles. There was a televised live broadcast in the presence of party group spokespersons. It was certainly no coincidence that there were different representatives from the EPP group (Pöttering) and the Green group (Patricia McKenna) than in 1995, and these representatives were considered more 'reliable'. Whereas Mp may have feared a too EU-friendly spokesperson, M wanted to avoid the embarrassment of the last election when the then EPP group chairman, Wilfried Martens, made known that the EPP programme favoured a common defence, including a nuclear component. However honest, this undiplomatic declaration probably lost M and Kd votes. The latter narrowly failed to reach the 4 per cent threshold.

Since the media had focused on the voting behaviour of Swedish members of EP party groups and found some striking differences from the positions of national parties, the latter generally seemed reluctant to highlight the transnational partisan dimension of the election. This could have had negative effects both in the electoral arena, where vote-seeking takes place, and in the internal arena, where the goal is to keep the party together. Instead, the parties tended to downplay the transnational platforms to which they had contributed. Also, pro-EU parties sensitive to not politicising European issues therefore appeared defensive, especially when compared to the offensive anti-EU parties, who played this card to mobilise votes. Consequently, the parties ran campaigns on national platforms rather than the transnational platforms of European parties. Although it was hard to discern the transnational dimension of the election, it does not follow that the election was not 'European', in terms of the issues. As for the campaign, however, these elections are perhaps inevitably national, at least under the present circumstances.

The results

Turnout was 38.8 per cent, down from 41.6 per cent in the September 1995 extraordinary European election in Sweden. Shortly before the election of 13 June 1999, predicted turnout ranged from 52 to 58 per cent. Statistics show, however, that many people made up their minds rather late, often even on election day, and thus after the opinion polls had been conducted. Aside from a widespread disillusionment with (and hostility towards) the EU, a number of factors might have contributed to the low turnout. There is general discontent with politics *per se*, including domestic politics as indicated by distrust of politicians, decreasing party identification, change of party and electoral volatility, and, perhaps most important and worrying, abstention, not least among younger voters. Turnout also fell in the most recent national election. Some possible factors behind the low turnout concern the very procedure and timing of the election. As always in Sweden, the election was held on a Sunday, but this time in June rather than the third Sunday in September, as is the practice for national, regional and local elections. In many parts of the country, including Stockholm, the weather was sunny on 13 June. In addition, the public schools had already closed for summer vacation and many Swedes were on their way to the countryside or abroad. Overall, the date seems unfortunate for the Swedish electorate, even though votes could also be cast at post offices from 26 May. Some people had perhaps not yet got accustomed to the electoral procedure. A relatively new electoral procedure was used, with an element of a personal vote system within the traditional party list system.[6] According to this procedure, if a candidate gained a 'personal' vote of more than 5 per cent of the party's overall national vote, with all of Sweden as an electoral district and the threshold still at 4 per cent, then that candidate would advance on the list. In the 1999 election, several of the elected members won a sufficiently high vote to be elected, but most of them already had 'safe' seats. However, two of the 22 MEPs gained enough votes to advance above candidates apparently preferred by party managers. One of the elected candidates had been ranked as number five on the Kd list and the other as number two on the C list.

The candidate with the largest number of personal votes was the frontrunner of Fp, Marit Paulsen, an author and a former social democrat who only recently had converted to the Fp. She attracted a great deal of media attention and the so-called 'Marit effect' very much contributed to the three seats for Fp, up by two from the previous term. This success attests to the importance of personalities for voters' preferences, a process already well under way, and how strikingly different voting behaviour could be in European and national elections. The most illustrative example of this is provided by Fp, which experienced a result reminiscent of Mp's success in the 1995 European election (see Table 16.1).

Voters rewarded parties with a clear and uniform message. Whereas Fp has a distinctly pro-EU message, the attitudes of Mp respectively V are no less distinctly anti-EU. At the same time, however, the success of Fp must be seen in the light of the over-representation in the poll of voters

Table 16.1 Results of the Swedish EP election, 13 June 1999 (%)

Party	Seats	Gain/loss	1999	European election 1995	National election 1998
Social Democratic Party (SAP)	6	−1	26.0	28.1	36.4
Moderate Party (M)	5	0	20.7	23.2	22.9
Left Party (V)	3	0	15.8	12.9	12.0
Liberal Party (Fp)	3	+2	13.9	4.8	4.7
Green Party (Mp)	2	−2	9.5	17.2	4.5
Christian Democratic Party (Kd)	2	+2	7.6	3.9	11.8
Centre Party (C)	1	−1	6.0	7.2	5.1
Others			0.5	2.8	2.6
Valid votes		2 529 437			
Blank and invalid votes		59 077			
Total votes		2 588 514			
Electorate		6 664 205			
Turnout		38.84 per cent			

Sources: Riksskatteverket (www.rsv.se); Sveriges Riksdag (www.riksdagen.se). See also Gilljam and Holmberg (1998), 11; Miles (1997), 273; Miles and Kintis (1996), 230.

inclined to favour the EU and Swedish EU membership. Many of the volatile voters on the non-socialist side of the spectrum were apparently attracted to Fp. This applies especially to people who usually vote for M, whose election result was worse than expected, not least within the party, despite its clearly pro-EU message and widely respected leader. It was difficult for Bildt to participate in the campaign and work as the UN envoy to the Balkans as well. However, the party kept its five seats after all.

The SAP suffered particularly from the low turnout and massive abstentions of disillusioned and Euro-sceptic social democrats. A large number of blue-collar workers and members of the Swedish Trade Union Confederation (*Landsorganisationen*, or LO) abstained, as did many young people. The pattern of participation reveals a clear social and class divide, although there was no significant difference in turnout between men and women. The poor turnout may have repercussions for the ongoing debate about the state of Swedish democracy and constitutional change.[7] Although the prime minister insisted that the election would have no domestic implications, it was bound to have. As Karlheinz Reif (1984, 245) has pointed out, 'European elections nevertheless are events of domestic high politics with potential political repercussions on the balance of power between political parties in the national arena.'

An immediate effect was that the alliance parties of Mp and V, increasing their combined share of the votes since the alliance was entered into after the September 1998 election, called for a decision in the near future about a referendum on full Swedish participation in EMU. Except for EMU, the electoral result may also have implications for the Swedish positions in view of the forthcoming IGC, perhaps making the government even more reluctant to raise and discuss questions about institutional reform. In fact, the Prime Minister's immediate comments after the election targeted the EP and its allegedly 'deep crisis of legitimacy'. Thereafter, he sought to evade questions about the role and legitimacy of 'his' government and party.

As for the balance of power in the European arena, the most notable effect of the result was that the Swedish EP contingent would contain more pro-EU members than before. Then, half of the 22 Swedish MEPs were anti-EU campaigners, at least when confronting domestic audiences. However, the Swedish delegations within party groups remain numerically small so their members are unlikely to make much impact, in general terms, considering the logic of numbers and games in the EP arena. Yet a social democrat, the Euro-sceptic mentioned above, was appointed chairman of the Committee on Women's Rights and Equal Opportunities. This was the first chairmanship ever of a Swedish MEP. Perhaps more importantly, however, a 'moderate', Staffan Burenstam Linder, became a vice-chairman of the expanding EPP group, bound to be the main broker of deals between party groups in the increasingly powerful EP.

Concluding reflections

Against the background of the evolution of the EP, the 1999 elections should have been of a different order than previously. In many ways, however, the second European election in Sweden also displayed similar characteristics to other EU member states and to the first one in Sweden, in September 1995, when it was commented: 'The most likely implication arising from the election result is that the anti-EU parties have not yet relinquished the idea of possible withdrawal from the Union' (Miles, 1997, 274). Also the 1999 election can be interpreted as a referendum on full EU membership. At the same time, anti-EU parties' rhetoric must be seen as primarily symbolic and for domestic consumption, aiming at mobilisation and maximisation of the vote. There is a reservoir of Eurosceptic voters for whom the choice is one between abstaining altogether or voting for anti-EU parties. Through the behaviour and 'out of EU' message of those parties, voters are (mis)led to believe that withdrawal is a reasonable option, whereas it is unlikely to happen for a number of reasons. The same parties are certainly aware of this and 'their' MEPs display different behaviour in the European arena of politics, seemingly accepting the fact that Sweden, after all, is a member.

As for the transnational dimension of the election in Sweden, it was more evident than in 1995 and the major Swedish political parties have taken part in the drafting of the election manifestos of European parties, including ELDR, EPP and PES. Thus there were transnational platforms alongside the national platforms. Seemingly, however, this had no obvious effect for the mobilisation of the vote. Overall, there was apparently too little stimulus for the electorate to find it meaningful to turn out to vote or to create an interest in the election. The nature and logic of the European elections and the campaign in Sweden attest to the primacy of domestic arenas. It is at the domestic level that parties have their constituencies and the main players, the parties and also the media, are essentially nation-bound institutions.

In conclusion, the conduct of the campaign was not predominantly national but instead had a strong European dimension. The sensitive debate about a European defence identity is of a European order. In this connection, pro-EU parties, including the governing SAP, engaged in issue avoidance, but as so often the parties could not fully control the political agenda and had to deal with unexpected events. For the main contenders of the left and the right, SAP and M, the tactics were to structure the campaign precisely along the left–right scale. Considering the second-order nature of European elections, these are difficult for a governing party. In fact, social democrats tended to behave as if they did not want this election and treated it as a non-event at best, with the leader and prime minister claiming it would have no consequences for domestic politics whatsoever. In many ways, the Swedish case is thus illustrative of the second-order thesis. The prospects for a successful result are, indeed, different for governing and opposition parties. The election was very much a mid-term test of popularity of the government. It was a protest vote insofar as voters either voted against the governing party or abstained in massive numbers. Naturally, the very poor turnout was the focus of post-election debate.

The low turnout reflects a widespread discontent with politics more generally. One should therefore be careful not to draw hasty conclusions about the balance of support for and opposition to the EU and the integration process. Instead, it might be argued that there is a crisis of legitimacy for domestic political institutions rather than for the EU and the EP, as the prime minister conveniently suggested. There is a 'democratic deficit' also in the way in which EU matters are handled within Sweden (cf. Petersson *et al*., 1999, ch. 4). One problem that could be remedied is to provide for further transmission belts and forums for exchange between parliamentarians at the European level and the national level. In the task of scrutinising the executive branch of the polity, MEPs should be partners rather than subordinates and rivals to national parliamentarians. At present, the role of MEPs as interlocutors or go-betweens, linking EU institutions to each other, is not sufficiently taken advantage of, and neither is the MEPs' knowledge of EU matters. The national and the European levels are inter-related and communication between them is of fundamental importance. Today, there is a mutual distrust due to a lack of communication, and ameliorating this situation requires that the number of links and degree of communication between party representatives, parliamentarians and officials at the two levels increase along with increased contact and exchange, bilaterally as well as multilaterally, vertically as well as horizontally (Johansson, 1996, 223). The democracy problem related to European elections and to the EU more generally should be dealt with as primarily a communication problem. In terms of democracy, there is a need for more rational forms of dialogue both between the political class and the electorate in the domestic arena of politics as well as with representatives in the European arena.

Notes

1 I have also drawn upon the speeches and interventions by participants at a conference, organised by me, on the theme of 'European Elections in Sweden: European or Swedish?' at the Swedish Institute of International Affairs, Stockholm, on 10 May 1999. The participants included Juliet Lodge, Olof Petersson, Tom Bryder, Yvonne Sandberg-Fries, Charlotte Cederschiöld and Cecilia Malmström. I gratefully acknowledge the input of their expertise and experience. I also would like to thank Hans Hegeland and Ingvar Mattson, deputy secretaries in the Riksdag, for friendly assistance and conversations.
2 The social democratic network, or faction, in favour of the EU is called 'Social Democrats for the EU', and the one against is 'Social Democratic EU Critics'.
3 It is noteworthy that two of the party leaders were absent from this debate, namely Alf Svensson (Kd) and Carl Bildt (M). Again, the latter was on a foreign mission. For this reason, he was also absent from the party leaders' debate which was televised live on 17 May.
4 The election seemed to be of little interest to the average voter. Figures reveal that a large number of people switched off or over to another channel instead of watching the live televised broadcast with questions put to the two top candidates of each one of the parties represented in the Riksdag. Furthermore, the live televised programmes on the evening of the election attracted strikingly few viewers.
5 I myself was present both at the PES congress in Milan and at the EPP congress in Brussels.
6 The procedure had been used in the 1995 European election and in the 1998 national election. It was used for the first time, as an experiment, in some local elections in 1994.
7 One probable effect of the low turnout is that the proposal to move the national election day from the autumn to the spring is unlikely to materialise.

References

Dagens Industri, 14 June 1999, 'Medierna vann valet'.
Gilljam, Mikael and Sören Holmberg (1998) *Sveriges första Europaparlamentsval* (Stockholm: Norstedts Juridik).
Jerneck, Magnus (1993) 'Sweden – the Reluctant European?', in Teija Tiilikainen and Ib Damgaard Petersen (eds), *The Nordic Countries and the EC* (Copenhagen: Copenhagen Political Studies Press), 23–42.
Johansson, Karl Magnus (1996) 'The Nature of Political Parties in the European Union', in George A. Kourvetaris and Andreas Moschonas (eds), *The Impact of European Integration: Political, Sociological, and Economic Changes* (Westport: Praeger), 201–32.

Johansson, Karl Magnus (1999a) 'EU och partiväsendet', in Karl Magnus Johansson (ed.), *Sverige i EU* (Stockholm: SNS Förlag), 150–68.

Johansson, Karl Magnus (1999b) 'Europeanization and its limits: the case of Sweden', *Journal of International Relations and Development*, 2(2), 169–86.

Johansson, Karl Magnus (1999c) 'Tracing the employment title in the Amsterdam treaty: uncovering transnational coalitions', *Journal of European Public Policy*, 6(1), 85–101.

Johansson, Karl Magnus and Tapio Raunio (1999) 'Partisan responses to Europe: comparing Finnish and Swedish political parties', manuscript prepared for publication.

Miles, Lee (1997) *Sweden and European Integration* (Aldershot: Ashgate).

Miles, Lee and Andreas Kintis (1996) 'Appendix: The New Members: Sweden, Austria and Finland', in Juliet Lodge (ed.), *The 1994 Elections to the European Parliament* (London: Pinter), 227–36.

Petersson, Olof *et al.* (1999) *Demokrati på svenskt vis*, Demokratirådets rapport (The SNS Democratic Audit) (Stockholm: SNS Förlag).

Reif, Karlheinz (1984) 'National Electoral Cycles and European Elections 1979 and 1984', *Electoral Studies*, 3 (3), 244–55.

List of Swedish party abbreviations used

C   Centre Party
Fp   Liberal Party
Kd   Christian Democratic Party
M   Moderate Party
Mp   Green Party
SAP Social Democratic Party
V   Left Party

17
The United Kingdom

Janet Mather

Introduction

The UK has always demonstrated an ambivalent attitude towards membership of the European Union. On the one hand, its political class has embraced each aspect of European integration as it has presented itself. On the other, UK leaders of the two main political parties have been reluctant to activate potential Euro-enthusiasm, largely because of fear of the reaction of their main opponent. Although the Maastricht Treaty went through the Commons under Prime Minister John Major's Conservative government, he declared that his party would not venture towards the next stage of integration – economic and monetary union – in the foreseeable future. This line was obviously insufficient to save his government, and in May 1997 an electoral landslide gave a Labour government its largest ever majority, 179 seats overall.

When Labour took office, it appeared that a change of attitude towards Europe had taken place. The new government immediately announced its decision to sign up the UK to the Social Chapter, whilst its leader, Tony Blair, confirmed his intention to place Britain 'at the heart of Europe', and began to talk about the prospect of signing up eventually to the EU's single currency. A successful, although not particularly spectacular, UK presidency of the EU was undertaken during the first half of 1998. During this, the process of further enlargement of the EU was begun, and negotiations on the Agenda 2000 proposals on policy reform and future financing of the EU were launched (European Council, 1998). The successor to the Maastricht Treaty – the Treaty of Amsterdam – was signed in June 1997, to be ratified, without noticeable dissent, in May 1999.

The Labour government too, however, indicated some wariness towards further integration. None of the key Euro-events was publicised with undue emphasis, and speeches made by Labour leaders were more pro-European when delivered outside the UK than they were within the country. Some progress was made on the single currency issue. In October 1997, the UK's Chancellor, Gordon Brown, outlined five criteria which would need to be met before the UK's economy could be said to be sufficiently in line with that of euroland, so that Britain's economic interest would be served by embracing the euro. In the meantime, a euro-awareness campaign was launched (subsequently affirmed by the Britain in Europe campaign launched in October) and a referendum promised on the issue, following the next general election. William Hague, Major's successor as Conservative leader, noted that the Government's position on the euro had changed from 'if' to 'when' (*Guardian*, 3 November 1998).

For the 1999 EP elections, the UK Government finally introduced a system of proportional representation, bringing it into line with the other fourteen member states. The system chosen was based upon the d'Hondt system of counting votes with a regional 'closed' list. In this system, each party provided a prioritised slate of candidates, so that the voter voted for a party rather than a person, except where an independent candidate was standing. Winning candidates were those heading the parties' lists allocated in accordance with total voter party preferences. The measure was regarded as controversial, and the European parliamentary Elections Bill was rejected five times by the House of Lords on the ostensible grounds that its members saw closed lists as undemocratic. Nevertheless, it was enacted in January 1999 just in time for the necessary changes to electoral procedures to be undertaken. The electoral boundaries chosen followed those of the newly-created Regional Development Agencies in England, giving nine regions (including London), with separate regions for Wales and Scotland. Each region was allocated seats in accordance with its population. The largest was the South-East, with eleven seats; the smallest was the North-East, with four. Northern Ireland's three MEPs continued to be elected by means of the single transferable vote system (STV).

The main contenders in the Euro-elections in England, Scotland and Wales were the UK's three main political parties: Labour, Conservative and Liberal Democrat. These put forward a full list of candidates for each region. In addition, hoping to benefit from the new electoral system, the UK Independence Party (advocating withdrawal from the EU), the Green Party, the pro-European Conservative Party, the Socialist Labour Party and the Natural Law Party also provided candidates within every region. The British Nationalist Party contested each region except Wales. In Scotland, the Scottish Nationalist Party (SNP), and in Wales, Plaid Cymru (the Welsh Nationalist Party, PC) put forward full lists.

In Northern Ireland, the Democratic Unionist Party (DUP), the Social Democratic and Labour Party (SDLP), the Ulster Unionist Party (UUP) and Sinn Fein (SF) were the main contenders. Candidates were also put forward by the Alliance, the United Kingdom Unionist Party (UKUP) and the Progressive Unionist Party (PUP). The Natural Law Party was the only UK-wide party to put forward a candidate.

Three issues underlay the 1999 election campaign and its results. First was the prospect of joining the single currency. Emphasis on the EU was unusual in UK Euro-elections: traditionally fought on domestic issues, the ensuing results reflected voter preferences for the domestic political parties rather than their candidates' policies on the EU. Second, media attention focused on the war in Kosovo, which wound, apparently successfully, towards its close as the polls opened. Third, concerns were voiced quite early in the campaign about potential electoral apathy, said to be a combination of Euro-scepticism and voter fatigue resulting from elections which had taken place in May for the new Scottish Parliament, Welsh Assembly and the UK's local councils. Turnout had been unimpressive with 58.7 per cent for the Scottish Parliament, 40 per cent for the Welsh Assembly, and an average of only 29 per cent in England. Each had also shown a swing from Labour to existing alternatives, so that in Scotland and Wales the expected overall Labour majority had become 'no overall control' and in England the Conservatives regained 1300 council seats previously lost in 1995. The main issue in Northern Ireland continued to be the peace process, concentrating in particular upon the fate of the Belfast Agreement.

The campaign

State funding in the UK is restricted to the provision of one free piece of electoral material per party per household and broadcasting airtime for those parties fielding a complete slate of candidates, commensurate with the number of seats already held by that party. For the 1999 Euro-elections, the amount which could be spent within each electoral division was limited by the Secretary of State (using powers granted under the European Parliamentary Elections Act 1999) to £45 000 per candidate, but this restriction did not apply to money spent centrally by a political party. Voting is not compulsory and UK electors have habitually demonstrated a disinclination to turn out for any other than national elections. The 1999 Euro-elections posed the additional problem of acclimatising voters to a new electoral system (except for Northern Ireland).

The government's campaign

The government took two steps to reduce voter confusion over the electoral procedure. First, it launched a campaign directed at the general public to inform them of the changed voting procedures. Second, it arranged training for electoral returning officers and their staff to enable them to assist any perplexed voters. The mass campaign took the form of television advertisements and the delivery, to every door, of a Home Office leaflet, printed in the EU's distinctive blue with its twelve yellow stars, depicting a mock ballot paper and providing rudimentary information about the EP and its role. The leaflet also contained the address of its website (www.homeoffice.gov.uk/epelections) and of an ITV Teletext page (348) and offered to supply the reader with a copy of the leaflet in any of twelve languages, including Welsh and Punjabi, or a Braille version or audio cassette. The training package – 'The European Election Project' – was devised by SOLACE Enterprises Limited and the Association of Electoral Administrators at the request of the Home Office. It worked with the Home Office's Election Training Advisory Panel to help develop and co-ordinate the programme. The Project involved a dedicated helpline and a range of training and distance learning materials: for example, video and audio cassettes with regular bulletins about developments and legislation.

The Labour campaign

The UK Labour Party began from a strong basis in EP seats, from which it would normally expect a decline. In the 1994 elections, with a 36.4 per cent turnout, Labour had won 62 seats, with 18 Conservative, two Liberal Democrat and two Scottish Nationalist seats. Officials estimated that Labour EP seats would fall to 42–45 purely as a result of the new voting system. In 1999, three other problems jeopardised Labour's EP seats. First, apathy among both its own activists and the electorate was likely to benefit Labour's opponents. Second, the involvement of key Labour actors in the Kosovar War distracted attention from the elections. Third, local and regional election results showed a switch from Labour to alternative parties. The Labour Party did not respond well to these challenges. Criticisms of the campaign included the charges that it was unduly focused upon the Labour leader (whom Labour saw as a certain vote winner), tenuously connected to European issues (for instance, the adoption of the Social Chapter was rejected as a campaign issue), and too low key to make much mark upon a disengaged electorate.

However, for the first time the Labour Party adopted the PES manifesto, indicating its wish to be depicted as a European-wide party. This manifesto was essentially social democratic with an emphasis upon the social, rather than the market, dimension of EU. It referred to the challenges of democratisation, economic growth, policy, institutional and budgetary reform and enlargement. The manifesto also emphasised the importance of the success of the single currency ('It is in the interests of all Member States, whether members of the single currency or not, that the Euro is a success'). It was an integrationist manifesto ('We have an exciting opportunity to build a Europe that is united'), but even if it had not been, it held a symbolic significance as a European Socialist attestation in the absence of a specifically British Labour manifesto.

By contrast, Labour's leaflets, under the heading 'Taking a lead in Europe' were domestic and adversely criticised. They featured both photographs of and statements by Tony Blair, pictures of him seated with the Labour candidates for a particular regional constituency with a 'What has Europe ever done for me' headline, and pictures of 'ordinary voters' giving their reasons for voting Labour, with the emphasis upon the dangers of 'letting the Tories in'. EP candidates were listed in preference order, but their responses towards the EU gave way, in some cases, to a description of their hobbies and interests. The intention was to humanise the candidates, but the presentation left the misleading impression that Labour was fielding political lightweights. This generally domestic image was not helped by Labour's election broadcasts, which also portrayed the elections as an internal party contest ('vote Labour and keep the Tories out'). The week before the election, the Labour Party claimed that an interview with an Italian neo-fascist indicated that the British Conservatives would be involved in a reconstruction of the European right, if elected. Mysteriously, the last televised broadcast before the election featured the qualities of leadership and the way in which the Labour leader epitomised them.

The verdict on the Labour campaign was an unfavourable one, but may be explained by the particular problems that faced the governing party. First, Labour was reluctant to indulge prematurely in a debate upon the single currency, knowing that it would probably have to address this during the next general election campaign and ensuing promised referendum. This meant that positive EU campaigning was problematic and, inasmuch as it occurred, counter-productive. In the week before the Euro-elections, the euro fell again because of the Kosovar crisis. During election week, a poll reported by the *Guardian* (8 June 1999), showed that 61 per cent of UK citizens opposed joining a single European currency. Second, a large part of Labour's team, including its leader, was preoccupied with the Kosovar War and its presentation. Finally, the Labour Party clearly had difficulties in mobilising activists to fight the campaign. MPs were advised to spend time in their constitutencies, but those who did not were not put under particular pressure to do so. Other activists had recently fought local elections, and were reluctant to face the electorate directly again, particularly as there was a degree of discontent about the 'closed list' system. In addition, now that new regulations meant that voters could receive unaddressed election mail directed from party headquarters, activists' participation in addressing envelopes and so on was no longer essential. The campaign, unavoidably, was largely a media-led one, but directed by media which also had a focus elsewhere.

The Conservative campaign

These comments are equally applicable to the Conservative and other main parties. However, the Conservative Party had more to gain and less to lose. Having held only 18 seats in the previous EP (which represented its lowest ever point) Conservative fortunes were beginning to look up after its improved showing in the local elections. In addition, the general mood of the country indicated some animosity towards further European integration, which could benefit a Euro-sceptic party. The Conservative Party, however, suffered from a long-standing split on European issues. In 1999 it needed to formulate a campaign which, whilst not being Euro-sceptic enough to alienate some Conservative key supporters, provided a radical alternative to the 'when the conditions are right' integrationist policy of its main opponents. The Conservative leadership persuaded its domestic representatives to present a united front based upon the premise that entry into a single currency was debarred for the near future. This meant that Kenneth Clarke and John Redwood could share the same platform. However, Conservative MEPs have usually been, or have become, more pro-European than their national counterparts. In 1999 some Conservative MEPs seeking re-election – for example, Edward Macmillan-Scott, outgoing leader of the British Tories and a Vice-President of the EPP – had difficulty in reconciling their personal views with those of their party. This made it more difficult for the Conservative Party to promote highly Euro-sceptic views and present a radical alternative to Labour. One consequence of its attempts to do so, however, was the rise of the pro-European Conservative Party.

Unlike the Labour Party, the Conservative Party published its own election manifesto: 'In Europe, not run by Europe', demonstrating its more tenuous links with sister parties on the European mainland. It claimed that it was a manifesto of a 'proudly British party', favouring widening (admitting more countries), but not deepening (integrating further). It stressed, in particular, the dangers of a single currency ('Conservatives recognise that membership of the single European currency would involve huge risks for Britain') and of left domination of the EU ('In the last few years, Europe has been lurching to the left... They seem more motivated by left-wing ideology than by common sense'). The manifesto also adopted the European Commission's maxim (ironically, given the degree of criticism of the Commission the document contained): 'do less, and do it better', using terms like 'flexibility' and 'cutting red tape'. Reforms proposed involved taking more decisions at national level and vetoing harmonisation proposals. Defence should depend upon NATO, rather than an incipient European defence force, and the war in Kosovo was presented only in the context of the failure of the EU's leaders to agree. The manifesto concluded with a pledge to oppose entry into the single currency at the next election as part of the Conservative manifesto for the next Westminster Parliament. The official Conservative leaflet bore the same heading as its manifesto. Like the Labour leaflet, it included a substantial feature on the party leader, presenting his views upon Europe (including his reluctance to enter a single currency), but kept the candidates' biographies to age, location, professional background and political achievements. Conservative election broadcasts, although somewhat simplistic and dependent upon criticisms of the opposition, also put forward the anti-single currency case.

The unusual circumstances of the 1999 UK Euro-elections held on Thursday 10, June paradoxically meant that the most positive campaigning was conducted on the most negative presentation of a key issue. Whilst Labour was reluctant to be drawn into displaying its pro-integrationist credentials in advance of the 'real' contest of the future national general election, there was nothing to prevent a more sceptical opposition from capitalising upon perceived popular anti-Europeanism to promote a campaign with a distinctive anti-European message.

The Liberal Democrat campaign

In 1994, the Liberal Democrats won two seats. In 1999, they confidently expected to win more, given an electoral system more favourable to a party with diffuse support. The Liberal Democrats started from a more convincingly Europhile position than Labour, and maintained this throughout the campaign. Like the Conservatives, they published their own untitled manifesto under the slogan: 'local where possible, European where necessary'. It projected the Liberal Democrats as 'an unashamedly internationalist party', interested in 'active British engagement in Europe'. Positive on the euro, it presented the non-member UK as 'left standing vulnerably on the sidelines, our politicians still without influence and increasingly marginalised, while our currency is at the mercy of instability and speculation'. However, it too cautiously promised an early referendum on, rather than early membership of, the single currency. Its official leaflets named but did not describe the candidates, emphasising instead the advantages of the new electoral system and the impact of a Liberal Democratic vote on the particular region. Leaflets were headed 'Liberal Democrats: Ambitious for Britain'. Circulated in the regions, these focused on domestic issues (health, crime, schools, pensions, the environment, and 'A strong voice for our area') and called for 'A Europe that listens to people', a written EU constitution and a referendum on the euro.

Other parties (England, Scotland and Wales)

The first nation-wide application of proportional representation drew in a number of small parties eager to make an impact, although not necessarily expecting victory. Of those which won seats, the UK Independence Party unsurprisingly presented a firmly anti-European stance, under the slogan of 'What's the European Union ever done for us?' with photographs of Edith Cresson, Jacques Santer, Romano Prodi and, confusingly, Neil Kinnock. Green Party promotional literature stressed the significance of proportional representation ('This time your Green vote will count'). The Scottish National Party campaign focused upon Scottish fishing waters and the high price paid for fuel in Scotland. Fresh from its successes in the Scottish Parliament elections, it based its manifesto on 'Standing up for Scotland' (i.e., Scottish independence in Europe). Plaid Cymru, although less ambitious, also presented itself in a triumphalist mood after the recent encouragement provided by the Welsh Assembly elections. No other party actually returned an MEP to the EU, but the British Nationalist Party, disguising its racist roots, fought a distinctive campaign to 'Save our sterling', in the hope that proportional representation would enable it to win a seat.

The campaign in Northern Ireland

The Belfast Agreement was made the basis of the campaign by the Democratic Unionist Party (DUP), with Ian Paisley presenting himself as the only sitting Northern Irish MEP who had not supported it unconditionally. In the circumstances the other parties had little choice other than to fight upon an issue that was clearly of more urgent significance to the electorate than who should occupy seats in the EP.

The European Parliament

The EP itself did not play a significant part in the campaign, although it did launch a website on the elections that could be reached through the EU's Europa site. In mid-May it launched an air balloon which was seen in some parts of the country.

Overall, this was a lacklustre campaign, in which the only EU issue was the matter of the single currency. Earlier in 1999, the EP did steal the limelight when it helped bring about the EU Commission's resignation. However, although referred to from time to time, it was eclipsed by the focus instead on examples of EU mismanagement and fraud. Other important EU issues, such as enlargement and agricultural policy, were similarly marginalised, although the Belgian food crisis unhelpfully provided fuel for Euro-sceptics on the grounds that EU (as well as British) officials had acted with insufficient alacrity and decision, appearing to put the UK's citizens in a danger which they would not otherwise have faced.

The results

The overall turnout was exceptionally low; and the results indicated a decided swing against the governing party. Turnout in the UK as a whole fell to 24 per cent (from 36.4 per cent in 1994), which was the lowest turnout across the EU in 1999, and the lowest ever recorded in the history of EP elections. While turnout rose in Northern Ireland from 48.7 per cent to 57 per cent, this reflected concern about the peace process rather than European issues (see Figure 17.1).

In England, turnout in each region showed a similar trend to 1994, with areas showing the highest and lowest turnouts in 1994 replicating the pattern in 1999, although comparatively turnout in the West Midlands and London fell less (see Figure 17.2).

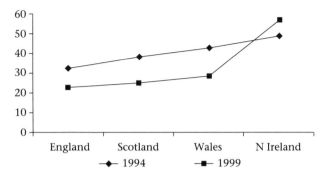

Figure 17.1 UK European election turnout, 1994 and 1999 (%)
Sources: Nugent (1996), 179; *Guardian*, 15 June 1999 (adapted).

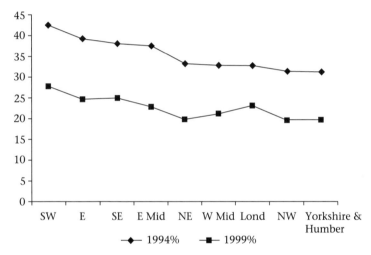

Figure 17.2 English regions: EP election turnouts, 1994 and 1999
Sources: *Guardian*, 14 June 1994 and 15 June 1999 (adapted).

When considering the data, one must remember that the regional boundaries changed between 1994 and 1999; figures have been adjusted accordingly.

Conservative gains (up 18 to 36 seats) were largely at Labour's expense (down from 62 to 29). Liberals rose from two to ten seats, and the UK Independence Party, Green Party and Plaid Cymru won seats for the first time. More information can be found in Tables 17.1–17.4.

How can the the low turnout and the swing from the governing party be explained, given the fact that general support for the government

Table 17.1 European election results in the UK, 10 June 1999

	1999			1994		Group in EP 1999
	Votes	%	Seats	%	Seats	
Conservative	3 578 217	33.51	36	26.85	18	EPP/ED*
Labour	2 803 821	26.25	29	42.67	62	PES
Liberal Democrats	1 266 549	11.86	10	16.13	2	ELDR
UK Independence Party	696 057	6.25	3	–	–	EDD
Green Party	625 381	5.86	2	3.12	–	Verts/ALE
SNP	268 528	2.50	2	3.08	2	Verts/ALE
Plaid Cymru	185 235	1.71	2	1.03	–	Verts/ALE
DUP	192 762	1.80	1	1.03	1	Ind
SDLP	190 731	1.79	1	1.02	1	PES
UUP	119 507	1.12	1	0.84	1	EPP/ED
Others	752 764	7.08	–	4.23	–	–
Total	10 679 552	100.00	87		87	

Seats: 18
Turnout: 24% (36.4% in 1994)
Electorate: 44 126 427
Votes cast: 106 792 552
* The UK Conservative delegation is allied to the EPP/European Democrats, but is not a full part of it.
Sources: Nugent (1996), 178; *Guardian*, 15 June 1999; *Irish Times*, 15 June 1999 (adapted).

Table 17.2 European election results in Scotland

	1999			1994	
	Votes	%	Seats	%	Seats
Labour	283 490	28.77	3	42.5	6
Scottish Nationalist Party	268 528	27.24	2	32.6	2
Conservative	195 296	19.81	2	14.5	–
Liberal Democrats	96 971	9.84	1	7.2	–
others	141 393	14.34	–	3.2	–
Total	985 678	100.00	8	100.00	8

Seats: 8
Turnout: 24.78% (37.9% in 1994)
Electorate: 3 979 845
Votes cast: 985 678
Sources: Nugent (1996), 179; *Guardian*, 15 June 1999 (adapted).

Table 17.3 European election results in Wales

	1999			1994	
	Votes	%	Seats	%	Seats
Labour	199 690	31.88	2	55.9	5
Plaid Cymru	185 235	29.57	2	17.1	–
Conservative	142 631	22.77	1	14.6	–
others	98 869	15.78	–	12.4	–
Total	626 425	100.00	5	100.00	5

Seats: 5
Turnout: 28.33% (40.6% in 1994)
Electorate: 2 211 162
Votes cast: 626 422
Sources: Nugent (1996), 179; *Guardian*, 15 June 1999 (adapted).

Table 17.4 European election results in Northern Ireland

	1999			1994	
	Votes	%	Seats	%	Seats
DUP (Paisley)	192 762	28.40	1	29.16	1
SDLP (Hume)	190 731	28.10	1	28.93	1
UUP (Nicholson)	119 507*	17.61	1	23.84	1
Sinn Fein (McLaughlin)	117 643*	17.33	–	9.86	–
others	58 166	8.56	–	8.21	–
Total	678 809	100.00	3	100.00	3

Seats: 3
Turnout: 57.04% (48.9% in 1994)
Electorate: 1 190 160
Votes cast: 678 809
* Northern Ireland's STV system allocates seats on a quota basis, with a candidate needing to achieve at least 25 per cent of the total votes cast to win. In 1999, only the top two candidates did this on the first count. The UUP candidate, Jim Nicholson, was elected as a result of the third count, when he received 184 789 (including transferred) votes to SF's Mitchel McLaughlin's 119 384.
Sources: Nugent (1996), 179; *Irish Times*, 15 June 1999 (adapted).

seemed to be holding quite well at a mid-term point? First, it was suggested that Labour's campaign was insufficiently high profile to encourage voters, although none of the campaigns was very stimulating. Second, it was argued that Labour had 'neglected' its traditional core voters so that they were not motivated to turn out. This, however, does not explain the relatively constant level in Labour support shown in opinion polls. (In a survey conducted by MORI in June 1999 after the Euro-elections, 51 per cent indicated their intention of voting Labour in a general election.) Third, against this, it was suggested that a large proportion of the electorate was too contented to be motivated to vote. At the same time, presented with a 'second-order' election, immediately after another set of 'second-order' local and regional government elections, voters were said to be growing weary of voting. This does not explain the exceptionally low turnout in 1999, since local and Euro-elections always occur within a month of each other in the UK. Fifth, it was pointed out that the UK's electoral performance was not unique: it was not just in the UK that voting levels declined, and it was not only the UK that experienced a swing against its government. However, turnout fell more steeply than in other EU states and the UK was the only country with a consistently popular government that was hit.

While these explanations help to explain turnout and while voters have a mixture of motives for voting, the most compelling explanation relates to the campaign issues. The most significant topic – the single currency – was unpopular with a high proportion of the electorate. A MORI poll for the *Sunday Telegraph* during the election in June reported that only 24 per cent would vote for a single currency should a referendum be held, with 57 per cent against and 33 per cent with no opinion. Only 12 per cent thought that they would 'strongly support British participation'. In an earlier May MORI for *The Times*, 39 per cent suggested they would vote to withdraw from the EU. This suggests that some Labour supporters remained at home to avoid choosing between voting against their party or against their personal views. The unexpected three-seat win by the UK Independence Party alongside the more noticeable decline in voting in traditional Labour areas (reflecting previous Euro-election turnouts) supports this hypothesis.

Conclusion

The results suggest that the EU rather than the Labour Party needs to be perturbed by them. Labour may need in due course to translate its 'soft' support into votes; but there is no suggestion at present that it is likely to lose the general election on the strength of its EP electoral performance. What are the consequences for the EU? First, low turnout questions the legitimacy of MEPs. A more powerful and more effective EP did not impress UK voters sufficiently to motivate them to vote for its candidates. This, replicated (although to a lesser degree) in the other member states except Portugal and Spain, may make the EP's quest for increased powers harder to sustain. Second, a larger contingent of Conservative, nationalist and anti-EU MEPs will affect the balance of power within the EP and the way in which its parties develop their identity. Conservative MEPs have agreed to remain allied members of the EPP (making it the largest single grouping in the EP). However, under the Malaga Declaration (July 1999) instigated by William Hague and co-signed by the German, Italian and Spanish leaders of their centre-right parties, the EPP parliamentary group has agreed to add 'and European Democrats' to its title. This acknowledges that the centre-right group in future will represent the two trans-European parties: the EPP and the European Democrat Union. This in turn suggests that the new centre-right parliamentary group may become a looser confederation. A letter dated 16 September 1999 from Hans-Gert Pöttering, Chairman of the EPP/European Democrat group to Edward Macmillan-Scott, Chairman of the Conservatives in the EP, confirms that allied membership of the Group does not mean, and has never meant, acceptance of the EPP transnational party manifesto, and presumes no obligations towards the group whip where EPP/European Democrat policies conflict with the UK Conservative European Manifesto. Third, the swing against Labour may make the government more circumspect as it pursues its European strategy. For example, after the elections, the Prime Minister took the decision to lead the all-party Britain in Europe (BiE) campaign providing its aims were widened beyond the original campaign for a single currency. In the meantime, British representatives in the Council of Ministers will also bear in mind expressed public opinion when considering future EU developments, particularly in relation to the next IGC.

References

European Council (1998) 'Conclusions of the Presidency' (Cardiff).
Nugent, N. (1996) 'The United Kingdom', in J. Lodge (ed.), *The 1994 Elections to the European Parliament* (London: Pinter), 271–82.

List of UK party abbreviations used

DUP	Democratic Unionist Party
PC	Plaid Cymru
PUP	Progressive Unionist Party
SDLP	Social Democratic and Labour Party

SF   Sinn Fein
SNP   Scottish Nationalist Party
UKUP   United Kingdom Unionist Party
UUP   Ulster Unionist Party

Part III
Overview

18
The Party System of the European Parliament after the 1999 Elections

Tapio Raunio

When comparing the party system of the fifth EP (1999–2004) with the groups formed after the 1994 elections, the observer is struck by two major developments: the emergence of the EPP as the largest party group, and the changes among the smaller groups. While the latter feature has characterised group formation after every Euro-election, the former constitutes a major shift in the party system. The electoral triumph of the EPP ended the social democrats' (PES) 20-year reign as the biggest party in the EP since the first direct elections of 1979.

This chapter analyses (i) the role of transnational party groups in the EP; (ii) the evolution of the party system and the internal EP rules for establishing groups; (iii) the importance of party groups for parliamentary organisation; (iv) parliamentary decision-making, both within and between party groups; (v) ideological differences and similarities of the Euro-parties' manifestos in the 1999 elections; and (vi) the party system of the 1999–2004 legislature.

Background

Examining briefly long-term trends in the party system, the number of political groups has remained rather stable since the first Euro-elections, despite the fact that six new member states have joined EC/EU during that time. PES and EPP have been the two dominant groups in each EP, controlling over half of the seats after each Euro-election. After the 1999 election, EPP and PES controlled 66 per cent of the seats. The third core group is the ELDR. Social democrats, Christian democrats and Liberals have been present in the EP since 1953 when party groups were first officially recognised in the standing orders of the European Coal and Steel Community Common Assembly. There is notable discontinuity among the smaller groups. The Communists, or to use a more modern term, the radical left, have formed a group since 1973, and the Greens since their electoral breakthrough in 1989. The conservative party family has been represented by both the European Democratic Group (EDG), a group formed around the British Conservatives, which joined the EPP in May 1992, and the Gaullist European Democratic Alliance (EDA), which was founded back in 1965. To the annoyance of the other groups, the extreme right parties were also able to form a group after the 1984 and 1989 elections.

Turning to the situation after the fourth Euro-elections, Table 18.1 shows the seat distribution between party groups in the 1994–99 EP. The PES was the largest group in the 1994–99 EP with just over one-third of the seats. The EPP made substantial gains during the legislative term, first in November 1996 when the Portuguese PSD joined the group from ELDR, and then in June 1998 when Forza Italia MEPs left the UFE to enter the EPP group. The ELDR lost its usual third slot to UFE for three years from 1995 to 1998. The UFE was formed in July 1995 from the merger of EDA and Forza Europa, and its three main national parties were the Gaullist RPR, Forza Italia and Fianna Fail. The fragile ideological foundation of the UFE was suggested by the fact that prior to the formation of the group Forza Italia had applied for membership in both the ELDR and the EPP, but had been rejected by both of them. The Confederal radical left group EUL–NGL functioned, as its name implies, more as an umbrella for various Communist and left parties than as a unitary group. Their influence depended primarily on co-operation with the social democrats. The same applies to the Greens. The emergence of the European Radical Alliance (ERA) was a by-product of the electoral success in France of the Energie Radicale (MRG) list headed by Bernard Tapie, with representatives from various regionalist parties making up the numbers needed for group recognition. The Europe of Nations was formed around the French l'autre Europe, two Danish anti-EU movements and the Dutch Calvinists, the unifying factor behind group formation being their opposition to the Maastricht Treaty. The non-attached members consisted almost exclusively of extreme right MEPs, with representatives from the Austrian Freedom Party, French Front National, Belgian Vlaams Blok and Front National, and Italian Allianza Nazionale.

The final column in Table 18.1 shows the changes in group strength during the 1994–99 legislative term. As in the previous parliaments, the groups underwent substantial restructuring over the five-year period. Two groups disappeared altogether, and when comparing the seat distri-

Table 18.1 Party groups in the 1994–99 EP

Groups	1994	1995	1997	1999	Seat change (1995–99)
PES	198	221	214	214	−7
EPP	157	173	181	201	+28
ELDR	43	52	43	42	−10
EUL–NGL	28	31	33	34	+3
Forza Europa	27	29		0	−29
EDA	26	26		0	−26
V	23	25	28	27	+2
EN	19	19	18	15	−4
ERA	19	19	20	21	+2
UFE		0	56	34	+34
NA	27	31	33	38	+7
Total	567	626	626	626	

List of Party group abbreviations
V   The Green Group
EN   Europe of Nations
NA   Non attached
After 1995, EUL–NGL included the Nordic Green Left.
Dates: 1994 = after the Fourth Euro-elections; 1995 = situation in January 1995 after Austria, Finland and Sweden had joined the Union; 1997 = situation at mid-term in January 1997; 1999 = situation at the end of the fourth legislature.
Seat change: the number of seats at the end of the legislature compared with seat distribution in January 1995 after the entry of Austrian, Finnish and Swedish delegations to the EP.
Source: The European Parliament.

bution at the end of the legislature with the situation in January 1995 after the latest enlargement, the combined difference was 76 seats. Such lack of stability is usual for the EP, with whole national party delegations or individual MEPs switching from one group to another during the legislative term much more frequently than occurs in the national legislatures of EU member states. Smaller groups are particularly vulnerable to such defections, with the two largest groups attracting MEPs from their ranks. European elections are also difficult hurdles for the smaller groups. Should a key national party lose all or most of its seats in the elections, the group may fail to meet the numerical criteria needed to gain group status, and will cease to exist unless replacements can be found (Bardi, 1996). For example, the survival of the EUL–NGL and the Greens continues to depend in each election on the performance of a small number of national parties. While there are still considerable differences between the electoral rules of the fifteen member states, all countries applied for the first time in the 1999

Euro-elections some variant of proportional representation. In the previous elections the single member constituency first-past-the-post system used in Great Britain had led to substantial disproportionality in seat distribution between the groups. For example, in the 1989 elections the Green Party won nearly 15 percent of the votes without gaining any seats, and in 1994 Labour won just under 43 per cent of the votes, but 71 per cent of UK seats.

Group formation is regulated in Rule 29 of the EP's Rules of Procedure (Fourteenth edition, June 1999).

1 Members may form themselves into groups according to their political affinities.
2 A political group must comprise Members from more than one Member State. The minimum number of Members required to form a political group shall be twenty-three if they come from two Member States, eighteen if they come from three Member States and fourteen if they come from four or more Member States.
3 A Member may not belong to more than one group.
4 The President shall be notified in a statement when a political group is set up. This statement shall specify the name of the group, its members and its bureau.
5 The statement shall be published in the Official Journal of the European Communities.

An important change from the 1994–99 EP is that mono-national groups, such as FE in 1994–95, are no longer permitted. A party group must include MEPs from at least two member states. The EP's Rules provide a further incentive for group formation through the allocation of considerable material and procedural benefits. Such strategic calculations explain the emergence of almost purely technical groups, like the Technical Group of Co-ordination and Defence of Independent MEPs in the 1979–84 and the Rainbow Group in the 1984–94 EP. Material benefits include, for example, office space, staff, and money for organising meetings and distributing information. The sum each group receives depends primarily on the number of members in the group. Turning to procedural rights, the party groups have over the years strengthened their position at the expense of individual MEPs (Williams, 1995). Appointments to committees and intra-parliamentary leadership positions, and the allocation of reports and plenary speaking time are based on the rule of proportionality between the groups. Non-attached MEPs are thus to a large extent marginalised in the EP.[1]

Party groups and parliamentary organisation

Appointment of committee seats and chairs and EP Presidency are all controlled by the groups. The President of the EP is elected for 2 1/2 years. Since 1989 the two largest groups, PES and EPP, have shared the Presidency. In the 1994–99 legislature, for example, the first President was Klaus Hänsch, a German SPD representative from the PES, with José María Gil-Robles Gil-Delgado from the Spanish Partido Popular replacing him at mid-term in January 1997. Smaller groups had protested against the informal alliance. For example, in January 1997 the smaller groups aligned to support the candidacy of Catherine Lalumière (European Radical Alliance, or ERA), who received 177 votes. Therefore it came as quite a surprise, at least for the dumbfounded social democrats, when a centre-right coalition elected Nicole Fontaine (EPP) as the new President in July 1999. Fontaine won 306 votes with the rival PES candidate Mário Soares coming second with 200 votes. Imitating the deals between EPP and PES, the EPP and ELDR struck what they termed as a constitutive agreement, according to which the Liberals would support Fontaine and the EPP would in turn back the candidacy of ELDR group leader Pat Cox at mid-term in January 2002. Fontaine also received support from the other right-wing groups.

Committee assignments are decided in the first session of the newly elected EP held in July. The number and size of the committees are decided first, followed by the appointment of committee members and substitutes. Table 18.2 shows the names, membership and chairmen of EP committees at the start of the 1999–2004 term. Committee membership lasts for 2 1/2 years, with seats reallocated half-way through the electoral term. The majority of representatives are full members of one committee and substitutes in another one. Due to the limited number of places available, the party groups often cannot accommodate members' wishes, and MEPs are forced to accept assignments to other committees. Moreover, members with seats in two committees often focus on work in one of the committees and pay much less attention to the work of the second one. The role of the substitutes should not be under-estimated. Members may be highly active in the committee in which they are substitutes, especially if they were not given full memberships in their priority committees. Substitutes usually have full speaking and voting rights, and it is rather common for them to be allocated rapporteurships, especially if they are recognised as policy experts. According to the EP Rules of Procedure (Rule 152/1), 'Members of committees and temporary committees of inquiry shall be elected after nominations have been

Table 18.2 Committees in the 1999–2004 EP

No.	Committee	Seats	Chairman
I	Foreign Affairs, Human Rights, Common Security and Defence Policy	65	Elmar Brok (Germany: EPP/ED)
II	Budgets	45	Terry Wynn (UK: PES)
III	Budgetary Control	21	Dietmut Theatro (Germany: EPP/ED)
IV	Citizens' Freedoms and Rights, Justice and Home Affairs	43	Graham Watson (UK: ELDR)
V	Economic and Monetary Affairs	45	Christa Randzio-Plath (Germany: PES)
VI	Legal Affairs and the Internal Market	35	Ana Palacio Vallelersundi (Spain: EPP/ED)
VII	Industry, External Trade, Research and Energy	60	Carlos Westendorp y Cabeza (Spain: PES)
VIII	Employment and Social Affairs	55	Michel Rocard (France: PES)
IX	Environment, Public Health and Consumer Policy	60	Caroline Jackson (UK: EPP/ED)
X	Agriculture and Rural Development	38	Friedrich-Wilhelm Graefe zu Baringdorf (Germany: Greens/EFA)
XI	Fisheries	20	Daniel Varela Suanzes-Carpegna (Spain: EPP/ED)
XII	Regional Policy, Transport and Tourism	59	Konstantinos Hatzidakis (Greece: EPP/ED)
XIII	Culture, Youth, Education, the Media and Sport	35	Giuseppe Gargani (Italy: EPP/ED)
XIV	Development and Co-operation	34	Joaquim Miranda (Portugal: EUL/NGL)
XV	Constitutional Affairs	30	Giorgio Napolitano (Italy: PES)
XVI	Women's Rights and Equal Opportunities	40	Maj Britt Theorin (Spain: PES)
XVII	Petitions	30	Vitalino Gemelli (Italy: EPP/ED)

Source: The European Parliament (situation in September 1999).

submitted by the political groups and the Non-attached Members. The Conference of Presidents shall submit proposals to EP. The composition of the committees shall, as far as possible, reflect the composition of Parliament.' The distribution of committee seats in the 1989–99 EP was indeed highly proportional, even among the smaller groups who often are over-represented.

Committee chairmanships are highly influential positions and heavily contested by individual members. Apart from the prestige associated with the position, the post of the chairman carries also a lot of potential political weight. She or he is in charge of the committee meetings, speaks on behalf of the committee in the plenary debates (particularly on politically sensitive matters), and is a major broker in drafting the committee agenda. Active chairmen, such as Ken Collins (PES), the Chairman of the Environment, Public Health and Consumer Protection Committee in 1979–84 and 1989–99, have received quite considerable attention outside the EP, both among the EU institutions and in the media. The d'Hondt system is used in allocating chairmen and vice-chairmen between the groups. As with committee memberships, the chairmanships are reallocated at mid-term. The selection mechanism is eloquently described by Westlake (1994, 180):

> the election of committee office-holders is almost always the result of a prior arrangement between all the mainstream political groups. First, on the basis of their d'Hondt-calculated strengths, the political groups 'choose' their committees. Second, on the basis of their numerical strengths, the national contingents within the political groups choose their preferred committees. Lastly, the national contingents choose preferred chairmen from within their ranks.

During the first half of the 1999–2004 EP, the EPP has eight, PES six, ELDR, Greens, and the EUL one chair. Party group co-ordinators are responsible for leading, or at least co-ordinating, the work of their groups in the committees. Together with the committee chairman, the co-ordinators negotiate the committee agenda and the distribution of reports between the groups. Once a group has been assigned a particular rapporteurship, the co-ordinator allocates the report to a member of her or his group. However, the co-ordinator must take into account the wishes of the group leadership and especially of the national parties. While there is naturally great variation between particular policy questions and party groups, group positions are influenced, or even determined, by the group co-ordinator and members sitting in the respective committees. This has increased party groups' need to have some control vis-à-vis the committees and individual members (Bowler and Farrell, 1995).

The rapporteur is responsible for drafting a report on an issue handled by the committee. The distribution of rapporteurships between political groups is a well-established bargaining process. While there are differences between the committees, the process goes as follows: each group receives a quota of points out of the total point tally based on its share of the seats in the committee.[2] Co-ordinators and the committee chairman then decide on the value of each report, and the co-ordinators identify their groups' priority reports and bid on behalf of their groups in specific co-ordinators' meetings. Average reports normally cost 3 or 2 points and opinions 1 point. The auction-like system means that the more expensive reports, such as important pieces of legislation under the co-decision or budgetary procedures, go almost without exception to the EPP and PES. Within party groups, national parties are again in a predominant role, but policy expertise is also an asset.[3] When drafting the report, the rapporteur must be prepared to make compromises in order to accommodate the views of the other groups. Such compromise-building is necessary in order to ensure the smooth passage of the report in the committee and then in the plenary, especially on legislative reports falling under the absolute majority requirement. Rapporteurs normally consult a variety of actors when preparing the draft report: party groups, external experts, interest groups, and often also the Commission. It is common for the MEPs to rely on the committee staff to carry out the background work, including consulting relevant outside actors, and even to produce the initial text which the rapporteur then scrutinises. For individual MEPs, rapporteurships offer the opportunity to gain recognition as policy experts. Indeed, certain MEPs have managed almost to monopolise reports on a particular issue area. The rapporteur system means that individual members – and not the committee chair – are the key persons in the passage of individual pieces of legislation. The rather chaotic decision-making structure of the EP strengthens the autonomy of the committees *vis-à-vis* the whole chamber, and the rapporteur often has a considerable informational advantage over the other members, especially in technical matters.

Party groups control the EP's agenda and internal organisation through the Conference of Presidents and the Bureau. The Conference of Presidents consists of the EP President and party group Chairmen. The non-attached MEPs can delegate two members to the Conference of Presidents, but they have no voting rights. Decisions are based on consensus, failing which matters shall be put to a vote with votes weighted according to the sizes of the groups. The duties of the Conference include taking decisions on parliamentary organisation and on matters related to legislative planning; drawing up the draft agenda of the plenary; deciding the composition and competence of committees and temporary committees of inquiry and of joint parliamentary committees, standing delegations and *ad hoc* delegations; authorising the drawing up of own-initiative reports by committees; and submitting proposals to the Bureau on administrative and budgetary matters relating to the party groups.[4] The Bureau consists of the President and the fourteen vice-presidents of the EP,[5] and is primarily responsible for internal administrative and financial matters.

The strong position of party groups is the result of gradual, piecemeal changes introduced since the first elections held in 1979. The chief motive has been to rationalise parliamentary work in order to enable the EP to make more effective use of its hard-won legislative powers. But while the party groups may control proceedings in the committees and in the plenary, their influence is undermined by internal structural weaknesses. This is particularly evident in the passage of legislation.

Party groups and parliamentary decision-making

Decision-making inside the EP is based on the interplay between party groups and committees. The following description is simplified for reasons of space, and presents the basic stages of how an average legislative bill is processed.[6] When the bill arrives in the EP, the legislative co-ordination unit normally takes the decision, announced in the plenary by the President, on which committee is responsible for producing the report on the issue, assigning simultaneously one or more committees as opinion-giving committees. The Conference of Committee Chairmen is also involved in planning forthcoming legislative work. In case of competition between two or more committees, the dispute may be brought to the Conference of Presidents. In the overwhelming majority of cases the allocation of reports is unproblematic. Once the responsible committee has been chosen, the committee chooses – according to the system described in the previous section – a rapporteur from among its membership whose task is to produce the draft report. The rapporteur consults the party groups in order to ensure the support of the majority of committee members. Party group co-ordinators, or specially nominated spokespersons, also keep their group informed of the contents of the bill and transmit the opinions of their groups to the committee, particularly on politically controversial matters. The draft report is then voted upon in the committee. Groups may debate the bill in their own meetings, but more usually the MEPs of the group seated in the committee convene to agree, if possible, on a common stand. When adopted, the report is then sent to the plenary. Before the final debate and vote the groups decide on their positions: what amendments to propose, and whether they support the report or not. National (party) delegations can hold their own meetings prior to the whole group meetings. While there is substantial variation between the national parties, delegations are well organised, (especially the largest) electing their own leaders and spokespersons. Finally the report is voted upon in the plenary.

Party groups penetrate all aspects of Parliament's work, including the passage of legislation. However, despite the impressive 'partyness' of EP, the transnational party groups are still organisationally weaker than their counterparts in national member state legislatures. The House of Representatives of the US Congress provides in many ways a better point of comparison: the work of both legislatures, especially the processing of legislative bills, is structured around the committees, and individual representatives enjoy much freedom of manoeuvre. The main organisational weakness of the EP party groups is the lack of clear hierarchy: group whips and leaders have precious few sanctions available against troublesome MEPs or national party delegations, particularly as the party groups are excluded from the process of candidate selection.

The internal organisation of party groups is similar to the structure of party groups in national legislatures. The groups elect their chairmen who normally serve the whole five-year legislative term. The chairman is the figurehead of the group and represents the group in the Conference of Presidents. The number of vice-chairmen varies between the groups. The Bureau, composed of the chairman, vice-chairmen, heads (and possible additional members) of national party delegations and the treasurer, is primarily responsible for organisational and administrative issues, but also prepares policy decisions for the group meetings. Group meetings are important in intra-group decision-making. The groups convene regularly in Brussels prior to the plenary week as well as during the plenaries in Strasbourg. The meetings in Brussels constitute a 'Group week', usually lasting two to three days, during which groups primarily discuss matters appearing on the agenda of the next plenary. When MEPs feel they cannot follow the group position, they may use the forum offered by group meetings to explain the reason behind their decision (Bay Brzinski, 1995). Party groups, in particular the more established ones, also set up working groups to deal with specific policy questions. The EPP, for example, has four working parties at the beginning of the 1999–2004 legislature to co-ordinate the work carried out by group members in the seventeen committees (see Hix and Lord, 1997, 77–166; Raunio, 1997, 43–78).

Candidate selection provides the key to understanding the strong position of national parties. MEPs depend on their national parties, especially on national party leadership, for nomination as candidates in future national or European elections. Therefore MEPs have a powerful incentive to listen to the wishes of their national parties. When members receive conflicting voting instructions from their EP party group and from their national party, they can in the majority of cases be expected to obey the orders of the latter.[7] As the EP groups are in practice completely excluded from candidate selection, the group leaders have few, if any, sanctions to wield against troublesome representatives.[8] While national parties have so far largely refrained from giving voting instructions to their MEPs, such mandating is likely to increase in the future.

The Parliament can arguably be categorised as a strong policy-influencing legislature (Scully, 1999). Following the implementation of the Amsterdam Treaty, the co-decision procedure will (according to various estimates) apply to between two-thirds and three-quarters of EC legislation. This substantial increase in legislative powers has potentially far-reaching consequences for party group behaviour. The Council is composed of ministers from the same national parties which are represented in the Parliament. Considering the political significance of post-Amsterdam EC legislation, it is probable that not only representatives of same national parties (MEPs and ministers), but also of same party families, will seek to co-ordinate their behaviour in the Council and the EP to a greater extent than before. This applies not only to matters of national interest (such as the distribution of regional policy funds), but also to wider European matters, many of which are traditional left–right matters concerned with macro-economic policy, employment, and (de)regulation of the internal market. Second, we can expect more partisan politics in dealing with the Commission. The vote on Jacques Santer's nomination as the new Commission President in July 1994 revealed how MEPs from government parties supported him, while those in opposition back home voted against him. The new Parliament has made it clear that it will subject the Commission to tighter scrutiny than before. Partisan tensions are bound to surface between the centre-right led EP and the more social democratic Commission. Moreover, Commission President Prodi has promised to listen carefully to the Parliament, and individual Commissioners have agreed to resign if so required by Prodi 'in the light of any unacceptable circumstances'.[9]

Roll call analyses of the directly elected Parliament indicate that the majority of EP parties have achieved rather high levels of cohesion (Attinà, 1990; Hix and Lord, 1997; Raunio, 1997; Kreppel, 1999). When internal conflict surfaces within the more established groups, it is usually confined to specific votes, with the more cohesive party groups achieving unitary behaviour in an overwhelming majority of votes. The performance of the smaller groups has varied a lot, with groups dominated by a single party often achieving very high levels of cohesion, while the more technical groups have not even tried to reach unitary behaviour. Group unity and coalition behaviour are linked to the EU's legislative procedures which require the Parliament to muster absolute majorities (314/626) to amend or reject legislation. Party groups thus have an incentive to achieve unity during voting periods in order to increase their influence in EC legislation and to put pressure on the Commission and the Council on non-legislative matters. Until the 1999 elections the manufacturing of legislative majorities was primarily based on informal co-operation between PES and the EPP. Co-operation between these two large groups can also be regarded as a sign of 'maturity', as the Parliament has needed to moderate its resolutions in order to get its amendments accepted by the Council and the Commission (Kreppel, 1999). After all, the overwhelming majority of national ministers represented in the Council and members of the Commission are either social democrats or Christian democrats/conservatives. The influence of smaller groups has in turn depended on their ability to align with one or both of the two large groups. Broad or over-sized coalitions are very common in the Parliament. Such coalition formation, facilitated by the rules on inter-institutional legislative procedures, also explains in part the ideological convergence between the Euro-parties. The ideological moderation implied by coalition building also makes it difficult for Euro-parties to claim credit for policy achievements.

Party groups and Euro-elections

The institutional environment of the EP party groups differs significantly from the standard European version of majoritarian parliamentary democracy. In the EU member states the result of general elections is the determining factor in government formation. Once appointed, the government must enjoy the confidence of the parliamentary majority. The multi-level nature of the European polity, together with the separation of powers system, mean that the policy influence of the EP depends on the preferences of the Council and the Commission. While this institutional set-up limits the powers of party groups, it also provides multiple opportunities for Europarties to influence European and national policy.

Officially the main EP party groups are organs of their Europarties. EPP was established in 1976. Following the inclusion of the Party Article in the Maastricht Treaty, the socialists and liberals turned their federations into parties: PES was established in November 1992 and ELDR in December 1993. The Greens set up a European Federation of Green Parties (EFGP) in June 1993. While qualified and even simple majority voting is nowadays applied in all four Euro-parties, their internal decision-making is still better characterised as co-operation and policy co-ordination between the member parties. EP groups are represented in the party congresses and executive committees of the Euro-parties. The group Chairman reports to the congress on the work of the group, and sits in the Euro-party's executive committee (the name and functions of which vary between the parties). The exact policy influence of Euro-parties is practically impossible to measure, and depends primarily on the willingness of national member parties to pursue and implement the agreed policy objectives. However, the Euro-parties serve as an important arena for diffusion of ideas and policy co-ordination, and such networking has undoubtedly become more important since the early 1990s (Ladrech, 1997; Bell and Lord, 1998).

The formulation of Euro-parties' manifestos for the EP elections is shaped by national parties, the executive bodies of the Euro-party, and the EP party group. All of them appoint special working groups, with often considerable overlap in terms of personnel.[10] The draft programme is circulated among the national parties, and then submitted to the Euro-party's congress for final adoption. The congress can still propose amendments to the programme. For example, in the ELDR congress in Torquay in December 1993 the delegates tabled over 400 amendment proposals. Drafting of the liberals' 1999 election manifesto began in January 1997 under the chairmanship of Pat Cox, the current leader of the ELDR group in the EP. The national member parties were consulted extensively in order to ensure a less cumbersome adoption process in the forthcoming party congress. Despite wide consultation rounds, nearly 100 amendment proposals were tabled at the Berlin congress in April 1999, with many of them producing extremely tightly contested votes: for example, on institutional reform and the division of competencies between the EU and the Member States (Sandström, 1999).

Euro-parties' election manifestos have almost without exception been described as vague, bland or uninspiring, with disagreements among member parties leading to the adoption of lowest common denominator programmes. Such criticism is arguably misplaced: election manifestos are intentionally brief, outlining the basic policy objectives and priorities of the party. Each Euro-party has additional, much longer and more detailed sectoral programmes which contain clear policy objectives and strategies for achieving them. This reflects the widespread practice in most European countries: election manifestos are brief, designed to appeal to a broader audience, while concrete policy objectives are explained in sectoral programmes. The following is a brief overview of the election manifestos of the four Euro-parties. More specific sectoral programmes are not examined here.[11]

The EPP election manifesto was the shortest one, and contained only very general, broad objectives. Individual freedom and solidarity are both emphasised, as are full employment, education, and protection of the environment. Enlargement is welcomed 'as soon as possible' and CFSP should become a reality. 'Radical institutional reform' is envisaged, but no actual reforms proposed (EPP, 1999). The PES manifesto lists 21 commitments, which reflect the party's overall ideology of 'yes' to a market economy, but 'no' to a market society. A European Employment Pact is singled out as the top priority, accompanied by closer economic co-ordination and improved dialogue between the social partners. Other commitments include the adoption of a European Charter of rights, the promotion of equal opportunities, effective CFSP, and sustainable development. Interestingly the manifesto is silent about institutional reform, merely stressing subsidiarity and the need to extend qualified majority voting in the Council (PES, 1999). The Liberals, on the other hand, argue that employment policy should be decided at the national level, and the EU 'should focus on developing the macroeconomic conditions and the framework of the single market in which appropriate national measures can more fully realise their potential'. Enlargement, fraud prevention, human rights, and effective CFSP are emphasised. On institutional questions ELDR supports granting the EP equal legislative status with the Council, transnational lists for Euro-elections, reserving the unanimity rule in Council for clearly defined constitutional issues, and the adoption of a Bill of Rights (ELDR, 1999). The Greens' manifesto includes a whole host of concrete policy objectives. These include the application of environmental and social standards in economic decision-making, combating tax competition, the introduction of eco-taxes, the adoption of a charter of civil rights, and the establishment of a European-level Pact for Sustainability and Employment to create jobs and to achieve 'environmentally and socially sustainable economic development'. Enlargement is prioritised, a pan-European security system should be developed, and decision-making should be democratised through granting the EP full co-decision powers with the Council and the right to initiate legislation and to dismiss individual Commissioners (EFGP, 1999).

The Euro-parties are in broad agreement over several issues: increasing transparency and openness in decision-making, strengthening civil rights at the European level, promoting sustainable development, subsidiarity and enlargement. The policy profiles of the four Euro-parties have become increasingly similar since the 1980s, especially since the PES and the EFGP both changed their attitude to European integration (Hix, 1999). This reflects the overall policy convergence, especially in economic policy, between the national parties belonging to the Euro-parties. With the exception of marginal fringe parties, basically all national parties are considered fit to govern, with national coalition cabinets often including parties across the ideological spectrum (Mair, 1997). Institutional matters remain problematic within all four Euro-parties. More interestingly, from the point of view of the new EP, employment and European-level regulation arise as crucial issue for all parties (Marks and Wilson, 1999). Whereas the PES and EFGP advocate EU action to fight unemployment and to improve social and environmental standards, EPP and ELDR emphasise closer economic co-ordination and national-level action. The salience of this classic left–right cleavage can have important consequences for coalition behaviour in the 1999–2004 EP.

The new EP party system

Reflecting the agreement struck between EPP and ELDR over the sharing of the Presidency, the new party system of the 1999–2004 EP looks substantially different from the PES/EPP duopoly which has dominated parliamentary decision-making since the late 1980s. However, such a move to the right is moderated by two main factors: the numerical threshold of the legislative majorities, and the shifting coalition formation between party groups across policy sectors. Moreover, the first part-session of the newly elected EP, held in July 1999, indicated that groups themselves are again prone to internal factionalism.[12]

Group formation in the EP is a far more complicated process than in national legislatures, involving a lot of uncertainty and bargaining among the approximately 100 national party delegations. EPP became (for the first time since the 1979 elections) the largest group with 233 members. The German CDU/CSU is the largest national party, with 53 seats. The conservative wing of the group has strengthened over the years, and the title European Democrats was added to the group name in order to appease the British Conservatives. The Conservative Party MEPs held a meeting in June on which group they should join, but eventually decided to continue in the EPP as this gives them more influence and office benefits than sitting in a smaller alternative conservative group. In the end both the British Conservatives and the French RPR–DL joined the EPP, and this may strain group cohesion, particularly on institutional questions. The PES group lost 34 seats. The largest party delegation among the 180 MEPs is the German SPD with 33 seats. Both EPP and PES comprise members from all member states.

The liberals gained eight seats, bringing their seat tally to 50. The largest party delegation is the British Liberal contingent with ten members. The accommodation of Nordic centre parties has proved problematic for ELDR, especially on anti/pro-integration matters in which the Danish, Finnish and Swedish parties are considerably more Euro-sceptical than the group majority. The Green group has now 48 MEPs, but the group is in effect an alliance between 38 representatives from green parties and ten members from Basque, Flemish, Scottish, Welsh and Spanish regionalist parties. The regionalist parties of the European Free Alliance did not win enough seats to sit on their own and thus chose to join the Green group. The EUL–NGL also gained seats, and now has 42 MEPs. The only real question mark in the group formation stage was the French Trotskyist Lutte Ouvrière, but in the end the French Communist Party accepted them into the group. UEN is the successor of the Europe of Nations, and brings together parties opposed to further integration. The main national parties are the Pasqua–Villiers list (RPFIE) from France, the Irish Fianna Fail, and the Italian Allianza Nazionale. Interestingly the four Danish anti-EU movement members are not included among the 30 MEPs. Finally, the 16-strong EDD includes six MEPs from the French pro-hunting/defence of rural traditions group, the three members of the UK Independence Party, four anti-EU Danes and three Dutch Calvinists. The ranks of the non-attached include Italian Bonino list and Lega Nord MEPs and extreme-right representatives from the Austrian Freedom Party, Flemish Vlaams Blok, and the French Front National.[13] Table 18.3 shows the composition of party groups in September 1999.

Decision-making in the EP will continue to be based on compromise-building between the groups as no group controls the majority of the seats. However, the election result and the proceedings of the first part-sessions indicate that the bargaining power of the groups has changed to the advantage of the centre-right groups, and that winning coalitions will alternate depending on the issue voted upon. The agreement reached between the EPP and ELDR about the Presidency, and the vote on electing Fontaine, signal that the centre-right intends to extend this

Table 18.3 Party groups in the EP, September 1999

Party groups	B	DK	D	GR	E	F	IRL	I	L	NL	A	P	FIN	S	UK	Total
EPP	6	1	53	9	28	21	5	34	2	9	7	9	5	7	37	233
PES	5	3	33	9	24	22	1	17	2	6	7	12	3	6	30	180
ELDR	5	6		3			1	7	1	8			5	4	10	50
GREENS/EFA	7		7		4	9	2	2	1	4	2		2	2	6	48
EUL/NGL		1	6	7	4	11		6		1		2	1	3		42
UEN		1				12	6	9				2				30
Independent	2				1	6		12			5				1	27
EDD		4				6				3					3	16
Total	25	16	99	25	64	87	15	87	6	31	21	25	16	22	87	626

Member states: B = Belgium, DK = Denmark, D = Germany, GR = Greece, E = Spain, F = France, IRL = Ireland, I = Italy, L = Luxembourg, NL = The Netherlands, A = Austria, P = Portugal, FIN = Finland, S = Sweden, UK = United Kingdom

co-operation to legislative matters. However, the success of such a non-socialist alliance depends on ideological cohesion and on the level of absenteeism. First, while the centre-right groups (EPP, ELDR, EDD, possibly UEN and the non-attached) may find agreement on internal market legislation, their ideological unity is not likely to extend to social policy, human rights or constitutional matters. Second, while absenteeism has declined over the years, most notably as a result of the introduction of the co-decision procedure and the legislative majority requirement, EP plenaries are still marked by relatively poor levels of attendance (Scully, 1997; Kreppel, 1999). Provided that the centre-right groups can agree on a common position, the non-left alliance still needs even for minimum winning coalitions the backing of about 85–95 per cent of its MEPs in votes falling under the legislative majority requirement. The centre-right coalition therefore needs additional allies, and the Greens might be the answer. The Greens are now seated right in the middle of the hemicycle, between the PES and ELDR.[14] The Greens have become significantly more pro-integrationist and market-friendly since the 1994 elections, and their ideological moderation facilitates co-operation with the centrist groups, notably the EPP and ELDR. PES may to an increasing extent align with EUL–NGL and the Greens. Leftist groups control 270 seats between them, and therefore they need to coalesce with other groups to produce winning coalitions. The new balance may mean that the ELDR and Greens have influence way beyond their pure numerical strength.

Previous studies of coalition behaviour in the EP have confirmed that winning coalitions are usually broad and over-sized, often including all the main groups, and that alignments vary across policy sectors (Hix and Lord, 1997; Raunio, 1997; Kreppel and Tsebelis, 1998). It is very likely that the same tendency will continue, not least because the EP's legislative influence depends on making amendments which are acceptable to both the Commission and the Council. At the same time the new EP is more bipolar than before, with the new legislative powers of the EU/EP increasing the salience of the left–right dimension. The smaller groups are also very unpredictable, both in terms of coalition behaviour and group cohesion. The party system of the 1999–2004 EP may not constitute a radical break from the past, but it is more competitive, and also more interesting.

Notes

1. For detailed information on such material and procedural rights, see the EP Rules of Procedure and Corbett, Jacobs and Shackleton (1995), 56–61, 86–9.
2. The exception is Budgets Committee, where the point totals have been determined by the share of the seats in the whole Parliament.
3. The major reports are often distributed well in advance. For example, in the Budgets Committee the key reports are allocated at the start of the five-year legislative term. While some committees operate a highly hierarchical points system for the various types of reports, in other less powerful committees, such as the Petitions Committee, the decisions are largely consensual without the need to resort to the points system. Occasionally rapporteurships are divided between two members. Examples from the 1989–99 EP include reports on the annual EU budget in the Budgets Committee and on the IGCs in the Institutional Affairs Committee.
4. See Rules 23 and 24 of the EP Rules of Procedure (Fourteenth edition, June 1999).
5. The group affiliation of the fourteen Vice-Presidents in the first half of the 1999–2004 Parliament is PES 6, EPP 5, ELDR 1, EUL–NGL 1, Greens 1.
6. For more detailed information on the processing of legislation, see Corbett, Jacobs and Shackleton (1995), 105–244.
7. While the importance of national party delegations should not be underestimated, these party delegations are by no means unitary actors themselves. National parties in every member state are, to varying degrees, divided over Europe, and these cleavages are reproduced in the Parliament. Well-known examples are the two main British parties: both the Conservative and Labour contingents have experienced splits when anti/pro-integration matters have been voted upon in the plenary.
8. Available (but very seldom used) sanctions include not assigning MEPs to committees or delegations of their choice. Obviously members who regularly vote against the group line may end up marginalised within their group. In the run-up to the vote on Jacques Saunter's nomination as the Commission President in July 1994, the President of the EPP group, Wilfried Martens, indicated to his group that those dissenting from the group line could risk expulsion

from the group (Hix and Lord, 1996). However, this kind of threat is exceptional in the Parliament.
9 Speech by Mr Prodi to the Parliament, 21 July 1999. Responding to Prodi's speech, the EPP leader Hans-Gert Pöttering emphasised that, unlike what Prodi had claimed, the team was not 'well balanced', singling out particularly the two proposed two German Commissioners, both of whom represented leftist parties (PES, Greens).
10 All national member parties may not establish special working groups, but normally at least appoint one or more delegates to participate in the Europarty's manifesto working group.
11 The EPP adopted (at its Brussels congress in February 1999) the *EPP Action Programme 1999–2004*. The programme contains a long list of concrete policy objectives under ten sub-headings. PES had sixteen detailed sectoral programmes, while the ELDR had five (all available on the parties' web pages in September 1999).
12 I am grateful to Simon Hix for information on group formation in the July plenary.
13 Following the elections the Italian Bonino list and Lega Nord MEPs, the French Front National and the Belgian Vlaams Blok delegations set up a purely technical coalition entitled Technical Group of Independent Members (TGI). The Committee on Constitutional Affairs decided on 28 July that the group did not meet the first criterion required from groups: political affinity. Also, all the other party groups opposed the recognition of the TGI. The group was disbanded on Monday, 13 September.
14 The Dutch Green MEP Alexander de Roo commented that 'the new seating order is a good reflection of the new political landscape in the new Parliament which will allow for much more flexible alliances in the political decisions'. See 'New Seating Order in the Parliament's Hemicycle Reflects New Strength of the Greens/EFA', Greens/EFA group press release, 15 July 1999.

References

Attinà, Fulvio (1990) 'The voting behaviour of the European Parliament members and the problem of the Europarties', *European Journal of Political Research*, 18(4), 557–79.
Bardi, Luciano (1996) 'Transnational Trends in European Parties and the 1994 European Elections of the European Parliament', *Party Politics* 2(1), 99–113.
Bay Brzinski, Joanne (1995) 'Political Group Cohesion in the European Parliament, 1989–1994' in Carolyn Rhodes and Sonia Mazey (eds), *The State of the European Union, Vol. 3: Building a European Polity?* (Boulder, Colorado: Lynne Rienner), 135–58.
Bell, David S. and Christopher Lord (eds) (1998) *Transnational Parties in the European Union* (Aldershot: Ashgate).
Bowler, Shaun and David M. Farrell (1995) 'The Organizing of the European Parliament: Committees, Specialisation and Co- ordination', *British Journal of Political Science*, 25(2), 219–43.
Corbett, Richard, Francis Jacobs and Michael Shackleton (1995) *The European Parliament*, 3rd edn (London: Cartermill).

EFGP (1999) *Greening Europe – More Urgent Than Ever*, Common Green manifesto for the 1999 European elections.

ELDR (1999) *Unity in Freedom: The Liberal Challenge for Europe*, Electoral manifesto adopted by the ELDR Party Congress in Berlin on 29 April 1999.

EPP (1999) *Election Manifesto 1999* Adopted by the XIII EPP Congress, 4–6 February, 1999, Brussels.

Hix, Simon (1999) 'Dimensions and alignments in European Union politics: Cognitive constraints and partisan responses', *European Journal of Political Research*, 35(1), 69–106.

Hix, Simon and Christopher Lord (1996) 'The Making of a President: The European Parliament and the Confirmation of Jacques Santer as President of the Commission', *Government and Opposition*, 31(1), 62–76.

Hix, Simon and Christopher Lord (1997) *Political Parties in the European Union* (London: Macmillan).

Kreppel, Amie (1999) 'Coalition Formation in the European Parliament: From Dogmatism to Pragmatism', Paper presented at the ECSA Sixth Biennial International Conference, Pittsburgh, 2–5 June 1999.

Kreppel, Amie and George Tsebelis (1998) 'Koalitionsbildung im Europäischen Parliament', in Thomas König, Elmar Rieger and Hermann Schmitt (eds), *Europa der Bürger? Voraussetzungen, Alternativen, Konsequenzen* (Frankfurt: Campus), 295–327.

Ladrech, Robert (1997) 'Partisanship and Party Formation in European Union Politics', *Comparative Politics*, 29(2), 167–85.

Mair, Peter (1997) *Party System Change: Approaches and Interpretations* (Oxford: Oxford University Press).

Marks, Gary and Wilson, Carole (1999) 'National parties and the contestation of Europe', in Thomas Banchoff and Mitchell P. Smith (eds), *Legitimacy and the European Union: The Contested Polity* (London: Routledge), 93–113.

PES (1999) *Manifesto for the 1999 European Elections*, Adopted by the fourth PES Congress, Milan, 1–2 March 1999.

Raunio, Tapio (1997) *The European Perspective: Transnational Party Groups in the 1989–94 European Parliament* (Aldershot: Ashgate).

Sandström, Camilla (1999) 'Europeiskt partisamarbete och partiernas idémässiga utveckling', Paper presented at the conference of the Nordic Political Science Association, Uppsala, 19–21 August 1999.

Scully, Roger (1997) 'Policy Influence and Participation in the European Parliament', *Legislative Studies Quarterly*, XXII(2), 233–52.

Scully, Roger (1999) 'Parliamentary Democracy and Political Legitimacy in the European Union', Paper presented at the ECPR Joint Sessions of Workshops, Mannheim, 26–31 March 1999.

Westlake, Martin (1994) *Britain's Emerging Euro-Elite? The British in the Directly-Elected European Parliament, 1979–1992* (Aldershot: Dartmouth).

Williams, Mark (1995) 'The European Parliament: Political Groups, Minority Rights and the "Rationalisation" of Parliamentary Organisation. A Research Note', in Herbert Döring (ed.), *Parliaments and Majority Rule in Western Europe* (Frankfurt and New York: Campus and St Martin's Press), 391–404.

List of European party abbreviations used

EDA	European Democratic Alliance
EFGP	European Federation of Green Parties
ELDR	European Liberal, Democrat and Reform Party
EN	Europe of Nations
EPP	European People's Party
ERA	European Radical Alliance
EUL-NGL	Confederal Group of the European United Left
PES	Party of European Socialists
UEN	Union for the Europe of Nations
UFE	Union for Europe
V	The Green Group

19
European Electoral Systems and Disproportionality

Mark Baimbridge and Darren Darcy

Introduction

The EP elections held in each of the EU's fifteen member states between 10 and 13 June 1999 followed the set pattern of being largely a collection of mini-referendums on the domestic records of national governments in which voters frequently took the chance to voice a note of dissatisfaction. First, the military achievement in Kosovo failed to translate into an electoral reward for almost any governing party. Second, although eight member states are run by centre-left parties with a further five possessing a mix of left and right, these elections signalled a general shift towards the centre-right. However, even where voters did not swing to the right (France, Portugal and Spain) this was still largely for domestic reasons. Perhaps the most noticeable feature was how little the voters seemed to care, with the 49.9 per cent turnout being the lowest since the first elected MEPs were returned in 1979. Possible explanations include voters distracted by Kosovo, uninspired by national campaigns and disillusioned with the EU following allegations of waste and nepotism which presaged the Commission's resignation.

However, these elections were a momentous occasion for numerous reasons. First, they were the largest pannational exercise in representative democracy with some 10 000 candidates contesting 626 seats across the fifteen participating countries. Second, under the 1991 Maastricht Treaty and the 1997 Amsterdam Treaty, the European Parliament currently possesses greater powers than at any previous period in its history. Third, the voting system adopted across the EU was based on proportional representation, albeit in different variants.

This chapter seeks to supplement the existing literature on the European Parliament elections, which have focused, *inter alia*, upon turnout, the concept of second-order elections, legitimacy and representation, politicians, federalism and EU integration and general summaries of European Parliament elections and its workings.[1] Thus whilst election systems will have been a feature within several of these themes, the analysis of disproportionality at the EU level has been absent from specific examination. This chapter seeks to analyse this topic since it is vital, at a technical level at least, to the degree of legitimacy accorded to the European Parliament itself.

Summary of voting systems

The analysis of elections is significantly influenced by the specific electoral system. We begin by summarising key features of the alternative voting structures within the fifteen member states. Table 19.1 illustrates the technical operation and chosen seat allocation method, together with the existence of thresholds and compulsory voting. Discussion at the European level had long centred on the idea of developing a 'uniform electoral procedure'. However, various debates had made clear the difficulty of reaching complete agreement, such that it was decided to look for a series of 'common principles' to be used for EP elections across all member states which were subsequently formally introduced by the 1997 Amsterdam Treaty. Thus when examining Table 19.1 it should be noted in July 1998 the outgoing EP adopted common electoral principles to ensure that the elections possessed a sufficient 'European' character. Hence, all fifteen member states used proportional representation, although in different variants. While the majority adopted the d'Hondt method for seat allocation, a further three systems were also used (STV, Hare – Niemeyer and modified-Sainte-Laguë). A similar degree of uniformity is also evident in terms of the threshold of votes required to win a seat where only five countries have such a rule (Austria, France, Germany, Greece and Sweden) and, to a lesser extent, regarding compulsory voting which is a legal requirement in Belgium, Greece, Italy[2] and Luxembourg.

Table 19.1 also provides *a priori* indication relating to the level of disproportionality associated with the electoral system of each country: first, this revolves around the three factors of electoral formula, district magnitude and ballot structure, second, the ranking of electoral formulae, with the following group order in descending proportionality based on theoretical assessments of likely electoral outcomes: (i) LR–Hare, Sainte-Laguë (ii) LR–Droop, STV, modified-Sainte-Laguë; (iii) d'Hondt, LR–Imperiali; third, the electoral threshold also influences disproportionality through reducing the opportunity for small parties to win seats. Moreover, even in the absence of legal thresholds, low district magnitudes possess similar effects.[3]

Analysis of disproportionality in the 1999 EP elections

This section focuses on the degree of disproportionality in each EU member. The analysis is undertaken for each party (i) between the share of vote (v_i) and the proportion of seats won (s_i) to determine how efficiently the wishes of the electorate were transposed into the final result. To examine the level of disproportionality, the familiar selection of traditional measurements is used: the Rae index, Loosemore–Hanby index,[4] least-squares index,[5] largest-deviation index (Lipjhart) and the regression coefficient index.[6] Although each of the alternative measurements possesses inherent advantages and disadvantages, space limitations preclude a discussion of their relative merits although we utilise them to examine the consequences of all fifteen members states having adopted proportional representation for the 1999 EP election.

In addition, we seek to supplement these traditional measurements through extending the concept of the effective number of parties[7] as it denotes the degree of efficiency the chosen election technique possesses in terms of the theoretical objective of accurately translating the wishes of the electorate into representation within the elected body. Our party representation index is thus calculated by subtracting the effective number of parties based on seat shares (N_s) from the vote proportion (N_v). The indicator can therefore either be determined from the two individual measurements as originally proposed or by using the consolidated formula of $(\sum s_i^2 - \sum v_i^2)/(\sum v_i^2 \sum s_i^2)$. Clearly, the 'ideal' score representing perfect efficiency is zero, whilst inefficiency is illustrated by increasing divergence from this base value in either direction.

Not only does this indicator of disproportionality provide an alternative to those previously utilised, but it also has an intrinsic political meaning beyond that of an abstract index. The integer score from this efficiency measurement directly refers to the number of parties who are either disadvantaged or advantaged as a direct consequence of the voting system. A positive figure ($N_v > N_s$) indicates the number of parties who are deprived of seats within the elected body (under-representation), whilst a negative figure ($N_v < N_s$) is suggestive of the voting system possessing bias towards parties attaining seats beyond that indicated by their vote shares (over-representation). Given that the operation of most electoral systems tends to result in a lower number of parties achieving representation compared to those gaining support in elections we would therefore usually expect to see positive values predominating. Furthermore, as with other measurements of disproportionality, this concept is equally applicable either when examining a single election or when comparing different elections (for instance, on the basis of alternative voting systems or intertemporal elections determined by either alternative or identical voting systems).

Table 19.2 illustrates the level of disproportionality measured by the six indicators derived from the results published by the EP's Directorate-General for Information and Public Relations.[8] The traditional measurements of disproportionality (Rae, Loosemore – Hanby, least-squares and largest-deviation) which comprise abstract indices raise difficulties when attempting to analyse the wealth of information presented in Table 19.2. Thus we adopt a ranking procedure to simplify the volume of information provided (see Table 19.3). In contrast, the regression coefficient and party representation indicators which possess intrinsic political significance permit discussion at this stage. The former measurement suggests that in thirteen EU member states a systematic advantage for larger parties was prevalent, with an advantage for smaller parties only apparent in Belgium and Italy. In relation to our party representation index, the hypothesised predominance of positive outcomes is evident for all fifteen countries. Examination of these positive values indicate that parties were excluded from gaining seats in ten member states, with no loss of representation in five (Austria, Greece, the Netherlands, Spain and Sweden) which are later shown to display amongst the lowest levels of disproportionality (see Table 19.3). Moreover, the problem appears particularly acute in Denmark, France and Ireland where two parties are indicated to have been excluded from the European Parliament. The final row of Table 19.2 reports the mean associated with each measure of disproportionality. These indicate, first, that the average level of disproportionality as illustrated by the four traditional indices is generally low and reflects the familiar inter-relationship between them. Second, the two measurements of disproportionality which possess intrinsic political interpretation indicate that a bias exists towards large parties and that on average at least one party is excluded from the European Parliament in each EU member state.

To simplify and facilitate understanding of the data presented in Table 19.2, we rank each EU member state, where 1 = lowest level of disproportionality, with respect to the six indices, and also indicate their overall position in terms of ascending disproportionality based on the average of the individual indicators. The examination of these relative

Table 19.1 Summary of electoral systems

Country	Voting system	Seat allocation method	Threshold	Compulsory voting
Austria	Electors vote for national lists, but may express preference for individuals on those lists.	d'Hondt	4%	No
Belgium	Proportional representation based on three electoral colleges. Electors vote for entire party lists or individuals.	d'Hondt	No	Yes
Denmark	Proportional representation by list. Votes can be cast for a party or individual candidates.	d'Hondt	No	No
Finland	Proportional representation with the whole country as a single constituency. Voters cast ballots for a party, electoral alliance or joint list, but are also able to prefer a particular candidate. Once it has been decided how many seats have been won by each party, alliance or list, then the candidates are elected according to how many preference votes they received.	d'Hondt	No	No
France	Electors vote for national party lists, each consisting of 87 names. The lists are ordered and voters cannot express preference for individual candidates	d'Hondt	5%	No
Germany	Proportional representation by party lists. Parties can submit either one federal list or lists at state level.	Hare – Niemeyer	5%	No
Greece	Proportional representation by national party list. No preference voting for individual candidates.	d'Hondt	3%	Yes
Ireland	The country is divided into four regional constituencies, each electing its own MEPs. Voters mark ballot papers in order of preference for the individual candidates. A candidate who reaches a quota dependent on the total number of votes cast is elected. Excess votes, and those of eliminated candidates, are redistributed on the basis of their supporters' second preferences.	STV	No	No

Italy	The country is divided into five constituencies, but a party must gain more than a specified quota of the vote at national level before it qualifies for seat allocation. This is done at constituency level, but surplus votes are redistributed at national level, helping smaller parties win seats.	d'Hondt	No	Yes
Luxembourg	Proportional representation on a national basis. Electors have six votes and may cast them en bloc for a whole list, for candidates from more than one list, or selected candidates from a single list. Party lists have a maximum of 12 names.	d'Hondt	No	Yes
Netherlands	Proportional representation on a national list system. There are 19 constituencies for administrative purposes only.	d'Hondt	No	No
Portugal	Proportional representation by national list.	d'Hondt	No	No
Spain	Proportional representation by national list.	d'Hondt	No	No
Sweden	Proportional representation at national level. Voters choose a party or one or more candidates. They can also change the order of party lists, and add or delete names. Candidates are ranked according to preference votes. To change the order of the list, candidates must obtain 5% of their party's votes.	modified-Sainte-Laguë	4%	No
UK	England, Scotland and Wales used proportional representation under a closed list system, while Northern Ireland employed STV. The UK was divided up into 12 electoral regions, electing a number of MEPs depending on the size of the electorate. Parties supplied a list of candidates ranked in order; electors cast their vote for the party or independent candidate of their choice.	d'Hondt & STV	No	No

Table 19.2 Disproportionality measurements

	Rae	Loosemore –Hanby	Least-squares	Largest– deviation	Regression coefficient	Party representation
Austria	1.5	4.5	3.1	2.7	1.13**	0.4
Belgium	1.7	10.2	4.9	3.8	0.85**	1.1
Britain*	3.0	11.9	7.8	7.1	1.27**	1.2
Denmark	2.6	10.2	6.7	8.0	1.5**	1.9
Finland	1.9	6.6	4.4	3.9	1.01**	0.6
France	1.2	6.0	3.8	3.3	1.26**	1.6
Germany	1.9	5.7	4.1	4.1	1.14*	0.7
Greece	2.0	7.2	4.3	3.3	1.05**	0.4
Ireland	3.4	11.8	7.3	6.6	1.02**	1.5
Italy	0.3	3.2	1.2	0.8	0.98**	1.4
Luxembourg	5.0	17.5	11.1	10.1	1.22**	1.0
The Netherlands	1.1	3.8	2.3	2.1	1.06**	0.3
Portugal	3.1	6.2	5.2	4.9	1.20**	0.7
Spain	0.8	3.7	2.6	2.4	1.07**	0.4
Sweden	1.2	4.8	2.8	2.2	1.07**	0.3
Mean	2.0	7.6	4.8	4.4	1.12	0.9

** = $p < 0.01$

Source: Calculated from EP Directorate-General for Information and Public Relations (1999).

Table 19.3 Ranking of disproportionality

	Rae	Loosemore – Hanby	Least-squares	Largest– deviation	Regression coefficient	Party repre-sentation	Overall rank
The Netherlands	3	3	2	2	5	=1	1
Italy	1	1	1	1	=2	12	2
Spain	2	2	3	4	=6	=3	3
Sweden	=4	5	4	3	=6	=1	4
Austria	6	4	5	5	8	=3	5
Finland	=8	9	9	9	1	=6	6
Greece	10	10	8	=6	4	=3	7
Germany	=8	6	7	10	9	8	8
France	=4	7	6	=6	13	14	9
Belgium	7	=11	10	8	10	10	10
Portugal	13	8	11	11	11	=6	11
Ireland	14	13	13	12	=2	13	12
Denmark	11	=11	12	14	15	15	=13
Britain	12	14	14	13	14	11	=13
Luxembourg	15	15	15	15	12	9	15

Source: Derived from Table 19.2.

This analysis solely focuses upon the outcome in Britain, thus excluding Northern Ireland which has operated under various forms of PR in European Parliamentary elections since 1979, together with possessing its own political party structure.

rankings indicates that the EU member states divide into several categories concerning, first, their overall ordering and second, their internal level of consistency across the indices of disproportionality.

Whilst the final column in Table 19.3 gives the appearance of a smooth ranking of countries, the mean positioning across the six measures of disproportionality reveals significant disparity. Hence, the member states disaggregate into four distinct groups under closer analysis. The first consists of the Netherlands, Italy, Spain and Sweden, where the mean ranking is between 2.7 and 3.8. This places these countries above the remainder who possess average rankings ranging from 5.2 to 13.5. The second group comprises Austria, Finland and Greece, with tightly clustered average rankings between 5.2 and 6.8, thus approximately occupying the middle zone relative to the number of EU member states, whilst the penultimate group possess mean rankings ranging from 8.0 for Germany to 10.0 for Portugal, indicating levels of disproportionality which place these countries beyond the sample midpoint. However, the final group of Ireland, Britain, Denmark and Luxembourg, with average rankings between 11.2 and 13.5, consistently demonstrate the highest relative levels of disproportionality amongst all fifteen member states. Indeed, Britain's ranking (even following its adoption of proportional representation) remains lowly, reaching eleventh on the party representation measure, twelfth on the Rae index, fourteenth according to the Loosemore–Hanby, least-squares and regression coefficient indicators, and fifteenth place with respect to largest-deviation. At an aggregate level this places Britain as displaying the equal second highest degree of disproportionality in the 1999 EP election. This analysis solely focuses upon the outcome in Britain, thus excluding Northern Ireland which has operated under various forms of proportional representation in EP elections since 1979, together with possessing its own political party structure. This analysis of countries' relative positioning shows that although the member states are theoretically harmonised in terms of the choice of voting system, significant variations continue to exist when the comparative level of disproportionality is considered. Thus whilst Table 19.1 indicates that the electoral systems of the Netherlands and Luxembourg nominally possess common characteristics regarding seat allocation method, thresholds and non-compulsory voting, the inherent proportionality of the outcome is far from uniform.

Table 19.3 also permits examination of the internal level of consistency across the measures of disproportionality for each country. This is similar to the identification of outliers, where the occurrence of a disproportionality ranking substantially different from the rest of the observations is seen to be generated by unique factors regarding the electoral system. The degree of divergence deemed noteworthy is arbitrarily determined to be deviations of more than four ranking positions relative to the member state's overall rank. This benchmark indicates that seven countries (Austria, Belgium, Britain, Denmark, Germany, Spain and Sweden) display lower levels of divergence than our selected value, suggesting that the comparative degree of disproportionality is stable. Furthermore, for seven member states only a single relative ranking significantly deviates, and for Finland, Greece, Luxembourg, the Netherlands and Portugal these are minor aberrations in comparison to their overall ranking. Differences are, however, more profound for Ireland and Italy, with one index suggesting relative disproportionality significantly at odds to their overall position. The voting system of Ireland on this evidence appears to result in a superior level of proportionality concerning the regression coefficient measurement relative to its overall performance. For Italy, the party representation measure is at least ten places above all other rankings, thereby ultimately losing it the opportunity to attain the top overall position. However, it is France which possesses the greatest degree of inconsistency based on the evidence presented with three of its relative rankings (Rae, regression coefficient and party representation) significantly deviating from its overall position. Moreover, further inspection reveals that none of its rankings bear any resemblance to its final position, which is thus derived from individual measurements of considerable variation. This second aspect of analysis, derived from Tables 19.2 and 19.3, suggests that substantial inconsistencies are evident from the 1999 EP election results concerning the relative ranking of disproportionality indices in relation to a country's overall position. In particular, for eight EU member states the level of discrepancy appears significant for at least one measurement.

An additional feature of the above discussion is the frequency with which inconsistency occurs in relation to specific indices of disproportionality. For instance, based on our chosen degree of variation of four ranking positions, one highlighted case occurred concerning the Rae index with four in terms of the regression coefficient and five regarding the party representation measurement. In contrast, the Loosemore–Hanby, least-squares and largest deviation recorded none. Thus although the six indices all purport to measure disproportionality, it appears that they are far from uniform in doing so.

Conclusion

This chapter has sought to examine the outcome of the 1999 EP elections with respect to the issue of disproportionality. However, such elections are generally perceived as atypical within the context of the usual domestic political process and as such are frequently described as second-order contests. The 1999 election displayed these customary features with the majority of governments losing support and the 49.9 per cent overall turnout being the lowest ever for direct elections to the European Parliament. Our analysis suggests, first, that in the majority of EU member states a systematic advantage for larger parties was prevalent. Second, in relation to our party representation index, the expected predominance of positive outcomes is evident for all fifteen countries, with at least one party being excluded in ten countries, with the problem appearing to be particularly acute in three member states where two parties have been excluded from the EP: in relation to the 111 parties gaining seats, a number equivalent to 12 per cent failed to get represented. Third, the relative positioning shows that significant variations continue to exist concerning the comparative level of disproportionality. In particular, the internal level of consistency across the measures of disproportionality suggest that for eight EU member states the level of discrepancy appears significant for at least one measurement. Although the outgoing EP adopted a series of common electoral principles to ensure that the elections possessed a sufficient 'European' character, the proportional representation voting systems utilised display significant variation when examined under the microscope of disproportionality. Thus the search for a series of common principles to be used for European Parliament elections which subsequently result in even quasi-equal levels of proportionality remains elusive.

Notes

1. For a review of the recent literature see, Blondel, J., Sinnott, R. and Svensson, P. (1998) *People and parliament in the European Union: participation, democracy and legitimacy*, Oxford: Clarendon Press; Pinder, J. (1999) *Foundations of democracy in the European Union from the genesis of parliamentary democracy to the European Parliament*, London: Macmillan; Lodge, J. (1996) *The 1994 election to the European Parliament*, London: Pinter.
2. Italy falls between the two positions in that its compulsory voting law was abolished in 1993. However, we follow Blondel et al. (1998) by incorporating Italy in the compulsory voting category on the basis that the obligation to vote remains part of the constitution and that there was little publicity surrounding

the change in electoral law. Hence, it is plausible to assume that voting is still regarded as at least quasi-compulsory in Italy.
3 Lijphart, A. (1985) The field of electoral systems research: a critical survey, *Electoral Studies*, 4 (1), 3–14; Lijphart, A. (1994) *Electoral systems and party systems*, Oxford: Oxford University Press; Rae, D. W. (1971) *The political consequences of electoral laws*, Yale University Press, New Haven; Taagepera, R. and Shugart, M.S. (1989) *Seats and votes: the effects and determinants of electoral systems*, Yale University Press, New Haven.
4 Loosemore, J. and Hanby, V. J. (1971) The theoretical limits of maximum distortion: some analytic expressions for electoral systems, *British Journal of Political Science*, 1, 467–477.
5 Gallagher, M. (1991) Proportionality, disproportionality and electoral systems, *Electoral Studies*, 10 (1), 33–31.
6 Lijphart (1994).
7 Cox, G. W. and Shugart, M. S. (1991) Comment on Gallagher's 'proportionality, disproportionality and electoral systems', *Electoral Studies*, 10 (4), 348–352. Laakso, M. and Taagepera, R. (1979) 'Effective' number of parties: a measure with application to West Europe, *Comparative Political Studies*, 12, 3–27.
8 Directorate-General for Information and Public Relations (1999) *EP elections – June 1999: results and elected members*, Brussels: Directorate-General for Information and Public Relations.

20
Making the EP Elections Distinctive: A Proposal for a Uniform E-election Procedure[1]

Juliet Lodge

Integration theorists and politicians for years argued that direct elections to the EP presaged a federal political union (preferably complete with a separation of powers, among the executive, legislature and judiciary). It was argued that direct democratic legitimacy would be conferred on the EP by the act of voting; that the EP would accordingly use its direct legitimacy to justify an accretion in its then very limited powers; and that it would eventually but inevitably usurp national parliaments for the allegiance of the voters and for the right to exercise effective political controls over EU level action by national governments. They had hitherto escaped effective parliamentary supervision at either the national level (where member parliaments' powers over the content and passage of EU legislation were negligible) or at the supranational level, where the EP had either no real authority or only nascent authority *vis-à-vis* the Council of Ministers. In short, direct elections to the EP were seen as a threat to national governments' supremacy; to national parliaments' position in a hierarchically conceived EU; and to national sovereignty. For some member states, by conferring direct legitimacy on the EP, voters were relocating their loyalty from the national to the supranational arena on a zero-sum basis, thereby eroding national sovereignty. The fact that this ignored the possibility of multiple loyalties and of people in effect having dual citizenship (both of the member state of which they were nationals, and of the EU) was conveniently forgotten in the flurry of political rhetoric that accompanied the arguments over the role and direct election of the EP over the years.

This chapter calls for changes in the electoral procedure, arguing that some flexibility in the implementation of a uniform electoral procedure (UEP) should be sacrificed in the name of distinctiveness. Advocating a system using personal direct electronic voting, aspects of a UEP designed to enhance the impact of the elections by underscoring their difference from national elections and their unique and distinctive features are outlined. It is suggested that *e*-politics be harnessed to the service of creating a participatory democracy close to the people.

Direct elections and mobilising the electorate

While the EP's powers have changed dramatically, notably since the time of the Single European Act, the matter of whether and by how much direct elections augment the democratic legitimacy of the EP (and indirectly of the European Union by demonstrating a willingness to exercise a political function in the polity of the EU) remains contested. Moreover, it has been argued that only if people are informed about the election of the EP (an event in which they may or may not be entitled to participate) and only if they understand the process and the point of European integration as developed by the EU institutions, and especially as shaped by the political forces in the EP, will they have an interest in voting. If they do not vote, they do not confer legitimacy on the EU. Therefore, member governments are better placed to insist on restricting the scope, pace and depth of European integration. Accordingly, low turnout can be interpreted as a signal to slow the pace of integration generally, and specifically to deny the EP greater legislative power.

Three things are inextricably linked: the 'event' of Euro-elections; the process by which MEPs are elected; and the impact on the EU's democratic credentials. For many reasons, it followed that efforts would have to be made to mobilise the electorate. The notion of electing representatives to a parliament lacking any legislative power and the power to turf out those purporting to carry out tasks normally associated with governments meant that at the first Euro-elections, a general information campaign had to be run (by the Commission and EP) to inform the voters about the event, the process, the purpose and the objectives of the Euro-elections. Though sensitive and provocative to Europhobes, anodyne and often rather dull information lacking party political bite had to be presented in a non-partisan manner (Herman and Lodge, 1978). Even so, turnout at the 1979 Euro-elections was higher on average than twenty years later.

In 1999, the precise purpose and nature of the EP and Euro-elections remained elusive. The use of the EU as a scapegoat for domestic politics obscured the reason for electing a supranational parliament. While the EU's purpose was often misportrayed by the media as being to constrain national governments from protecting, insisting on and acting in the best interest of their people, it proved particularly difficult, on the occasion of the elections, to convince them that the EP served a useful purpose, or had done a good job since the last elections, or was likely to do something genuinely useful in the foreseeable future. The stakes were unclear. Few knew why precisely they were voting, or whether they were voting for or aginst anything in particular. It is hardly surprising, then, that voters used to voting in domestic elections against those in office should think that voting was of any consequence. The spirit underlying Euro-election voting derives from the principle of democratic parliamentary government being created for and by the people who have the power to usurp it if it fails to uphold democratic practice. This is embedded in the psyche of contemporary democratic politics in Western Europe and can be readily traced back to the Dutch experience in the seventeenth century (Kossman, 1999).

Democracy is largely taken for granted and rarely questioned in the member states. Only at EU level is there continuing scrutiny of what it means in practice. Democracy was believed to be satisfied, at least in reasonable measure, by the common political act of universal suffrage to the EP. While the democratic deficit is far more complex and apparent throughout the EU member states' polities, the idea that EP elections would democratise the EU persisted until well after the first Euroelections. A more nuanced understanding is required but does not erase the idea that universal suffrage is a key democratic right, and that a basic democratic requirement is satisfied by *voluntary* voting in EP elections.

Turnout and legitimacy

Turnout was seen as important to counter the argument that the EP (and by inference the EU) lacked democratic legitimacy and that therefore the decisions and policies made by and in their name need not be upheld. The idea that an EP elected by a minority of the people lacks legitimacy and credibility is potentially damaging to the nascent democratic (sometimes termed pre-democratic) EU (Pinder, 1999). The issue of getting voters to participate in the election of the EP is conflated, however, with providing and ensuring that EU citizens have access to (and are informed about) the decisions taken in their name and on their behalf by the EU. This is not the place to investigate the process of augmenting transparency in the EU, vital though transparency is to democratic practice. The right to vote in EP elections is a political right universally bestowed on member states' nationals, subject to certain qualification. It complements the economic and social rights entrenched in the founding treaties. Its exercise confirmed the EC's transition from a largely elite-led, neo-functionalist organisation geared towards maximising economic benefits for selected economic actors to a neo-federal and political community. The confluence of changes in the scope of the EC's competence is quite marked. The attendant equation of the exercise of universal suffrage with the transfer of popular sovereignty from the national to supranational domain was also apparent but hotly contested and refuted by member governments. What survives is the idea that direct participation in supranational political life via direct elections is an element of EU citizenship which confirms the direct link between the EU citizen and the EU 'sovereign' without an intermediary. EP elections are a potent source of direct legitimacy and turnout is interpreted for a variety of purposes, but above all it is seen as symbolising public participation and acceptance of the EU more generally.

It is this that gives concern over turnout at EP elections particular significance. Turnout above around 45 per cent is seen as desirable and necessary for the above reasons. Getting a relatively high *voluntary* turnout is important because it is seen to confer legitimacy on them. If it was simply a matter of boosting turnout at EU elections, then – arguably – that goal could be served by making voting compulsory. However, two distinct processes have been conflated: (i) the idea of boosting participation in the EP elections by ensuring that citizens are informed about them, and (ii) enhancing ongoing citizen awareness of EU affairs and how they are affected by them in order that they can judge performance and vote on it (somewhat imperfectly) at the next Euro-election. It is relatively easy to see why people might need basic information and education as to the role and purpose of the EP. It is less easy to persuade them to take an ongoing interest in EU affairs and to ensure that they inform themselves regularly about those affairs: there is no obvious connection with how the EU performs over a given period, or whatever dimension – socio-economic, political, etc. – is measured, and the outcome of the EP election.

Euro-elections: a failure of participation and a source of de-legitimisation?

Elections to the EP do not serve the same purpose as elections to national parliaments; they do not result in the selection of a 'government' of one particular persuasion over that of another. The EP is not the arena in which an EU 'government' has to show that it has acquired and retains the support of the people's representatives. The fact that the Commission is not the EU's pre-government, though it has to retain EP confidence, and the fact that member governments in effect question its legitimacy when they use it as a scapegoat, does not help. This is exacerbated by the dissonance between the EP's growing influence and relative public invisibility as the seat of democratic accountability. MEPs therefore have sought to mobilise voters to support the EP's quest for more power and to endorse its efforts to exercise it in their interests. Public awareness and knowledge of national parliaments (as opposed to national governments' activities) is generally imperfect and low, but media attention is constant. By contrast, media attention to the EU and EP fluctuates. Some correlation has been presumed to exist between public awareness of the EP and intention to vote (see Eurobarometers). While additional variables have refined this and favourability towards the EU is also believed to correlate with propensity to vote, the basic premise of supposing that if armed with more information, voters would turn out, remains intact. Euro-phobes continue to use low turnout to justify opposition to increasing EU and EP competence.[2] This position is paradoxically more difficult to sustain after the all-time low turnout in 1999. This is partly because low turnout is no longer seen as the only indicator of democratic deficiency in the EU, but this should not excuse complacency. The democratic deficit in the EU, and in its member states, is real (Andersen and Eliassen, 1996).

Disillusionment with the act of voting cannot be dismissed as of little consequence for the maintenance and practice of democratic government. Disenchantment is not uniform across the EU; neither are those disinclined to vote in EP elections necessarily the same people who abstain from political affairs at national and subnational levels. Social movements' activities, tactics and leverage differ within and across member states (Marks and McAdam, 1996, 119) and the desirability and purpose of elections in upholding 'good government' and democracy has become obscure. The sense that by voting one could genuinely influence legislative as opposed to electoral outcomes has diminished at the same time as governments' ability to influence those factors which shape and condition real financial and policy choices too have been eroded. The impact of globalisation has not, however, translated itself into realistic electoral manifestos: politicians do not usually suggest that big choices are made outside their territories and escape their control.

This leaves the electorate in an invidious position, and especially so when called upon to elect MEPs. Euro-manifestos and campaigns have always been conducted with reference primarily to domestic political preoccupations. Making Euro-elections as much like nationally-based municipal, regional or general elections was therefore the short-cut to getting across the idea that Euro-elections are about voting people into the EP. These elections were based around a series of (often dull, poorly differentiated, weakly communicated or incomprehensible) political choices. Member governments failed to respond to the joint call of the national parliaments and EP to contribute to reducing the democratic deficit either by systematically publicising EU legislative proposals, or by making themselves and national parliaments fully accountable for their actions within the EC, later EU.

Not surprisingly, national parties simply used Euro-elections as yet another occasion on which to lambast the opposition, using topical domestic preoccupations and personalities regardless of real European issues. They were arguably encouraged to interpret repeated relatively low turnout as merely a reflection of electoral fatigue, the confluence of several national and or local or regional elections having dissipated energy, interest and depleted party funds. All this lulled national parties into the false belief that they need not significantly amend their Euro-election efforts to mobilise the electorate. In some states, too, this helped the Europhobes' cause and reassured national MPs that they were not (yet) redundant and remained superior to the upstart 'parliament' (the EP). Not only is this attitude destructive in terms of ensuring an improved quality in legislative outputs, since the quality of EU and national level legislation is, to a degree, interdependent [Kellerman, 1991, 30], but it inhibits the promotion of understanding of the nature and role of parliaments in the daily conduct of political life. If pseudo-Euro-elections are merely seen as yet one more occasion on which to vote on tedious and repetitive national issues and personalities, why bother to turn out? After all, another election is likely relatively soon, so missing one may not seem important or of much personal consequence.

As yet a political class that transcends the national psyche and national boundaries, and is able to conceive of itself and work as such, is in its infancy. Here lie the clues to electoral mobilisation in the future. National, regional and local politicians work together on the occasion of elections within their national boundaries to mobilise the vote and draw knowledge and ideas on how to mobilise the electorate at large from their own party followers. Unlike Euro-elections, these have understood ground rules, star performers, multi-media campaigns and often clear strategic, political campaign management with parties acting, to some extent still – while harnessing the media – as interlocutor between citizen and state.

The parties are less successful in Euro-election campaigns partly owing to continuing ignorance among their own activists as well as the public about the role, purpose and consequences for EU decision-making of the Euro-elections themselves; and partly owing to ignorance about, and even lack of interest in, what the EU does. However, the absence of many national political leaders in Euro-election campaigns reduces their impact on the electorate, downgrades their importance and fails to enable the public to participate in a nascent European *demos*, shaping and taking responsibility for decisions in civic society. It inhibits the development of an EU closer to its citizens.

National governments need to harness political parties to involve civil society in the process of European integration, in a process that is undertaken freely, democratically and consensually (Toulemon, 1998, 117). At the EU level, it means facilitating communication and mobilisation. Society's rules and values need to be transmitted and reinforced. At EU level, this is missing. Voters are neither shown nor understand the link between solidarity, the common good and liberty and giving effect to sustaining collective goals and collective responsibility for their attainment. Just as democracy as a way of life needs to be learnt at the national level, so it needs to be experienced at EU level. That is why in the EU, participating in one aspect of democratic practice – the Euro-elections – is so important (Lodge, 1978). A common European identity cannot be resolved simply by affording EU citizens common political rights, obligations or symbols of identity, and cannot therefore be overcome simply by holding Euro-elections. However, the Euro-elections are an occasion for maximising awareness of participating in a common venture. If a sizeable proportion of the electorate voluntarily votes, it demonstrates a capacity for EU participation and acceptance of its underlying democratic principles. Any system losing such support compromises its ultimate source of legitimacy and power.

Low turnout as a symptom of the lack of Euro-election distinctiveness

The first *five* sets of direct elections to the EP have been largely contested on national issues. There are many explanations for this. While voters' EU knowledge remains low, there might be a greater tendency to adhere to pre-conceptions, to screen out contrary views and (depending perhaps on what importance individuals attach to the impact their vote will have *nationally* on the outcome) to vote either as they might in a national election, or to experiment in voting for a minority, unconventional or high-salience single policy issue party. At the same time, it is illogical to expect citizens to bother to vote in EP elections on national issues when they have many opportunities to vote on national issues in national elections (including local, municipal and regional elections within member states). In such elections, moreover, they can have far greater impact and one that appears meaningful to them because such elections result in the confirmation or change in the governing party/parties and normally in some directly relevant and perceptible change in political priorities informed by a more or less meaningful ideology. This suggests that the EP political parties need to reappraise the way in which they communicate with national politicians as well as with voters. There is a chicken and egg dilemma here: attracting sustained media attention is difficult partly because of the way the EP organises its business and partly because of the nature of much EU business. Making the EP visible and *relevant* to the public is difficult, moreover, for MEPs who, if they are not well-known domestic political figures, have to earn their status in the public eye.

Citizens need to have a political voice and European parties need to ask and get answers to questions concerning the agenda, priorities and expenditure on a range of public issues; to make political choices visible and understandable; and to check that the decision-makers are accountable, are effectively scrutinised and held in check, and are open and answerable for decisions they take in the name of the public. Citizens need to be able to see that the people (MEPs) they elect do these tasks effectively in their own right on their behalf. They need to be able to vote them out of office if necessary and on the occasion of meaningful and genuine European elections.

It has been repeatedly claimed that because the Euro-elections are not about electing a European government, the electorate finds it difficult to discern the stake in the outcome of, or reason for, voting. The politicisation of the Commission has been seen as a step towards making clear their ideological affinities, proclivities and likely political choices. Closer links with the EP and its party groups have been seen as instrumental to achieving this. Even so, it has only been possible gradually for the EP, largely through the medium of cross-examining nominated Commissioners prior to the EP's endorsement of their formal appointment, to make an institutional interdependence between them apparent to the public at large. MEPs have also had to assert that this extends beyond the one-off appearance to generally retaining their trust and confidence. It is a moot point whether MEPs are sufficiently well-organised and disciplined to achieve this. Their own political party groups themselves need to reform how they manage their business and present their activities to the public (Pridham and Pridham, 1981; Bardi, 1994; Bowler and Farrell, 1993). Much remains to be done.

Euro-elections will remain for the foreseeable future the only transnational occasion on which EU citizens participate in a common political event. How might this event be made both more visible and relevant as a uniform event geared towards a common, shared purpose? This is not the place to rehearse either the deficiencies inherent in the existing national Euro-election procedures which distort the representation of voters' choices in the EP's final composition, or to suggest how politicians may make progress before the EU enlarges, vital to the success of enlargement as such progress is. Instead, here are a few ideas on how to make EP elections distinctive, more user-friendly, Euro-focused and geared towards underpinning a European *demos*.

Euro-elections: a process of ongoing reform

The purpose of Euro-elections is to determine the composition of the EP. Article 21 of the Treaty of Paris and Article 138 of the Rome Treaty prescribed the EP's election by a uniform procedure. Article 137 EEC states: 'The Assembly, which shall consist of representatives of the peoples of the States brought together in the Community, shall exercise the advisory and supervisory powers which are conferred upon it by this Treaty.' Article 138 (i) EEC goes on: 'The Assembly shall consist of delegates who shall be designated by the respective Parliaments from among their members in accordance with the procedure laid down by each Member State.' Although the Rome Treaty prescribed the election of MEPs by direct universal suffrage (Article 138, 1: new articles 190–5) in accordance with a uniform procedure (Article 138(3)), governments' fears that this heralded a federal Europe, would deprive them of power, and ultimately challenge their supremacy as policy-makers led them to delay holding direct elections for 20 years. In line with treaty provisions pending approval and implementation of a UEP drafted by the EP itself (and unanimously approved by the Council: see Article 7 of the 1976 Act, *Official Journal* L278, 8 October 1976, p. 1) MEPs were appointed from among the membership of national parliaments. As such, they held a dual mandate in two parliamentary bodies. The Council failed repeatedly (and still fails) to agree on MEPs' proposals for a UEP. The first such proposals date from the Dehousse Report of 17 May 1961. Others followed in 1963 and 1969. More adventurous states stepped back from holding their own national direct EP-election as this would have contravened the treaty. In 1973, new proposals were prepared under the rapporteurship of the Dutch Socialist, Schelto Patijn. The institutionalisation of EC summits as European Councils in 1974 led to the argument that this intergovernmental element be balanced, following the EP's widened budgetary competences in 1973, by direct elections. In 1975, the EP presented a Draft Convention which broke the requirement on uniformity by permitting each member state to use its own procedures for the first Euro-elections, and awaiting EP proposals thereafter. The EP adopted a further four resolutions on this in 1976. In the summer of 1976, the Council decided to advance that goal by the Act of 20 September, subsequently incorporated into the EC Treaty (Article 190). The Act on MEPs' election entered into force on 1 July 1978, several weeks after the initial election date of June 1978 (Council Decision 76/787/ECSC, EEC, Euratom in *Official Journal* L278, 8 October 1978). While prescribing the first such elections to be held in 1978, delay in implementing the necessary enabling legislation in the member states resulted in the first Euro-elections' postponement until 1979, since when they have occurred every five years. The task of drafting a UEP falls to the EP. After 1979, it accordingly adopted on 10 March 1982 the Seitlinger report, rejected by the Council of Ministers on 24 May 1983. Subsequent attempts to get a UEP accepted have failed and the EP's coupling of minimalist and maximalist reform strategies in the 1980s and 1990s led it to focus on the acquisition of real power in the meantime.[3]

Continuing failure, primarily by the member governments, to agree on a UEP (drafted on several occasions by the EP) resulted in major and minor distortions in the representative character of the EP. (We ignore here the distortions produced by the ratio of voters to seats which discriminates against and among the big states). The UK's insistence on adhering to first-past-the-post simple majority voting (in Great Britain but not Northern Ireland which had proportional representation *ab initio*) until the 1999 Euro-election in particular seriously distorted the relative power of the social democratic and conservative/Christian democratic party groups. Extreme proportional representation in Italy, coupled with the lack of a minimum electoral threshold of around 5 per cent (common in many states), by contrast, permitted the representation of tiny – and sometimes extremist – minorities. While the EP has systematically reformed its Rules of Procedure and those relating to the number and composition of groups of MEPs seeking recognition, the status and attendant privileges of forming a party group, member governments have eschewed the adoption of a UEP. They have maximised their right to elect MEPs (and to fill any vacancies occurring within the legislative period) in line with their own preferred electoral methods. This, coupled with differences over voter and candidate eligibility, party funding, advertising, polling days and electoral procedures, detracted from the notion of a common, single European election for a common, single purpose of electing representatives to serve in a common, single (ill-understood and misunderstood) institution, the EP.

The Amsterdam Treaty represents a modest attempt to cut through the interminable arguing over the precise details of a UEP by advocating agreement on common principles (Articles 190–5, formerly Article 138 EEC). This allows member states to combine maximum flexibility over the way in which they wish to interpret and implement those common principles with minimum concessions to the principle of uniformity. Agreement on a *single* procedure implemented in a uniform and identical way in each member state is not crucial, but acceptance of a common procedure acknowledging EU states' political cultures and traditions is. Flexibility must be bounded and not allow EU states to evade their responsibilities. It is a first step towards a system that accommodates, adapts to and draws strength from diversity without abandoning adherence to a common set of rules and principles. These would enable the EU to enlarge without additional obstacles being inferred from a requirement to adhere to a uniform system: if its construction proved tortuous in the past, even among states whose recent political past conformed more or less to liberal, democratic ideals, how much more difficult could it prove with new applicants?

Reforming the election of the EP: towards a uniform electoral procedure

Subscribing to a set of principles enables members, including nascent democratic regimes, to re-examine questions of accountability and democracy within their national settings. This is more important than might be thought at first sight. Democratic legitimacy implies that government is seen by the people as lawful: if it acts in a just and lawful way, its authority is seen to be acceptable and accepted. This is neither something that can be taken for granted in some applicant states, and nor is it something that can be deduced merely from the convening of Euro-elections. It may be that broader based political disenchantment, scepticism and ignorance have had an adverse impact on levels of participation in Euro-elections. Why is this relevant to the Euro-election reform proposal suggested below? If people do not understand in broad terms how their own government is elected, why and on whose behalf it takes decisions; why, how, when and to what end the Parliament is elected; and what the role of MPs is, then it is very difficult to explain to voters what the purpose is of holding elections to the less visible, and less comprehensible and seemingly irrelevant EP. This proposal does not make recommendations as to the need for wider civic education and the roles and opportunities for the expression of a public voice in the emerging European 'public space', but it does attempt to query past assumptions and advocates changes capitalising on the rapid development of and access to the digital technology.

Little is gained from suggesting that voting for MEPs is a process which is either familiar to voters, or akin to voting in local, regional or national elections. It is not (or not yet) about electing a government. However, it starts from the premise that in the EU democratic politics rests on basic principles of party governance [Lane and Ersson, 1996, 119]. While the stakes need to be explained to voters by genuine European parties, and by politicians able to communicate with, inform and educate voters in a relevant and intelligible way, the distinctiveness of the Euro-election could be celebrated and exploited. Accordingly, this proposal seeks to marry distinctiveness to the flexibility advocated by the Amsterdam Treaty.

Towards common principles

Any reform must address confusion over: (i) the Euro-elections' purpose; (ii) their lack of distinctiveness; and (iii) the nature and role of the EP in EU decision-making. While the following suggestion is by no means comprehensive, it attempts to address some of the issues. Its recommendations aim:

(a) to make the Euro-elections distinctive;
(b) to facilitate maximum, *voluntary* participation by eligible voters;
(c) to enhance democratic legitimacy by giving voters the sense of their vote being valued and a sense of ownership of distinctive elections in a distinctive organisation of which they are a vital part;
(d) to enhance and awaken public interest in shaping the EU polity and political process on a continuing basis;
(e) to enable citizens to see that they have a genuine role in determining outcomes;

(f) to encourage European political parties and movements to act as genuinely transnational and supranational political forces that aggregate, express, represent and follow through common interests via coherent strategies and mature political processes that facilitate democratic follow-up action.

It is designed to help combat voter ennui and contribute to the debate about how the EP electoral system might be reformed to encourage participation and make the EP elections genuinely European.

Distinctiveness

The process should be distinctive: polling day, polling outlets and voter eligibility should be changed. The reforms suggested require all member states to change their procedures but allow them to conduct the election campaigns in line with their respective political traditions and cultures. By exercising flexibility over the precise day on which the vote is cast, how and when it is cast, the UEP would have unique and uniform elements which would be instantly recognisable by voters regardless of nationality.

However, as regards common principles, so far member states have held EP elections on a traditional voting day, within a four-day period in the second week of June. The various states vote on different days: for example, Denmark, Ireland, the Netherlands and the UK traditionally vote on Thursday; Germany, Belgium, France, Italy, Luxembourg, Greece, Spain and Portugal on Sunday. Results are announced after the polls close in the state voting last. Time zone differences produce further delay and disjunction between the act of casting a vote and hearing the outcome. The returning officer for each constituency is responsible for the count and announcing the result, but the results are verified in line with different member states' practices. By abolishing this flexibility, all the actors and voters would participate in the major political democratic act of electing the EP on the same day. To underline the distinctiveness of this election, it is proposed that voting take place on a single, non-traditional voting day throughout the EU – Tuesday – once every five years.

Casting the vote

Uniformity would enhance Euro-election distinctiveness if inflexibility applied to voter eligibility and the minimum voting age was set at 17 years old in all EU states for Euro-elections only. All other aspects of eligibility may be flexibly interpreted and fixed in accordance with local practice, providing that the principle of permitting EU citizens resident in a member state of which they are not nationals to vote and to stand as a candidate in EP elections is honoured [Directive 93/0109/EC]. Voting should be voluntary but only for Euro-elections. This may pose particular problems in states where voting is compulsory (such as Belgium and Greece). The political sensitivity of this, however, may mean that it is not feasible and failure to adopt voluntary voting should not be allowed to postpone the introduction of the UEP.

Distinctiveness could be enhanced by making voting easy, for example, by facilitating: (i) voting by electronic means in non-traditional outlets (e.g., supermarkets), and also via net and digital interfaces in addition to traditional ballot stations; (ii) voting during normal working hours, for example, 07.00 to 20.00; and (iii) speedy announcement of the results (electronic voting facilitates this) on the same day. Voting would not only be relatively easy and easily accomplished while undertaking a necessary task (shopping for food), but it should help boost voting among those groups disinclined or too busy to make a special trip to a polling station. It might boost voting by women, pensioners, and the unemployed. Voting while shopping need not trivialise the elections. (Voting in church or school halls, for example, does not confer gravity on the process.) The EP election would become more newsworthy by dint of the distinctive, novel procedure. Other appropriate electronic outlets could be licensed to record votes, such as banks, post offices and public libraries. Electronic voting might boost social inclusion by encouraging participation by certain disadvantaged social groups if technology and security requirements were sufficiently advanced to permit participation via television and the Internet. In some countries, such as the UK, this would complement the nascent development of 'electronic governance' whereby citizens are already encouraged to use the net and e-mail to contribute to the discussion of policy options. However, it must be recognised that there are disadvantages to electronic voting directly from the home since secrecy of the ballot could not be as readily protected as it is in ballot stations.

Ideally, all votes should be cast electronically for EP elections. In practice, the availability of the requisite technology might be patchy at first. Consequently, while electronic voting would be a distinctive feature, voting by traditional means might have to complement it. Moreover, in order to reduce the possibility of electoral fraud, member states would have to have the flexibility to determine how they validate each voter's identity. Depending on the electronic methods used, verification could proceed automatically through the correct double entry of the a personal identification number, PIN (which might, for example, be the existing PIN number used by those with bank cards), or iris recognition, etc. Alternatively, if votes were cast at supermarkets, tellers could verify details of individual voters in the same way as they do when checking out identity in ballot stations, or when checking identities before permitting telephone financial transactions. If direct voting from home were permitted, the physically disabled and/or elderly voters would be able to use touchscreen televisions. It might also ultimately allow all voters perhaps to phone in their vote, entering their PIN as required at the start and end of the process. Mechanisms need to be in place that make the casting of the vote simple but unique to each voter and which guard against fraud [Commission of the European Communities, 1997, 503]. While the Commission has examined fraud and the impact of electronic commerce, it has yet to extend this survey to electronic political processes. Making voting simple and easy would help to convey to voters the idea that their vote mattered to those in power in that the latter had made it simple and easy to vote in terms of both access to and the location of voting outlets (polling opportunities/voting booths/ screens, etc.). By 2004, the next generation of new voters (many of whom those are seen to be among the least politically interested and motivated) is likely to be computer literate, and is likely to find on-line/ screen and virtual voting easy. Therefore new voters should readily accept and see it as an extension of other activities carried on by electronic and digitised means. This also applies increasingly to older cohorts of voters, including informed pensioners.

Impact

The impact of having speedy results should not be over-estimated but that would certainly be a bonus to candidates and politicians. Greater visibility for the elections would result directly from the employment of digital and electronic means of casting a vote. Much public debate and political discussion, not least among the candidates and parties contesting the Euro-elections, is bound to result from the suggestion that *all* member states lower the voting age to 17 for this election. Political leaders would, moreover, be likely to want to be seen as favouring participation by young people entitled to vote. Often absent from the Euro-campaigns in the past, it would be inconceivable that they would absent themselves from the public debate about Euro-elections conducted by electronic means. The digital age's ramifications are so wide, electronic commerce is growing apace, and the public is increasingly aware on a daily basis of the benefits as well as the pitfalls of new technology.

There are obvious disadvantages to the system. Setting aside electricity strikes and the usual valid concern over hackers and computer fraud, and focusing on the issue of mobilising participation, it is possible that information technology phobics would need help in voting. Member governments would have to judge whether to make electronic voting the single and only means of voting in the Euro-elections. An information campaign and set of instructions would be needed to guide voters on the mechanics of voting. This need not be more difficult than the touch-screen operations of automatic banking. It would be feasible, in theory, simply to change the mechanical aspects of the voting procedure without changing the electoral procedure for the next Euro-elections. Moreover, member states who felt unable to countenance personal direct electronic voting (PDEV), for example, because they had yet to introduce machine readable identity cards for all residents (a political problem in the UK) might either opt for a system that allowed individuals to choose whether to vote in the traditional manner or to use PDEV. This is a feasible alternative since individuals could make such a choice when they join the electoral register.

Satisfying the requirement of producing a UEP that has as many uniform features in it as possible is arguably a separate issue from that of making the Euro-elections distinctive. It would be a missed opportunity, however, to focus purely on the traditional elements of the UEP. While flexibility has been essential in the past in overcoming domestic political opposition to the principle of electing the EP directly, care now needs to be exercised to inhibit the proliferation and maintenance of avoidable distortions resulting from certain national practices. In the past any variant of PR was acceptable. Reconciling the ideal of genuine European elections in the long term with the short to medium-term requirements of introducing practicable but flexible provisions that could be revised later could be facilitated by adopting a system of proportional representation in constituencies (whether national constituencies or, where a member state so wishes, subdivided into smaller regional constituencies)[4] where voters have two votes. One vote could be cast explicitly for an official EP party group already represented in the EP and headed in each state by (i) the leader-designate of the group and backed or seconded by (ii) the leader-designate of that state's delegation to that group. The second vote could be cast within the state for a national (or where appropriate regional) party list of candidates. Flexibility would be served by states choosing uni-national party lists or party lists subdivided according to regional constituencies. Uniformity would be served by having a 5 per cent minimum electoral threshold in all member states. This electoral system would cut distortions arising out of 'extreme PR' and reduce party splintering and excessive factionalisation in the EP. The onus would be on the national parties to get out the vote; to present themselves effectively within EU states; and to co-operate with MEPs. The EP party groups would have a direct stake in the campaign in mobilising the vote, and in the outcome by virtue of having to present themselves for election in each EU state. Vote splitting between the EP group and a national list could be possible[5] even if unusual or tactically used as in national elections. At subsequent elections, the EP party groups might try to present their record in a perhaps more convincing, credible and – to the voter – relevant way than in the past. This might encourage the EP, its party groups, candidates and national parties to become closer to the electorate. Non-engagement would mean invisibility at the EU level. Arguably, the system would discriminate against entry into the EP by newer parties and minorities. This might encourage smaller groups to pool their endeavours and inhibit independent candidates. They would not be entitled to any special treatment or prerogatives in the EP. The current EP system favouring official party groups should be retained. Accordingly, uni-national parties may not qualify for recognition as an official EP party group. Over the longer term, the electoral procedure could be modified to permit the development of transnational regional constituencies and genuine transnational party lists.

It would be desirable but not necessarily essential to eliminate flexibility over candidate eligibility criteria. It is proposed that EU *citizens* over the age of 18 may contest the elections in any member state, though several EU states stipulate 21 as the minimum age. Multinational party lists might be facilitated[6] if local parties sought to Europeanise their campaigns by allowing young eligible adults to stand in any EU state. Practice over the nomination of candidates diverges. A common principle of not disadvantaging candidates by having potentially punitive 'candidate deposits' might be explored. Being an MEP would be incompatible with holding political office in any other EU institution or holding major political posts in the member states. Existing non-eligibility and incompatibility criteria and unacceptability of dual mandates would remain and should be harmonised and uniformly applied. Candidates elected to the EP would be required to declare any vested interests (financial or otherwise) in socio-economic ventures and organisations to their EP party group. The EP group would be responsible for logging these interests and, in line with agreed rules, making the register open to public scrutiny.

Discrepancies exist over public funding of elections. Ideally, maximum expenditure should be specified by a joint EP and Council decision. EU funds could be provided for (i) an information campaign on voting and (ii) EP party group manifestos and one short (i.e., 2-minute) television slot relayed throughout the EU. All campaign finance – and its sources – should be publicly declared eight weeks before polling day. Similarly, the official campaign period would have to be officially determined. An official campaign opening one month before polling day seems reasonable. Media coverage remains increasingly difficult to contain territorially. Many states apportion time according to parties' strength. Coverage is complicated by whether or not political advertising is permitted. Subsidiarity may need to be applied here.

It has been argued traditionally that publishing opinion polls, and more recently exit polls, influences turnout and voting behaviour. Accordingly, states have differing restrictions. Some prohibit polls and forecasts by law, some by custom, for given periods before the ballot: these range from less than a week (Spain and Germany) to a total ban for the campaign's duration (Portugal). Some (e.g., Denmark, Ireland and the Netherlands) have no restrictions. Uniformity might be necessary, for example, to allow publishing opinion poll data up to 48 hours before the polls open and prohibiting the publication of interim or final exit polls until after 18.00 on polling day to minimise their effect on the results. Easy access to websites, satellite and cable television compromises this already and, if anything, makes uniformity even more desirable.

Conclusion

A UEP would enhance political engagement and communication. This, in turn, should lead to a more wide-ranging debate about the EU and complement, if not eclipse, parochial preoccupations chosen in the past by national politicians pretending that voting – at any level in any election – is simply an expression of economic self-interest. The proposal places a premium on parties' ability to communicate with, educate and discuss with voters the election's purpose and what roles they might play in shaping the nature and sustainability of good democratic representative government in the enlarging EU. The right to participate in Euro-elections is, after all, one of the first and arguably intrinsically most important political rights linked to the status of EU citizen. It must not be allowed to atrophy through neglect. It is time to harness the potential of *e*-politics to its service and to the common good of an EU close to the people.

Notes

1 This draws heavily on my paper for the European Journal of Law Reform (April 2000). I thank Frank Emmert for allowing me to do so.
2 See the 1990 Colombo Report on the Constitutional Basis of European Union, where para.21 states: 'The citizens of the Union shall participate in the political life of the Union... in particular in local and European Parliament elections'; *EP Resolution* A3-0301/90. The Amsterdam Treaty, Article 8d, states: 'Every citizen of the Union may write to any of the institutions or bodies referred to in this Article or in Article 4 in one of the languages mentioned in Article 248 and have an answer in the same language.'
3 See reports from the Seitlinger Report of 1982 and the subsequent Bocklet Report onwards, in particular EP Working documents Resolution A3-0152/91, Resolution A3-0186/92, and A4- 0212/98.
4 Eleven (Austria, Denmark, Germany, Finland, France, Greece, Luxembourg, the Netherlands, Portugal, Spain and Sweden) define their constituency by the territory of the state: each state has one constituency electing a variable number of MEPs according to a ratio of population to seats formula which gives small states with small populations advantages over large ones. Belgium, Ireland, Italy and the UK have several constituencies. German parties may present lists at either *Land* or national level, and Finns may do so at electoral district or national level.
5 Electronic voting would facilitate the participation of eligible voters temporarily away from their place of work or residence. Eligible voters voting from outside the EU's territory could have two votes, one to be cast for an official EP party group and one for a party list (or independent candidate) standing in their normal state of residence. Eligible voters resident in another EU member state for more than one year prior to the Euro-election could opt to vote in either their normal state of residence *or* in their host state of actual residence, so allowing maximum inclusion and participation.
6 Italy alone let other EC nationals contest EP elections on the same basis as its own in the 1989 EP elections.

References

Andersen, S. and Eliassen, K. (eds) (1996) *The European Union: How Democratic is it?* (London: Sage).
Attinà, F. (1990) 'The Voting Behaviour of European Parliament Members and the Problem of Europarties', *European Journal of Political Research*, 18(4), 557–79.
Bainbridge, D. and G. Pearce (1996) 'EC Data Protection Law', *Computer Law and Security Report*, 12, 160–8.
Bardi, L. (1994) 'Transnational party Federations, European Parliamentary Party Groups, and the Building of Europarties', in R.S. Katz and P. Mair (eds), *How*

Parties Organise: Adaptation and Change in Party Organisations in Western Democracies (London: Sage).

Bowler, S. and D. Farrell (1993) 'Legislator Shirking and Voter Monitoring: Impacts of European Parliament Electoral Systems upon Legislator-Voter Relationships', *Journal of Common Market Studies*, 31, 45–61.

Centre for Finnish Business and Policy Studies (EVA) *Finnish EU Opinion*, Autumn 1998.

Commission of the European Communities (1997) *Ensuring Security and Trust in Electronic Commerce*, COM (97).

European Parliament (1992) *The European Community in the Historical Context of its Parliament*, 40th Anniversary Proceedings of the Symposium, Strasbourg.

European Parliament (1993) *The Powers of the European Parliament in the European Union*, Working Papers, Political Series E-1.

Final Declaration of the Conference of European Parliament and National Parliaments, adopted 30 November 1990 [known as the Rome Assizes]. Text reprinted in European Parliament (1992) *Documents on Political Union* (Dublin).

Herman, V. and J. Lodge (1978) *The European Parliament and the European Community* (London and New York: Macmillan).

Hix, S. and C. Lord, (1997) *Political Parties in the European Union* (London: Longman).

Jacobs, F., R. Corbett and M Shackleton (1995) *The European Parliament* (London: Longman).

Kellermann, P.A. (1991) 'Proposals for Improving the Quality of European and National Legislation', *The European Journal of Law Reform*, 1, 7–30.

Kossmann, E.H. (1999) 'Republican Freedom against Monarchical Absolutism: The Dutch Experience in the Seventeenth Century', in J. Pinder (ed.), *Foundations of Democracy in the European Union* (London: Macmillan).

Lane J.-E. and S. O. Ersson (1996) *European Politics: An Introduction* (London: Sage).

Loader, B. (1997) *The Governance of Cyberspace* (London: Routledge).

Lodge, J. (1978) 'Loyalty and the EEC: The Limitations of the Functionalist Approach', *Political Studies*, 232–48.

Lodge, J. (ed.) (1985) *Direct Elections to the European Parliament* (London: Macmillan).

Lodge, J. (ed.) (1990) *The 1989 Election of the European Parliament* (London: Macmillan).

Lodge, J. (ed.) (1996) *The 1994 Elections to the European Parliament* (London: Pinter).

Lodge, J. and V. Herman (1982) *Direct Elections to the European Parliament: A Community Perspective* (London: Macmillan).

Marks, G. and D. McAdam (1996), 'Social Movements and the Changing Structure of Political Opportunity in the European Union', in G. Marks *et al.*, *Governance in the European Union* (London: Sage).

Morgan, D. (1999) *The European Parliament, Mass Media and the Search for Power and Influence* (Aldershot: Ashgate).

Pridham, G. and Pridham, P (1981) *Transnational Party Cooperation and European Integration* (London: Allen & Unwin).

Radunski, P. (1995) 'The Election Campaign as a form of political communication', in J. Thesing and W. Hofmeister, *Political Parties in Democracy* (Bonn: Konrad-Adenauer Stiftung).

Reif, K.-H. and H. Schmitt (1985) *Ten European Elections 1979–81 and 1984: Campaigns and Results* (Aldershot: Gower).
Sasse, C. et al. (1981) *The European Parliament: Towards a Uniform Electoral Procedure* (Florence: EUI).
Tsatsos, D. T. (ed.) (1992) *Parteienfinanzierung im europäischen Vergleich* (Baden Baden: Nomos).
Toulemon, R. (1998) 'For a Democratic Europe', in M. Westlake (ed.), *The European Union beyond Amsterdam* (London: Routledge).

21
A Union Fit to Govern? The Role of the European Parliament and the Commission

Juliet Lodge

> We have laid the foundations of a political union with shared institutions and a directly-elected European Parliament.
>
> Commission President-designate, Romano Prodi, to the European Parliament, 22 September 1999

Twenty years after the first Euro-elections, the European Parliament has come of age as a parliament, but remains essentially a parliament in search of a role. It has established and consolidated its position as a *bona fide* institution of the EU, having for decades found that its status and legitimacy were contested, not least by hostile national governments and national parliaments. It remains in search of a role, however, in that it remains committed to continuously redefining its role as it strives to shape the EU polity in a way that corresponds to western understanding of liberal, democratic representative, open government. This chapter suggests that: (i) the relationship between the EP and Commission has become of increasing importance; (ii) this in turn heightens the prospect for institutional and constitutional change and further development of the EP's role following the implementation of the Treaty of Amsterdam (TEA); and (iii) a newly elected European Parliament exercising a modified role and having a wider role conception will have significant implications for the EU in the run-up to the next IGC and EU enlargement.

Background: the EP and Commission on the eve of the 1999 Euro-elections

In 1999 for the first time ever, all the member states used a system of proportional representation for the EP elections. All the member states contested the elections on ostensibly national platforms, with national campaigning methods, varying degrees of advertising, polling, and variable amounts of input from the European transnational parties and from national political leaders. As in the past, European election manifestos did not over-ride national concerns but complemented them. Different emphasis was accorded to specific issue areas both by the transnational parties, their national counterparts and EP candidates and by the different member states, but there were signs of an emerging shared and common agenda. This reflected, broadly speaking, the common concerns of the Commission, MEPs and member governments across the EU. It did not necessarily correspond to or effectively address the concerns of the broader electorate. As ever, the electorate remained somewhat bemused, at best, and bored at worst by the EP elections. Voters may understand that elections are about a contest for seats in the EP among different parties with whom they are broadly familiar. Beyond that, however, their purpose is unclear. The EP elections are neither about electing people of a particular ideological persuasion to power, nor are they about who holds power or exercises executive power. They are, therefore, unfamiliar in terms of their goal, and mysterious in terms of their impact.

As usual, turnout differed according to whether voting was compulsory or not, the effectiveness and persuasiveness of the election campaign, relative voter fatigue in view of adjacent national and municipal elections, voter interest and motivation. All this was predictable and could be more or less taken for granted. But the 1999 Euro-elections differed from the past four EP elections in several crucial respects:

(a) the legitimacy of the European Parliament is no longer disputed;
(b) the right of the European Parliament to demand and to exercise genuine legislative power is no longer contested;
(c) the democratic legitimacy of MEPs is no longer a major subject of debate;
(d) the democratic legitimacy of the present distribution of political authority among the various EU institutions is a subject of domestic political concern;
(e) the EU itself has become broadly speaking internalised in a way which suggests that domestic political administrations regard it as the key parameter and frame of reference for policy-making;
(f) the member governments have already implicitly accepted that further institutional reform is absolutely essential to enable the EU to act efficiently and democratically in the next century (something that in the past was often a major source of dispute);

(g) the member governments have accepted the need for EU level political leadership;
(h) the member governments anticipate and are prepared to discuss the desirability of moves towards a constitution.

While it would be interesting to ascertain the extent to which the European Parliament can take credit for this positive transformation in national governments' attitudes towards European integration in general, that question is outside the scope of this chapter. Instead, the argument proceeds on the premise that by the time of the 1999 elections, the political context of European integration had changed significantly since the last Euro-elections in 1994. System transformation was accelerated by the imperative to respond to external factors that required common responses on the part of the member states. The changes in the regional milieu since 1989, coupled with two major treaty revisions embodied in the Maastricht and Amsterdam treaties, EMU and the euro, and impending enlargement impelled not merely a reappraisal of the appropriate scope of integration but acceptance of the need to act in common and even to try to attain integration in the taboo fields of internal and external security.

The fact that the European Parliament's role no longer appears to be contested, however, cannot be explained merely by the imperatives of the external milieu: the EP still has limited power in respect of pillars II and III. Rather, the general problems of governance and the appropriate and lawful exercise of power by specific institutions suggest that there is renewed recognition of the desirability in principle and practice of one institution (preferably one enjoying the consent of and elected by the people) being responsible for checking the potential abuse of power. Traditionally, this is a function of an elected parliament. Traditionally, too, it has been argued that while the EP was the ultimate check against the abuse of power by the Commission, it shared with it the mission of promoting integration for the common good. When particularism appeared to take hold in a weakly-led Commission apparently lacking adequate financial controls and devoid of a mission, an over-arching goal, methods, procedures, and a timetable for achieving that goal, the EP had to wield its authority and seek its removal. Tactically and politically, it was fortuitous that this coincided with the preparatory phase in the run-up to the first Euro-elections under the TEA. The TEA came into effect on 1 May 1999 and opened up a range of interesting scenarios for the future development of relations between the EP and the Commission that presage further system transformation.

The Commission's resignation

The Commission's resignation justified the European Parliament's perennial claim that the EU needed a strong, independent Commission President in order to *sustain* European integration. This lesson was one that the member governments learned the hard way, but it is a principle that has recurred in MEPs' subsequent arguments. They link strong leadership with a raft of other ideas central to their sense of what the requirements are of a functioning, liberal democratic polity charged with sustaining European integration, namely:

(a) strong, independent Commission Presidential leadership;
(b) effective, policy-relevant links between the EP and the Commission (in short, answers to the question of who defines political priorities);
(c) accountability: Commission accountability to the EP, EP accountability to the public and, in view of co-decision, in effect Council accountability to the EP;
(d) institutional reform informed by the need to assert and sustain democratic values, behaviour, credentials and legitimacy of the EU as a democratic polity in its own right.

Strong, independent Commission Presidential leadership

The framers of the Paris and Rome treaties in the 1950s had asserted the need for apolitical (that is, party-free) technocratic, independent leadership of the process of European integration. Such leadership was not going to be value-free but it was supposed to be objective, consensual, able to withstand 'capture' by individual parties and to transcend individual parties, and to distil common interests and advocate and ensure the implementation of the common good. Therein lay the rationale behind the establishment of an independent Commission. Over the years, this principle has not been erased but has been subsumed under the welter of pressure to accommodate hostile member governments and the vagaries of national politics. The consequences were sharply outlined when the Santer Commission appeared to lose direction (and stagnate, as some governments no doubt hoped) and engage in seemingly haphazard initiatives. This highlighted the need for a clear programme of action and set of fulfillable priorities. Whereas, in the past, it had been largely left to the Commission to determine the broad direction of policy, latterly informed by the European Council, following the Commission's resignation new possibilities surfaced, as will be shown below.

288 *The Role of the EP and the Commission*

The Commission's resignation led MEPs to set out principles and lay down markers as to what they expected from a new Commission in terms of *leadership, accountability* and *credibility*. The process for appointing the Commission under the TEA (Article 214), replacing article 158 of the TEU, led to an examination of a range of procedural possibilities which were rehearsed alongside principles of good government.

Procedures and principles

At the end of March 1999, before confirming the appointment of Mr Prodi as the next Commission President, MEPs emphasised that they would not have confidence in a new College in which certain discredited Commissioners remained in post (Committee of Independent Experts, First report, 15/03/99). For some – though not for Mr Prodi – wholesale Commission replacement was essential in order to give citizens a clear signal of a fresh start. Some MEPs advised against the reappointment of certain Commissioners (e.g., both Germans), and suggested that Germany follow the practice of other states with two Commissioners by appointing one from the opposition parties. Some MEPs also advocated a new motion of censure to ensure that the Santer Commission could not remain in place until after the elections. At the beginning of April, seven of eight political groups (PES excepted) planned to present a motion of censure against the Santer Commission[1] on the basis of Article 144, which allows the EP to give its views on a censure motion only three days after it has been submitted; adoption requires a two-thirds majority and a majority of all MEPs and, if carried, all Commissioners must resign as a body. However, it proved difficult to insist on their immediate removal. Commission President-designate Mr Prodi argued in favour of a lengthy period of consultation to allow him to select a strong team for the College in consultation with the heads of government.

Confusion over, on the one hand, how to proceed between March and the Euro-elections, and, on the other, the time when the new Commission would take over led to different scenarios being mooted. Some politicians were anxious to see an interim Commission in place. However, others suggested that this would not be the most sensible course of action: (i) the procedures had to be set in motion for appointing an interim Commission quickly; (ii) account needed to be taken of the likely negative subsequent political implications of not re-appointing the interim Commission after the Euro-elections and, in late 1999, to allow it to assume a full-term of office from 2000. As it was, the EP had to act swiftly to consider whether it should immediately endorse the designation of Mr Prodi (a person whose name was already being discussed by the governments in any event in preparation for the scheduled appointment of the new Commission).²

EP President Gil-Robles convened an extraordinary meeting of the Council of Presidents (EP President plus presidents of the political groups) on 7 April to ensure that the agenda of the 2–16 April session included the debate and vote on the European Council proposal to appoint Prodi. In the meantime, Socialist Group leader Pauline Green invited Mr Prodi to meet the PES on 8 April (he met the Liberal Group on 7 April, and the EPP on 8 April). An extraordinary summit of the heads of government and Mr Prodi convened for 14 April. However, the chances of appointing a new Commission *before* the end of the current legislature were slim because (i) Euro-election campaigning was getting under way; and (ii) MEPs only had until May to give the EP's view on the new Commission as a whole before the end of the legislature. Procedurally, this would have required the convening of hearings of the different candidates between 19 April and 3 May by MEPs; and agreement among the member governments on who their nominees would be. Politically, it presented the possibility of a wide divergence of views between that of an outgoing EP (whose make-up might differ significantly from the newly elected EP) and that of the EP to be elected in June. A legacy representing the views of the former would have had serious implications for the newly elected EP, and difficult repercussions inevitably on the member governments. Moreover, it could have been a serious liability for the newly elected EP given the importance attached to making the terms of the Commission and elected EP broadly coincide. It would, for instance, possibly have opened the door to the outgoing EP in 2004 facing a similar scenario. This would not have been conducive to establishing either a good precedent in terms of EP–Commission legislative linkage; or for the EU's image as an organisation committed to good, 'clean', democratic government. In that sense, it was probably inevitable that the heads of government in Berlin stated in their Declaration that the newly elected EP should give its approval to the president and people nominated as members of the new Commission. The Commission, which was due to take office in 2000, needed to be invested by the newly elected EP and, in the interim, to enjoy its tacit approval.

This did not entirely resolve problems, however, because principles of good government needed to be demonstrated at a time when (i) Santer had not resigned, even though he was campaigning for election as an MEP in Luxembourg; (ii) Prodi as President-elect could not allocate portfolios because the new Commission-designate was not known, and the old Commission remained in post; and (iii) the imminent entry into force of the Treaty of Amsterdam on 1 May 1999 required attention to the legislative programme because existing legislation needed to be adapted to the TEA. (Thirty pending legislative items formerly subject to co-operation now came under co-decision and the EP decided not to vote on them in second reading but to leave them to the newly elected EP: (Palacio Vallelersundi report, 1999). The most urgent measure involved enhanced co-decision given by the TEA to the EP.

Above all, MEPs needed to be mindful of the potential political ramifications of their reaction to the resignation of the Commission and the designation of a new Commission President. It was imperative to take a medium to long-term view rather than merely to address short-term, politically expedient issues. They had a chance to exercise political muscle and set down guidelines for future action and relations between the EP and Commission, as well as for the next IGC. There is a premium on effective policy-relevant links between the Commission and the EP. The procedural implications were that the EP approved the appointment of Prodi, leaving the newly elected EP to proceed with a 'vote of approval' (Article 214, TEA) on the President and other members of the Commission as a College. To prevent Santer staying in office until the autumn, Mr Friedrich (EPP) suggested bringing forward the inaugural session of the new EP normally scheduled for 20–3 July. Accordingly, the EP would organise hearings of Commissioners in its different committees (as it did for the first time with the Santer Commission, which got through by the skin of its teeth). In the meantime, the political groups would – and did – hold hearings with Mr Prodi.

Strong leadership and effective policy-relevant links

Strong independent leadership by the Commission has two main elements: independence of thought; and independence of action. As designate Commission President, Prodi clearly saw the two as linked to the calibre of the appointees. He set the tone for this when asserting himself both *vis-à-vis* the governments and *vis-à-vis* the EP. Using his rights, under the TEA, to exercise 'general political authority' (or the *Richtlinienkompetenz*) and shape the future reform of working methods and structures, he insisted that he wanted strong, competent individuals as candidates for Commission posts; that he would 'veto' any names that did not match up to this in terms of skill and integrity; and that he intended to establish a strong Commission to follow through a Commission programme geared towards the over-arching goal of transparent, efficient enlargement through the transparent execution of a limited set of priority actions.

Where the EP was concerned, he resisted pressure to take up office prematurely; and in his consultations with the EP's party groups, only hinted at what policy priorities might be. This was significant because he accomplished two key principles: (i) independence of thought, coupled with an intention to translate this into action by using the Commission's right of initiative; and (ii) independence of action coupled with the promotion of dialogue with the EP's party groups. The parties would ultimately be responsible for facilitating independent action in giving or withholding their approval to draft legislation. In that sense they potentially have a real right of 'policy-control' though one that falls short, just as it does in national parliaments, of determining the agenda, direction and detailed content of the legislative programme. Such a right of policy-control, however, would be one which could be developed to ensure that the Commission President was obliged to incorporate into the annual or multi-annual programme a limited number of genuinely major policy priorities agreed by the EP.

There were, perhaps, hints as to the possible arenas in which greater co-operation and mutual agreement between the EP and Commission might be developed in future, and specifically in respect of reforms by the next IGC. Both Prodi and the EP stressed the importance they attached to matters that broadly fall under the remit of external affairs and foreign policy. Pillar II gives the EP limited authority. The EP, moreover, continues to seek a formal right to endorse all international treaties. In addition, it is inconceivable that there could be further reform on a common defence without the defence accountability deficit being made good: indirect accountability of either the Commission or the Council via national parliaments would not be sufficient. A Mr CFSP has to be seen to be accountable to the EP; and to an EP with a role in approving defence related expenditure, whether categorised as humanitarian relief/aid or defence spending. The same applies to pillar III, and the realisation of a zone of freedom, security and justice, which is one of the key priorities for future action according to Mr Prodi. This also corresponds to Council priorities and to identified areas of reform in the TEA, such a pillar III.

Mr Prodi made clear early on that he attached over-riding importance to a Balkan programme, something pushed also by the EU Council groups of experts on the Balkans, in the form of a Stability Pact for South-east Europe, tied to a possible EU special representative for the region, but linked to the Organisation for Security and Co-operation in Europe (OSCE) [Agence Europe 29/04/99 Europe Documents]. (It is perhaps significant that the representatives of the Foreign Affairs

Committees of the parliaments participating in the Royaumont process, at their meeting in Ohrid at the end of April 1999, indicated a desire to promote interparliamentary co-operation, including the EP, and proposed an action plan on the parliamentary dimension of the process [Agence Europe 26–27/4/99, 17.] Mr Prodi also hinted at future institutional scenarios linking the Commission and EP more closely together when suggesting that the issue of a Commission President who was somehow 'elected', possibly from within the ranks of the EP, might be a desirable development. The EP is, arguably, (or should be), the repository of a 'common European soul'.

Prodi argued that EP control of, and daily co-operation with, the Commission was vital; that there should be neither winners nor losers; and that political responsibilities and accountability should be exercised by EU institutions, notably the EP, rather than outside bodies, even those comprising the very 'wise'. Many MEPs applauded this view and felt that the EP's Budgetary Control Committee rather than the Group of Experts should have looked into fraud [Agence Europe 10/04/99, 14]. While clearly giving signals as to the future ideal interinstitutional relationships he foresaw, Mr Prodi drew a line between the feasible and desirable; between aspiration and expectation. This was shrewd as it allowed MEPs to take the rhetoric at face value and to proceed pragmatically both in the immediate run-up to the elections and pending the investiture of the new Commission. At the same time, it was clear that a culture of transparency and Commission responsibility had to be fostered and that distinctions were to be made and upheld in the Commission's brief which in turn would have implications for its relations with the EP and for the role MEPs might expect to play themselves in defining priorities and taking decisions. As he said in his speech as Commission President-designate to the EP on 14 September 1999, the EP's vote (of approval) would 'mark a new beginning in relations' between the two institutions:

> For now it is time to get down to work. Together we can and must put Europe at the service of the people. We have to win back ordinary people's confidence in Europe and in a European vision which puts their needs first. I want our two institutions, and the Council, to work together, genuinely and efficiently serving the people of Europe. I am determined to transform the Commission into a modern, efficient administration which has learnt the lessons of recent experience and put its house in order.

Mr Prodi's speech to the EP echoed many guiding principles germane to EP arguments on institutional reform since the 1980s. He distinguished between the political and administrative role of the Commission. At an administrative level, he wants greater precision in defining the functions of the various components of the Commission. These extend beyond the normal administrative functions common to domestic civil services and incorporate political functions which need to be exercised in an efficient, accountable and transparent manner. This has implications for the internal organisation of the Commission, notably the position of the much-criticised 'Cabinets'. While they had acted as a Commissioner's antenna or 'eyes and ears', and continue to perform an essential but discreet political function in this respect, they can develop into fiefdoms if there is neither strong Commission Presidential leadership, nor a common over-arching goal which determines the general direction of all initiatives taken by individual Commissioners and excludes the cabinets from interfering in the running of the Commission. The various (and subsequently reformed) Directorate-Generals also need now to ensure that they co-operate and implement policy in line with clearly defined responsibilities.

A similar distinction can be drawn between political control and the exercise of accountability, on the one hand, and the technical control exemplified by EP demonstration of legislative control via effective scrutiny and legislative follow-up. It was significant that Mr Prodi stressed the importance he attached both to transparency and to making the Commission open and accessible to the public. Transparency, for him, is implicitly part of demonstrating accountability, openness and readiness to share information with the people it is supposed to serve: the citizens. So far, Mr Prodi's apparent policy agenda reflects contemporary political reality, the priorities of the European Council and the TEA. It also echoes the major electoral manifesto themes of the EP's parties contesting the Euro-elections. But ignoring politically expedient references, there was, as German Council President-in-Office Gunter Verheugen said, 'vision, realism and pragmatism' displayed by someone whom the governments see as a high calibre, experienced politician.

Institutional and constitutional reform

The Amsterdam Treaty strengthened both the EP and the Commission President in the EU's structure; underscored clear commitment to transparency and democratic legitimacy; and to giving effect to them across the range of EU activities. As far as further Treaty reforms are concerned, the Commission's resignation emphasised the need to reconsider and possibly alter existing rules possibly to censure (with or without enforced resignation) individual Commissioners. If this were the only aspect of the EP's powers to be changed for the next IGC, then the EP's actual authority and its potential influence across the board would be significantly increased. At the same time, it is hard to argue against universal co-decision. At the very least, all areas subject to majority voting by the Council of Ministers should be made subject to co-decision. Given the ambitious agenda and requirements of enlargement, co-decision must be expanded in the name of equity, efficiency, effectiveness, democracy, accountability and transparency. There can be little doubt that the precise manner or point in the policy making and policy implementation processes at which co-decision is exercised must vary. This is essential in respect of the highly sensitive areas under pillars II and III where secrecy is a pre-condition of operational effectiveness and where intergovernmental arrangements can facilitate or constrain progress in unpredictable ways.

It is also important to consider what the potential impact on the public and outsiders alike might be of seeing a focused, strongly led Commission enjoying the support of an assertive and confident EP endowed with co-decision powers. The Commission would thereby enjoy a boost in credibility. The challenges posed by the CFSP, the euro, pillar III and enlargement need to be addressed systematically and coherently by a strong Commission that enjoys the respect, confidence and critical trust of a well-organised EP to whom it is accountable. The regular, visible and understandable demonstration of a critical exchange of ideas and views between the Commission and the EP is an essential element of democratic practice which has yet to be adequately exposed. Exacting EP scrutiny of draft legislation and equally exacting follow-up are needed to complement and sustain the idea of an EU imbued with robust political institutions that engage with the public, whether directly via the Internet [Commission of the European Communities, 1998, 476, final 29.07.98] or the media. Neither requires sweeping treaty reform. Both depend on changes in behaviour and resolution to act professionally as the custodians of European integration and its original goal: to preserve and sustain peace and stability.

The Commission and the EP share an interest in asserting the need for democratic responsibility and accountability. They may, however, imply slightly different things. For the EP, Commission responsibility means it must be responsible for the efficient and effective prosecution of policy that broadly enjoys the EP's support. Commission accountability means that it must be accountable to and retain the confidence of the EP. The difference in interpretation may lie in the procedures used to give effect to them. On the one hand, there is the matter of appropriate, institutionalised procedures to ensure effective consultation, responsiveness and accountability to the EP, not least for the new areas that have moved from pillar III to pillar I, and for those under pillar III and pillar II when EP control is rudimentary. On the other hand, there is the matter of the degree to which the majority view of the EP party groups should condition and determine the priorities announced in the Commission's legislative programme. The EP party groups have an interest in ensuring that the Commission, in drafting its legislative programme, reflects the political priorities of the parties themselves, or of the majority coalition. There is a fine line between that and a politicised Commission dependent on the EP or on a particular alliance within the EP. Over-dependence on one alliance could weaken the Commission's autonomy without necessarily increasing its accountability and responsibility to the EP.

However, the EP has already indicated how in future a Commission President might indirectly emanate from the ranks of the political parties. Elmar Brok suggested, on behalf of the Institutional Affairs Committee, that for Euro-election campaigns, parties should propose the candidate they would like to see appointed; member governments should take account of the outcome of those Euro-elections and, in nominating the Presidential candidates, take account of the parties' views. He also suggested choosing some Commissioners from within the ranks of sitting MEPs [Brok report A4-0488/98]. The over-riding suggestion is that the Commission has come to be seen as a political leadership in its own right with political functions that are legitimate and performed openly, democratically, accountably, effectively, responsibly and responsively alongside those of technocratic and administrative management. The EU clearly needs to accept and endorse the development of a new political and administrative culture when it reforms the institutions at the next IGC.

Ensuring EP capability to legitimise change

In theory, if the Commission is composed of people with political nous and an ability to pursue and implement independently and on their own individual responsibility strong, clear policies, then a sense of political direction for European integration should be restored. However, a strong independent Commission cannot operate in a vacuum: it is not self-sufficient, and neither is it meant to be. Its efficiency and its ability to act decisively does depend on it (i) enjoying the confidence of the EP; (ii) being able to initiate and see adopted a limited but clearly focused set of policy priorities; and (iii) having a co-operative legislative environment in which the EP and the Council work as two arms of a legislature.

A strong Commission implicitly assumes that there will be a strong EP. A strong EP would not simply be one that has been entrusted with a raft of additional powers, and neither would it be one that simply has the opportunity to exercise co-decision over a wider scope of policy than its predecessors. For the EP itself to be seen as strong, and for it to support Mr Prodi's aim of running a strong, independent Commission with strong and decisive leadership, the EP must organise its own business effectively. This puts a premium on the parties developing a relevant and coherent set of policy priorities and developing and demonstrating their ability to maintain party cohesion. Excessive splintering would inhibit this, reduce the EP's ability to act responsibly, accountably and credibly and enfeeble the Commission. A weak EP would not be able to lend the necessary support to the Commission. This would be a danger because it would weaken the overall capacity of the EU to act credibly both internally and internationally. In such a scenario, not only would the euro, stability, security and enlargement be jeopardised but the door would be opened to factions and national impulses not conducive to European integration. This in turn would seriously weaken the ability of the Commission to remain independent and sustain the momentum behind its programme.

Only if the EP is well-organised will it be able to follow through some of the interesting prospects for injecting into the Commission's programme policy priorities relevant to the people and ensuring that they are prosecuted as effectively as possible within the bounds of financial resources available. The EP and Commission President need to work together on this. If a Commissioner loses the confidence of the EP, it does not necessarily follow that he or she must curry favour with the political groups regardless. It does imply, however, either that the policy is unpopular and needs to be better explained, presented, argued, amended or abandoned or that the relevant Commissioner is incapable of acting responsibly and effectively. While it is true that neither the Commission President nor the EP will have the means to 'sack' individual Commissioners, together they can emasculate them. How? The Commission President can re-arrange responsibilities in such a way as to deprive Commissioners of control of important policy areas. That would send an important signal to all concerned, including the national governments who might then be persuaded to seek the individual's resignation and/or recall that person. The force of such 'reshuffling' would be significantly boosted if, in addition, MEPs indicated dissatisfaction. This could take the form of questions to the Commission and Council about 'reshuffling' and its consequences for the conduct of legislative business. Or MEPs might require the 're-shuffled' or 'failing' Commissioner to appear before a particular committee, or joint committee, for a hearing and cross-examination. In the event that MEPs were still dissatisfied, they might pass a resolution laying the blame for inadequate performance on an individual Commissioner. This would inevitably become a highly public event which member governments could not ignore. If, moreover, co-operation between the EP and national parliaments significantly improves, it is conceivable that national parliaments might either (i) also seek a hearing with the 'failing' Commissioner, or (ii) indicate, via an appropriate committee to the EP, their opinion on the EP's resolution.

When the TEA is revised at the next IGC, the relationship between the Commission and the EP could be affirmed. But the two must not be seen to be opponents of the Council of Ministers which has legitimate, national interests to project and protect. All could, however, be seen to be working together for the good of effective, democratic government. Thus, in the case of inadequate performance by an individual Commissioner, it might be desirable to make an *EP resolution of blame* subject to a form of co-decision whereby (i) the Council President or European Council has to respond, if such a resolution is passed by a sufficiently high majority of MEPs, and justify the Commissioner remaining in office; and (ii) the Commission President has to explain to the EP how he intends to deal with alleged failure. This would fall short of a procedure to oust a failing Commissioner but would show that the exercise of political authority in the name of good government is taken seriously.

Good government and closing the accountability deficit: the immediate problem of EMU

One area of vital concern is EMU. Here the absence of effective national parliamentary involvement in the scrutiny of EMU means there is a democratic gap, an accountability deficit, to be addressed. The realisation of the euro and EMU has substantial implications for a range of socio-economic policies to be implemented in the member states. The indirect consequences of EMU for national parliaments affect all EU member states regardless of their formal status and scrutiny powers in respect of EMU. They affect employment policy, fiscal policy, and measures linked to these designed to foster economic stability (including flanking programmes on education, lifelong learning and training, and measures to promote social inclusion). They concern the very areas most likely to affect individuals and families directly or indirectly. [Lodge and Lord, 1998; Lodge, 1999]. EMU is likely to unsettle publics across the EU. It is as yet unclear who is to articulate public views and convey them to those in a position to make, influence or execute policy. Who is the advocate and voice of the people? It would be consistent with parliamentarians' view of their representative function that they should assume responsibility in this respect. MPs cannot be effective alone, however; they need the EP. The MEPs and MPs could assume an obligation to act as the guardian of the EMU's conscience. They could then argue that they were justified in requiring the public accountability of the executive agencies responsible for EMU, and above all, justified in requiring interaction and dialogue in these policy areas with the Commission.

Addressing the EP in May 1999, Mr Prodi underscored the significance of the euro and EMU, and their importance for creating international monetary stability. There are technical and political sides to this. The latter are germane to the EP's deliberative role *vis-à-vis* the public and national parliaments. While recognising the direct impact on citizens in terms of transparency of prices, Mr Prodi opened up again the touchy issue of the full liberalisation of financial services, the removal of remaining barriers and reducing state aid. He acknowledged that closer economic and financial co-operation would have an implication for tax policy and combating unemployment. While emphasising his support for codes of conduct as far as capital gains and company taxes were concerned, he also felt an energy tax was an area where closer co-ordination could take place. Other plans for the EU's contribution to reducing unemployment included strong support for research and development and small firms through the fifth framework research programme, education and training. He pointed out that some 500 000 students had already benefited from the Erasmus student exchange programme to enable young people to study in another member state. Concluding, he once again emphasised the importance of maintaining peace and stability in Europe, as without this the integration process itself would be at risk, he felt.

The newly elected EP in the run-up to the new IGC and EU enlargement

The EP's short-term gain in confidence following the Commission's resignation needs to be extended and deepened. While the new EP had an immediate *raison d'être* in having to approve the appointment of the incoming Commission, it has (even in a purely technical sense) in line with its advocacy of transparency and democratic accountability, an obligation to ensure greater transparency and better and more effective communication with citizens. It should not ignore this on the somewhat specious grounds that until voters are better informed about the EU, they will not be much interested in the EP or its future election. With the Commission, the EP can contribute to closing the accountability deficit by endorsing and insisting on practices untainted by suspicions of maladministration, corruption or fraud. Reforms were put in place as the Euro-election campaign got under way: the Commission adopted a decision setting up a new anti-fraud office (OLAF/AFO) to replace the old one primarily concerned with agriculture-UCLAF – from 1 June 1999. The EP and Council had to complete their work on their own rules of procedures relating to investigations; and an interinstitutional agreement between the three was needed as AFO has the right to investigate all institutions. AFO is to be independent, able to instigate investigations on the initiative of its director, or at the request of the institution, body or member state concerned. A five-strong independent surveillance committee (comprising outsiders appointed jointly by the EP, Commission and Council) oversees investigations [Agence Europe 29/04/99, 7].

It is important to recall that AFO was the product of co-operation between the Commission and the EP which began in 1995. Acting Commissioner Anita Gradin noted that in 1995 the staff involved in anti-fraud work had been dispersed in several locations and that there were only 60 individuals employed in UCLAF. Now all the staff are under one roof and there are 141 employees in UCLAF investigating around 1000 cases of suspected irregularity, including 27 internal investigations in the Commission. She commended the close working co-operation between UCLAF and member states, noting that this had involved looking at around 5000 cases and would require a bigger budget. This, she concluded, sent a clear signal that the EU would display a common front in the fight against fraud.

The EP and policing 'squeaky clean' institutions

The Commission's resignation has implications for the new EP which have barely been discussed. First, there will be heightened expectations among the institutions, the member states and the public at large that the EP will systematically hold the Commission accountable for what it does. Second, this implies both that the EP must ensure that there is maximum transparency, and that the Commission is 'squeaky clean'. This is not simply a matter of improving links with the Ombudsman and ensuring optimum openness and availability of documents as guaranteed by the TEA (see A4-0476/98 – Lööw Report), and neither is it merely a question of Commission President Prodi making a register of his own correspondence available for public inspection in line with what is done in most open countries in Europe, nor of improving effective communication with the public by giving them open access to information. Rather, the purpose of given policies, and the reasons behind and costs and consequences of policy choices need to be explained. Third, the EP's own activities must be transparent and above suspicion. Not surprisingly, therefore, even before the Euro-elections, the EP clarified its own Rules of Procedure and tightened up rules governing the activities of groups of MEPs to ensure that they declare, from the outset, any support, no matter how seemingly trivial, from outside sources. Their activities must be above board. Their own credibility would be seriously tarnished if they were lax in respect of transparency, their own accountability or their own use of power.

A report by Shaun Spiers (PES, UK) seeking to alter the funding arrangements for groupings of MEPs and the way in which they are allowed to operate was adopted unanimously by the Committee on the Rules of Procedure in May 1999. This report recommended extending the rules on lobbyists (that currently apply to individual MEPs) to collective groupings of MEPs, thereby bringing them into line with individual MEPs. Specifically, it suggested that the chairmen of groupings of MEPs, both intergroups and other unofficial groupings of MEPs, should declare any outside support received, whether in cash or kind (e.g., secretarial assistance). The Quaestors would be responsible for keeping a register of such declarations and for drawing up detailed rules on the declaration of outside support by groupings. Further consideration is to be given to the introduction of a detailed set of rules on intergroups at a later date.

It would be disingenuous to believe that MEPs would be content merely with policing institutional probity. Political parties are about ideologies, ideas and their translation into relevant policies. How can the EP shape the policy agenda of a highly independent, strongly-led Commission?

It has several routes open to it:

(a) demonstrating the value of a mutually supportive, symbiotic relationship between the Commission and the EP in the adoption of legislation;
(b) acting as a conduit of information and ideas between national parliaments and EU policy-making levels, thereby performing via the parties and MEPs a communication and education function;
(c) discerning and articulating issues of common interest which might appear as primary or second-order issues in different member states but which potentially are important, though with differing priority, to the majority (playing a grand forum function);
(d) checking the lawful exercise of power by asserting respect for institutional balance;
(e) deliberating on and defining constitutional objectives for the next IGC;
(f) insisting on, and securing Commission support for, a right to participate on the same basis as the Commission and member governments in the next IGC;
(g) insisting on a right to ratify the outcome of the next IGC;
(h) deliberating on a common policy agenda to be presented for the next IGC which respects the need for focused development, matched by appropriate means and personnel, and which corresponds to the intention to create and sustain a functioning, liberal democratic, open, supranational polity, in short a constitutional pact before the next Euro-elections in 2004.

As early as the spring of 1999, Mr Prodi said that one of his priorities was 'to draw on the consequences of the single currency and create a political Europe' [Agence Europe 6&7/4/99, 4]. This might have been dismissed simply as part of his own politicking. He had suggested that there was no impediment to him running on the I Democratici list as it would be 'positive for a Commission President to pass some kind of democratic test', but he withdrew his candidature. That this was more than rhetoric was confirmed in his speech to the EP in September 1999 when he stated that it was essential to hold a major overall political debate at the beginning of each year and that he would present an 'Annual Political and Economic Report on the State of the Union' with a view to providing a unifying element in the policy formulation process, rendering it more coherent. He acknowledged that that policies do not come from the ether: someone is influencing the agenda, and someone is distilling areas where broad consensus over priorities exist. Transparency requires the EP party groups to play a role in ensuring, at the very least, that certain issues get aired and prioritised. If the Commission is not believed to be capable of carrying out its legal mandate and achieving its primary focused rather than diffuse goals, it will not be credible. A strong EP is a necessary but not a sufficient condition for Commission credibility. This must be supplemented, too, by participation in a political culture in which the principles of good government are embedded in a trusted political system whose ideals, behaviours, norms and values are shared, and regarded as lawful and open. Democratic legitimacy will not be enhanced unless the EP is able to sustain public engagement and persuade voters that they are heard: responsiveness, accountability, fairness and openness must remain EU leitmotifs.

The foregoing suggests that the EP has to reassess its role in the EU in terms of its responsibility to maintain, sustain and transform the institutional and political system, and in terms of its continuing search for a role. There can be little doubt that the EP has, since its inception, played the role of system transformer. The EU is in constant flux. It is sustained by institutions giving expression to shared, core political values entrenched in a constitution that evolves, responds to and is shaped by successive generations in line with those core political values.

Conclusion: the EP and democracy – searching for a role

Traditionally tension between passive acquiescence and an engaged civil society has been subsumed in the debate between the functionalists and neo-functionalists on the one hand, and federalists on the other. Crudely speaking, the latter are associated with the idea of a politically engaged citizenry partly because their over-riding interest has been in the creation of an integrated Europe based on a federal, liberal democratic constitution. Parliamentary democracy in shape and practice has been more important than functional representation of economically active interest groups, elites and employees (Andersen and Eliassen, 1996; 5.) (Pinder, 1999). However, democracy is more than a set of procedures and guiding principles: it is also the pursuit of democratic values involving the extension of popular participation in the political process (Archibugi and Held, 1995, 13). Therefore, the politicisation of a decision-making and legislative procedure once held to be largely technocratic has implications for the conduct of supranational government both among its institutional arms and *vis-à-vis* its citizenry. A symbiotic relationship is thereby implied between EU institutions which are separate from the EU's emergent civil society but central to developing and sustaining a democratic political order. Democracy may be 'concerned, on the one hand, with the reform of state power and, on the other, with the restructuring of civil society' (Held, 1996, 316).

Perhaps the problem lies more in the perceived need to give the public the impression of it being able to exercise (even indirectly via MPs and MEPs) some influence over binding public policy decisions taken by governments than in the reality. In practice, the more likely it is that effective solutions require EU level (or even international) action, the less likely it will be that effective democratic participation and control by the people will be feasible. It would be foolhardy to infer that EU decision-makers do not therefore require a democratic mandate and must be seen to be democratically accountable to the people. The EU must act as 'champion of the people'. Paradoxically, by facilitating a more participatory citizenry, the EU may stimulate greater interest in (and possibly opposition to) its agenda while simultaneously boosting its democratic legitimacy. The resulting reduction in the democratic deficit and enhancement of legitimacy would therefore reflect the idea that citizens had (and saw themselves as able to exercise) a just and rightful share in the process of supranational policy-making and 'government', possibly via local democratic mechanisms, such as a regional assembly, linked to higher levels (Martin, 1991, 18). Subnational units, by fostering participation, may help bring the EU closer to the citizen and define a European identity, cultural and social cohesion (particularly if the definition of citizenship were refined to include fundamental values such as social, environmental and cultural values), and enhance democratic practice (European Commission, 1997, 126, 297).

Supranational democracy, therefore, might be seen as a dynamic process mediated by multi-level networks (Ansell, Parsons and Darden, 347ff) embracing several variants of democracy. These include: direct and participatory democracy (derived from the ancient Athenian model where citizens directly engage in public policy-making); liberal, representative democracy (where citizens' interests are represented and articulated within the framework of the rule of law); deliberative democracy (where it behoves democratic institutions to resolve conflicts over political preferences): supplemented perhaps by constitutionalism (Weiler, 1997) and associational democracy (which prioritises freedom and argues that its effective pursuit rests on individuals working together: see Hirst, 1993, 112 and 132). The EU must work out how to maximise sufficient public engagement without compromising open, efficient, democratically accountable government.

This presents an interesting conceptual dilemma. It may be that optimal participation exists when policy communities normally associated with maintaining the system engage the expertise of epistemic communities and advocacy communities (Sabatier and Jenkins-Smith, 1993; Dudley and Richardson, 1996) and themselves adapt to permit wider participation. In the context of EU enlargement, this has implications for politico-economic, social and technocratic-bureaucratic elite communities, networks and interactions. Questioning the acceptability of such a widening may imperil how legitimate they are perceived to be and so widen divisions between included and excluded communities, however conceived. Applicant states' ability to behave politically in a liberal democratic manner with all that existing member states usually understand in terms of respect for the rule of law, justice, equality and liberty cannot be taken to mean that they comprehend, wish to uphold, value, share and will follow EU practice. The EU decided that a test of fitness would be their institutional and administrative capacity to implement the acquis (European Commission, 1997: 84). In a transformative phase, EU players need to explain procedures, practices and processes and the underlying normative beliefs and value system. The EP should fully exploit its implied socialisation role in this respect.

Defining and maintaining a European democracy: a millennium role for the EP?

The EP remains in search of a role. In the past, this search primarily focused on acquiring legislative authority and building a polity with appropriate built-in checks and balances. Theoretically, the EP is well placed to act as a voice and champion of the people and to scrutinise the executive and the legislative process effectively. In practice, internal organisational constraints inhibit it from performing as independent and supranational a role as might be desirable. Enlargement, with more MEPs and political groups, must not dilute its ability to sustain its credibility as a competent, responsive and responsible legislature able to insist on the democratic accountability of the 'government' in an open, just and acceptable manner. It must develop and sustain effective channels of communication internally within and among its own parties, Commission and Council; and externally with national and subnational elected bodies, notably parliaments and assemblies. To project a credible public persona, it must improve its relationship with the public. Herein lies the democratic deficit paradox. Paradoxically, the deficit grew with each step taken to reduce EP weakness in effecting control over the Council and the Commission, and with each increase in its visibility. It currently has a public, human face, and must now foster genuine participation.

Arguably, the EP remains in perpetual search for a role. How the EP evolves and the principles and values it practises in the name of liberal representative democratic government will inform any role it secures in contributing to further treaty revision and the next IGC. It must genuinely influence the EU's legislative agenda and so entrench a wider, traditional parliamentary role for itself in evolving, transforming, redefining and sustaining that system. The transformative role is arguably the more challenging, interesting and potent for the millennium. The EP signalled to both the public, but also, crucially, the member governments what it perceives to be key elements of democratic practice when it finally persuaded the Santer Commission to resign in March 1999. Since the EP's inception, it had been argued that the power to censure and force the resignation of the Commission *en bloc* was too nuclear a weapon to be used; that governments could theoretically reinstate the censured Commission (or most of its original membership); that the EP would probably never muster the requisite two-thirds majority to pass a censure motion; and that censuring the Commission would be merely symbolic, hitting the Commission rather than the real target, which is member governments and Council. The logistical obstacles were seen to be overwhelming. However, when the EP finally edged the Commission out, it achieved many things simultaneously and arguably established for itself a millennium role which had hitherto proved elusive. As custodian of the principles and practice of good government, it routed practices inimical to them. It established the need for a Commission capable of leading and enjoying EP support, not merely EP *confidence*. But does the EP enjoy public confidence? Mobilising high election turn-out may be one such indicator when it has the chance to rehearse its millennium role: to transform EU governance in line with accepted, if sometimes unarticulated, values and norms. These might embrace the following, derived from a more nuanced interpretation of its existing legislative, budgetary and control functions. Accordingly, the EP should:

(a) act as the custodian, guardian and transmitter of the supranational polity's normative beliefs, practices and values (to be upheld by EU political classes);
(b) promote the development of a EU participatory civil society and political culture derived from western liberal democracy;
(c) act as the voice and champion of the people, and be both responsible and responsive;
(d) communicate and educate; it should keep people informed and seek to be kept informed by others.

Custodian and guardian of normative beliefs and values

The EP should complement the Commission's 'guardian of the treaty' role designed to ensure compliance with EU law. It should act as a check against the abuse of values, practices, norms and political behaviour (by EU institutions) which compromise western liberal democratic values.

Advocate the development of a EU participatory civil society

In promoting a participatory supranational democracy and civil society, the EP could reform its own working methods to enhance the opportunity for mutually beneficial consultation with national parliamentary committees, the Committee of the Regions, and other democratically organised transnational groups, and develop an agenda-setting capability to reflect consent over major political priorities which it should help determine.

Communication and education

The EP should help to sustain and develop an open, participatory, democratic political culture by acting as a conduit for the transmission of information both among its own members, horizontally with other EU institutions, and vertically with national and regional bodies. This would reinforce effective legislative and control functions. In defining the agenda, the EP might accordingly stress appropriate use of information technology for civil society.

Voice of the people

The EP must continue to develop and act as the voice of the people, ensuring that it is responsive, accountable and responsible in its advocacy of EU level action, and worthy of public confidence and trust.

***

The foregoing implies that the EP will be a visible, intelligible and accessible forum in the future. The public needs reassuring over the purpose, rightfulness and scope of integration. Anxiety over potential abuse of power by EU (and national) level bodies – whether the Commission or national governments and their agents – has to be assuaged. Arguments about an appropriate division of power remain but one element in the construction of appropriate institutions for the EU. Traditional concerns about the conduct of democratic politics cannot be abandoned. They must be affirmed, articulated, advocated and practised. While information technology facilitates greater participation in policy discussion groups and networks, it does not eliminate the need for appropriate institutions to safeguard values and practices upon which liberal democratic government at all levels is founded. The EP must address this element of the debate about the appropriate organisational form that participatory, democratic government might take.

The EU as yet lacks the wherewithal to be as effective as it might be. Central to managing these issues is EU institutional reform. The old fudge over the f-word has given way to a more pragmatic agenda. Enlargement cannot proceed without institutional reform or without openness over the purpose of, and need for, transitional arrangements. The EU, at the next IGC, does not need to expand significantly the scope of its powers, but it does need to enjoy public confidence and to be able to exercise its authority, and it requires appropriate instruments to exercise its competences. It also needs to be sufficiently confident and open to enable it to make difficult choices, to justify them to the citizenry, to inform future decisions and help the authorities take informed decisions (Dyrberg, 1999). While internal organisational reforms must continue, and while they are essential, sight must not be lost of the bigger picture: facilitating and sustaining democratic good government. The EU needs in the millennium an EP that is the guardian, custodian and voice of good government based on liberal democratic values common to Western systems of representative government (a minority form of government in the international system). The EU needs an EP that enjoys public confidence, something that in turn needs to be sustained and not simply crudely measured by turnout in Euro-elections.

The millennium is a time when the traditions of west European liberal democratic practice should be applauded and refined in pursuit of inclusiveness: acceptance of them is an essential and necessary pre-condition both of EU membership and of the EU being able to act effectively, openly, accountably and credibly. The overall implication is that the EP retains multiple roles. They need re-prioritising to address the requirements of building an effective, credible, open, democratic supranational polity. Crucial is the EP's contribution to constructing a constitution for the wider Europe. The EP, if it performs the roles listed above, will be in an unparalleled position to distil the values which should guide and underpin the ever closer union of the new millennium.

Notes

1   A motion of censure against the Commission was tabled by Pauline Green (London North, PES) following Parliament's refusal to give a discharge to the 1996 budget. It was debated on Monday, 11 January, and the vote was on Thursday, 14 January. This was confirmed by the Conference of Presidents meeting on 7 January. A minimum of 10% of MEPs – or 63 – must put their names to a motion before it is debated According to the censure motion tabled on 17 December, the vote meant that the Commission was not assured of Parliament's confidence. The motion noted that the coming three months would be among the most complex in the EU's history and that, in these circumstances, it was essential for the Commission to operate to the best of its ability and in close co-operation with Parliament, which it could not do because of the uncertainty created by the vote. A double majority is required to carry a motion of censure; i.e., an absolute majority of MEPs and at least two-thirds of the votes cast. Since June 1974 motions of censure have been put to the vote only four times and have never obtained the necessary majority. If the motion is adopted, the Commission must resign collectively, as it is not possible to censure an individual Commissioner. The current Commission would continue to manage business until a new Commission was appointed. Various scenarios were discussed, and the appointment possibilities for the EP were:
(a) two procedures and 4 votes: current EP votes on confirming Prodi in April; votes on the College of Commissioners as a whole for the remainder of the Commission's term of office, until end of 1999, in May; new elected EP votes in June on Prodi as President of the Commission to take office in January 2000 for a 5-year term of office; and then votes on the Commission-designate;
(b) EP approves appointment of Prodi; newly elected EP proceeds with a vote of approval (Article 214 TEA) on President and other members of the Commission as a College;
(c) Present Commission stays in office until autumn 1999. To prevent this, Mr Friedrich (EPP) suggested bringing forward the inaugural session of the new EP normally scheduled for 20–3 July. EP would organise hearings of Commissioners in its different committees (as it did for the first time with the Santer Commission, which got through by the skin of its teeth). In the meantime, the political groups would hold hearings with Mr Prodi (Agence Europe 1/4/99).
2   Openness in the EU (A4-0476/98 – Lööw) Maj-Lis Lööw (PES) for the Committee on Institutional Affairs welcomed the Amsterdam Treaty provisions guaranteeing a public right of access to EU documents and stating that EU decisions must be taken as openly as possible and as closely as possible to the

citizen. She considered this to be a first step towards establishing a genuine principle of public access in the EU and called for full implementation of these provisions. She advocated proposals to improve a new code of public access to EU documents applicable to all EU institutions and bodies; greater public access to Council and Commission meetings; fuller use of the Internet; and more open administrative cultures within the various EU institutions and bodies.

References

Anderson, S. and K. Eliassen (eds) (1996) *The European Union: How Democratic is it?* (London: Sage,). Ansell, C., C. Parsons and K. Darden (1977) 'Dual networks in European regional Development Policy', *Journal of Common Market Studies*, 35 (3), 347–76.

Archibugi, D. and D. Held (eds) (1995) *Cosmopolitan Democracy* (Oxford: Polity).

Commission of the ECs, COM (1998) 476 final Brussels 29 July 1998, Communication from the Commission to the Council and the European Parliament, *Internet Governance: Management of Internet names and addresses – analysis and assessment from the European Commission of the United States Department of Commerce White Paper*.

Dyrberg, P. (1999) 'Citizens of the Union and the Union for the citizens: What does the Amsterdam Treaty bring?', *Cuadernos Europeos de Deusto*, 21, 28–52.

Dudley, G. and J. Richardson (1996) 'Why does policy change over time? Adversarial policy communities, alternative policy arenas, and British trunk roads policy 1945–95', *Journal of European Public Policy*, 3, 63–83.

European Commission (1997) *Agenda 2000: For a Stronger and Wider Union*, Bulletin of the EU, Supplement 5/97, Luxembourg.

Garrett, G. and G. Tsebelis (1996), 'An Institutionalist Critique of Intergovernmentalism', *International Organisation*, 50(2), 269–99.

Haas, E. B. (1971) 'The Study of Regional Integration: Reflections on the Joy and Anguish of Pretheorizing', in L. N. Lindberg and S. A. Scheingold (eds), *Regional Integration: Theory and Research* (Cambridge, Mass.: Harvard University Press).

Haas, P. M. (1992) 'Introduction: epistemic communities and international policy coordination', *International Organisation*, 46, 12–35.

Held, D. (1996) *Models of Democracy*, 2nd edn (Oxford: Polity).

Hirst, P. (1993) 'Associational Democracy', in D. Held (ed.), *Prospects for Democracy* (Oxford: Polity), 112–35.

Lodge, J. (1994) 'Transparency and Democratic Legitimacy', *Journal of Common Market Studies*, 43–68.

Lodge, J. (1996) 'The European Parliament', in S. Andersen and K. Eliassen (eds), *The European Union: How Democratic is it?* (London: Sage), 187–214.

Lodge, J. (1999) '1999 and All That: Why the euro makes parliaments matter', *The Parliamentarian*, LXXX, 171–5.

Lodge. J and C. Lord (1998) *Institutional Implications of EMU at the National and EU Levels* (Brussels: European Parliament).

Martin, D. (1991) *Europe: An Ever Closer Union?* (Nottingham: Spokesman).

Pinder, J. (ed.) (1999) *Foundations of Democracy in the European Union: From the Genesis of Parliamentary Democracy to the European Parliament* (London: Macmillan).

Sabatier, P. A. and H. C. Jenkins-Smith (eds) (1993) *Policy Change and Learning: An Advocacy Coalition Approach* (Boulder, Colorado: Westview Press).

Scharpf, F. (1994) 'Community and Autonomy Multilevel Policymaking in the European Union', *Journal of European Public Policy*, 1, 219–42.

Tsebelis, G. (1994) 'The Power of the European Parliament as Conditional Agenda Setter', *American Political Science Review*, 88, 128–42.

Weiler, J. H. H. (1997) 'The Reformation of European Constitutionalism', *Journal of Common Market Studies*, 35 (1), 97–131.

Wessels, W. (1996) 'The Modern European State and the European Union: Democratic Erosion or a new kind of Polity?', in S. Andersen and K. Eliassen (eds), *The European Union: How Democratic is it?* (London: Sage), 57–70.

Index

Note: references to illustrative / tabular matter are in **bold** type.

ABC 185, 194
abstentions in elections 5, 95, 121–2, 128, 151–2, 179
AD (Democratic Alternative: Portugal) 174
ADR (Pensioners' Party: Luxembourg) 46, **50**
African, Caribbean and Pacific (ACP) countries 27
Agalev (Anders Gaan Leven: Belgium) 52, **52**, 58
agriculture 139
Aguiar, J. 171
Ahern, Bertie 144
Ahern, Nuala 147
Aho, Esko 103, 109
Ahtisaari, Martii 76
Alema, Massimo d' 122, 149, 150, 155
Amsterdam Treaty 14, 15, 17, 18, 19, 21, 60, 91, 118, 162, 204, 214, 252
AN (National Alliance: Italy) 151, **153**, **157**, 232, 246
Anastassopoulos Report 8
Anguita, Julio 188, 194, 195
anti-Europeanism 5
  see also Euro-scepticism
Anttila, Sirkka-Liisa 111
ARC (Rainbow Alliance) Group 49
asylum policy 166
Australia 26
Austria 16, 18
  EU membership of 35, 43
  Euro-elections in 33–44; players in 34; players in a domestic context 34–7; party campaigns in 37–8; mobilising the vote in 38–9, 43; turnout in 38–9; issues in 39–41; results of 41–2, **42**; as part of series of elections 36; threshold of votes required

in 253; electoral systems
  in **256**, 259, 260
  and NATO 35, 37
  as net payer in EU
  neutrality of 35, 38, 39, 40, 42
  presidency of EU by 35
  security and defence policy of 35
Auto-Oil package 15–16
Aznar, José Maria 185

Barcelona Conference on the Mediterranean 28
Barcelona Process 28
Basque separatism 187, 189, 194
Bayrou, François 118
BE (Bloc of the Left: Portugal) 174, 178, **180**
Belfast Agreement 222
Belgian food crisis 222
Belgium
  Chamber of representatives election results 57, **57**
  'colleges' of electorate in 53
  Community Councils in 51
  constitutional structure and reforms in 51–2
  dioxin crisis 56
  electoral system in 53
  Euro-elections in 51–8; political situation in 45; obscured by national election 46; coincidence of elections in 54; background to campaign in 53–4; campaign issues in 54–5; results of 57–8, **58**; compulsory voting in 253; electoral systems in 260
  federal system in 51
  list order in 53
  political parties in 52–3, **52**
BE (Move Europe: France) 119

311

312 *Index*

Berès, Pervenche   123
Berlusconi, Silvio   151
Bertens, Jan-Willem   165
Bijlmer disaster (air crash)   165–6
Bildt, Carl   206
biotechnical inventions   16–17
Blair, Tony   122, 132, 214, 218
BNG (Galician Nationalist Bloc: Spain)   187, 192, 195, **196**, 197
Bodry, Alex   51
Bolkestein, Frits   160, 167
Borg, Sami   113
Borrell, José   193
Bowler, S.   271
Britain in Europe (BiE) campaign   227
British Nationalist Party (BNP)   216, 221–2
Brok, Elmar   19
Brown, Gordon   215
Buitenen, Paul van   167

C (Centre Party: Sweden)   200–1, 203, **208**
Cambio   192
CAP   141
Carneiro, Francisco Sá   172
Carrilho, Maria   182
Carvalhas, Carlos   174
CC (Canary Islands Coalition: Spain)   195
CCD (Christian Democratic Centre: Italy)   151, **153**, **157**
CD (Conservative Democrats: Netherlands)   162, **168**
CDA (Christen Democratisch Appèl: Netherlands)   160, 161, 162, 163, 165, 167, 168, **168**
CDS (Centro Democratico Social: Portugal)   **180**
CDU (Christian Democratic Union: Germany)   72, 81, 245
CDU (United Christian Democrats: Italy)   **153**, 154, **157**
CDU (Unitary Democratic Coalition: Portugal)   **173**, 178, 179, 180, **180**, **181**
CDU/CSU (Germany)   72, 73, 77, 79–80, 82–3, **83**, 85, **85**

Central Europe, democracy and human rights in   25
Centre Democrats (Denmark)   63, 65, **68**
Centre Party (Finland)   102, 106
CFSP   244
Chechnya   23
Chevènement, Jean-Pierre   119
China   26
Chirac, President   118, 122–3, 134
Christian People's Party (Denmark)   **68**
Ciampi, Carlo Azeglio   153
citizenship, nationality and immigration   7, 80
CiU (Convergence and Union: Spain)   186, 188, 191, 192, 195, **196**, 197
Claes, Willy   54
Clarke, Kenneth   219
Coalición Nacionalista (Spain)   195, **196**
co-decision procedures   14
 volume of   15
Cohn-Bendit, Daniel   119, 123, 124
Colla, Marcel   56
colonies, former   21
'comitology'   14
Commission of the European Communities *see* European Commission
Commonwealth of Independent States (CIS)   23
Communist Party (Portugal)   174, 178
compulsory voting   253
conciliation committee   14–16
Conservative Party (Denmark)   67, **68**
Conservative Party (UK)   127, 214, 215, 216, 217, 218, 219–20, 223, **224**, **225**, 245, 246
consumer protection   205
Cook, Robin   126
Cools, André   54
co-operation procedure   16
Council as council of member states   9
Cox, Pat   143, 146, 147
CPNT (Hunting, Fishing, Nature and Traditions: France)   119, 125, 131
Cresson, Edith   167, 221

crime, international, as election
  issue 4, 40, 61, 166, 221
CSA (Christian Social Alliance:
  Austria) 34
CSU (Christian Socialist Union:
  Germany) 72, 73, 83, 245
CSV (Chrëstlich-Sozial Vollekspartei:
  Luxembourg) 46–51, **50**
Cuba, association agreement with 24
Cuhna, Arlindo 182
Cuhna, C. 173
CVP/PSC (Christian Democratic Party:
  Belgium) 52, **52**, 55, 57–8, **57**, **58**

D66 (Democraten 66: Netherlands)
  160–1, 162, 163, 165, 166, 167,
  168, **168**
Dagens Industri 200
Danish People's Party 63, 66, 67, **68**
Dehaene, Jean-Luc 55, 56, 57
democracy and good governance 27
democratic credentials of Euro-
  elections 3
Democratici, I (Democrats: Italy)
  149–50, 151, 152, **153**, 156, **157**,
  158
democratic participation at heart of
  EU 7
Democrats for the Olive tree
  (Italy) 149–50
Denmark
  electoral system in 64–5
  Euro-elections in 60–71; focus on
    institutions rather than policies
    in 60; turnout in 60–1;
    attitudes to EU in 60–1; actors
    in 62–4; issues and campaign
    in 65–6; results of 67–71, **68**;
    volatility in 66–7; distribution of
    seats in 69, **70**;
    disproportionality in 255;
    electoral systems in 255, 259,
    260
  federalism in 63
  Maastricht referendum in 63
  Maastricht Treaty, exemptions
    from 65–6
  membership of EU, support for
    61–2, **62**, 62–3, 66, 67, 68–9, **70**–1

national decision-making, attitude
  to 61, **62**
'personal voting' in 64
Deprez, Gérard 52
de Rossa, Proinsias 147
development policy 21–2
  as foreign policy 22
Diáz, Rosa 191, 193, 194
DIKKI (Dimokratiko/Koinoniko
  Kinima [Democratic Social
  Movement]: Greece) 92, 93, 94,
  96, 97, **97**
dioxin scandal (Belgium and
  Netherlands) 130
Di Pietro, Antonio 149
disaster relief 22
DL (Liberal Democracy: France) 118,
  133
Doyle, Avril 147
DP (Demokratesch Partei:
  Luxembourg) 46–51, **50**
drugs as election issues 4
DS (Left Democrats: Italy) 151, 154,
  153, **157**
DUP (Democratic Unionist Party:
  UK) 22, , 216, **224**, **225**
Durao Barroso, José Manuel 172, 177,
  180, 181
Dutroux affair 54
Dybkjær, Lone 64–5

EA (Basque Solidarity: Spain) 187, 195
Eastern Europe, democracy and human
  rights in 25
Ecolo (Mouvement 'ecolo' – les verts
  [Greens]: Belgium) 52, **52**, 57–8,
  57, **58**
Economist, The 194
EDA (European Democratic
  Alliance) 156, 227, 232, **233**
education as election issue 221, 244
EDG (European Democratic Party)
  232
EFGP (European Federation of Green
  Parties) 243, 244, 246, 247, **247**
  manifesto of 244
EH (Basque Nation: Spain) 187, 194,
  195, **196**, 197
Eichel, Hans 78

EKA (Pensioners for the People: Finland) 101, **108**
ELDR (European Liberal, Democratic and Reform Party) 146, **146**, 147, **157**, 166, 197, 206, 231, 232, 235, 237, 243, 244, 245, 246, 289
 election manifesto of 244
elections, European *see* European elections
electoral procedures
 first-past-the-post 272
 public participation in 9
 uniform 8
Emma Bonino List (Italy) 150, 151, 152, **153**, 154–5, **157**, 158, 246
employment as election issues 4, 244
EMU 125, 201, 297–8
environmental protection as election issue 55, 61, 221, 244
EPP (European People's Party) 93, **146**, 156, 157, **157**, 175, 190, 206, 227, 231, 232, **233**, 235, 237, 238, 242–3, 244, 245, 246, 247, **247**, 289
 election manifesto 244
ERA (European Radical Alliance) 232, **233**, 235
Estaing, Giscard d' 118
ETA (Spain) 187, 191, 194
ethnic minorities 120
EU–ACP Joint Parliamentary Assembly 27
EU–Latin American Interparliamentary Conference 27
EUL-NGL (Confederal Group of the European United Left) 108, 232, **233**, 237, 246, 247, **247**
EU–Rio Group meetings 28
Eurobarometer 61, 175
Euro-Mediterranean Parliamentary Forum 27–8
European army 176
European balance of power 209
European Central Bank 75, 192
European Charter of Rights 244
European citizenship 177, 191
European Coal and Steel Community 231
European Coalition 187

European Commission (EC) 33, 277
 autonomy of 7, 10
 in CFSP activities 23
 Directorate General I and external relations 22
 election: of political heads of 8; of President of 17
 and environmental issues 23
 fraud in, as election issue 40
 Humanitarian Aid Office (ECHO) 22
 and law of the sea 23
 and Parliament, relationship with 7, 17
 and Parliamentary majority 20
 politicisation of 270
 Presidential leadership, quality required 287–93
 reform of 293–5
 resignation of 18, 48
 role of 284–309: prior to 1999 elections 284–6
 strengthening of, procedures for 288–90
 term of office 17
 trade policy powers of 23
 working of, as election issue 40
 see also Santer Commission
European Community (EC) 21
European constitution 124, 176, 192
European Council
 Cologne summit of 76
 composition of 241
European defence 204
European elections
 democratic credentials of 3
 direct elections in 264–5
 as direct participation in supranational political life 266: failure of 266–9
 as disaster for European socialists 86
 and disproportionality 252–62; analysis of 254–60, **258**; traditional measurements of 254, 255; ranking in 255–9, **258**; advantages for larger parties in 261
 effective number of parties in 254

enthusiasm for, lack of   5
funding of   280
information campaigns in   264
manifestos for   285
mobilising the electorate in   264–5
as outlet for party tensions   201
policy choices in, lack of   10
political parties' engagement in   10
proportional representation in   149, 234, 253, 279
as punishment for the government   140
purpose of   3
as referendum on domestic issues   43, 86, 252, 264, 268, 269–70
reform in   271–3
as transnational political event   271
turnout in   252, 264, 266, 285; as indicator of support for pace of integration   264; and legitimacy   265–6; as symptom of lack of Euro-election distinctiveness   269–71; *see also individual countries*
uniform electoral procedure (UEP): Council failure to agree on   271–2; Amsterdam Treaty and   273; Seitlinger report on   272; proposal for   273–80; common principle for   274–5; distinctiveness in   274, 275; time-zone differences in   275; voter eligibility in   275–6, 277; voting methods in   276–7, 278; PDEV in   278; impact of   277–80; regional constituencies in   278; candidate eligibility for   279–80
voting in, reasons for   265
voting rights in   265–6
voting systems in   253–3, **256–7**
see also individual countries
European Employment Pact   127, 244
European integration   7, 125–6, 133, 144, 152, 191, 201, 219
European Parliament (EP)
in accountability of executive   5
budgetary muscle of, in external policy   24
Bureau of   238–9
candidate selection in   240–1
character of newly-elected   299
Christian democrats in   231
and Commission, relationship with   7, 17, 20
committee assignments in   235–6, **236**
committee chairmanships   237–8
committee co-ordinators   237–8
Committee of Independent Experts of   18
committees of   **236**
Common Foreign and security Policy (CFSP)   21
in communication and education   306
Communists in   232
Conference of Committee Chairmen   239
Conference of Presidents   238–9
and Council, relationship between   9
credibility of MEPs in   19
as custodian and guardian of normative beliefs and values   306
customs union with Turkey, withholding assent from   24
as democratic control in the EU   21
in democratic oversight of external relations   28
election observers from   28
election of President of   235
and EU system, role in   19
evolution of   3
and federal political union   263
Greens in   232, 233
group co-ordinators   239
growth of authority of   4
as guardian of liberty and democracy   8
and IGC leading to Amsterdam Treaty   18
imbalance of power between Commission, Council and   192
influencing policy agenda   4
inter-parliamentary delegations of   27–9

316 *Index*

European Parliament (EP) (*cont.*)
  KEDO, blocking participation in 24
  legislative bill, procedure for 239
  legislative powers of 14–17, 25, 252, 264, 285
  in legislative process 4, 6
  legitimacy of 285; in implementing change 295–7; and direct elections 264
  and national parliaments, usurping 263
  non-statutory role of 25–9
  in nuclear energy accords 24
  in participatory EU civil society 306
  party groups: and Euro-election 242–5; internal organisation of 240; as organs of Europarties 242; and parliamentary decision-making 239–42; and parliamentary organisation in 235–9; roles of 240
  party system of, after 1999 elections 231–49
  pluralism brought by 19
  as policy-influencing legislature 241
  political groups in 231
  power to dismiss Commissioner 14
  progress (1994–99) 13–20
  radical left *see* Communists
  rapporteurships 237–8, 239
  ratifying foreign agreements of EC
  role of 284–309; in democratic Europe 302–8; prior to 1999 elections 284–6
  rules of procedure 234
  scrutiny of the executive by 17–18
  SiSo (Sign in Sod off) system in 166
  Social Democrats in 231
  statutory role of 22–5
  'unfinished business' of 3
  visits to, by international statesmen 28
  as voice of the people 307
  see also MEPs
European Radical Alliance 127, 132
European Socialist Party 126–7, 132
European taxation 176

European Union (EU) 21
  and constitution 286
  democratisation of 188
  desirability of 6
  development of 18–19
  eastern enlargement of 6, 22, 35–6, 37, 40, 42, 125, 244, 299
  as frame of reference for policy-making 285
  institutional reforms to 124, 285
  militarisation of 142
  policies of, presenting choices in 6
  political leadership required from 286
Euro-scepticism 5
Eurostat 172
Evert, Miltiades, 91
EVPN (Europees Verkiezers Platform Nederland) 162, **168**
Export Earnings Stabilisation Fund 27
external relations and foreign policy 21–22
  EP statutory involvement in 22–5

FD (Democratic Force: France) 118
FDF (Front democratique des bruxellois francophones: Belgium) 52, **52**, 57–8, **57**, **58**
FDP (Free Democratic Party: Germany) 72, 73, 77, 80–1, 83–4, **83**, **85**
Featherstone, K. 91
federal Europe 176
Fenner, C. 175
Fianna Fail (Ireland) 140, 141, 144, 146, **146**, 147, 232, 246
Figueiredo, Ilda 176
Financial Times 75, 83
Fine Gail (Ireland) 141, 144, **146**, 147, 190
Fini, Gianfranco 154
Finland 18
  agricultural sector, effect of CAP on 103
  coalition governments in 101
  combining national and EU elections 110–11
  constituency reform in 112–13
  EU membership referendum 102

European Council, Presidency of 106
European elections in 100–16; lack of European focus in 100; players in 100–1; players in domestic context 101–3; timing of 103; mobilising the vote for 103–5; EP information for 103–4; funding of 110; media coverage of 104; participation of party leaders in 104; projection of EU as intergovernmental organisation in 104; issues in 105–7; lack of political experience of candidates in 106; similarity of party views in 106; absence of transnational manifestos in 107; results of 107–10, **108**, **111**; electoral systems in 259, 260 National Coalition government of 106 parliamentary system in 105 rural–urban cleavage in 101
Fischer, Joschka 75, 82
Flesch, Colette 51
FN (Front National: France) 117, 119, 125, 128, 129, 130, 246
FN (Front National/Front voor de Natie: Belgium) 52, **52**, 53, 57–8, **57**, **58**, 232
Fontaine, Nicole 132–3, 235
food aid programmes 22
foreign policy *see* external relations
Forschungsgruppe Wahlen 83, **83**
Forza Italia 151, 152, **153**, 156, **157**, 232
Fp (Liberal Party) 201, 206, 208, **208**
FPOe (Austrian Freedom Party) 33, 34–43, **42**, 246
Frain, M. **173**
France 13, 252
European elections in 117–38; importance of 117; further integration as issue 120–1; campaign 121–3; issues in 124–8; results of 128–33, **129**; victory for 'Plural Left' government in 129; threshold of votes required in 253; disproportionality in 255; electoral systems in **255** proportional representation in 117 fraud prevention 244

GaL (Gréng a Liberal Allianz: Luxembourg) 49
Gaulle, Charles de 123
gender parity in lists 120
General Agreement on Tariffs and Trade (GATT) 25
Generalised System of Preferences (GSP) 22
Genscher, Hans-Dietrich 80
Germany 13
attitudes to EU in 74, 86
Bundesrat balance of power in 85
Bundestag election in 76
contributions to EU 75
divisions over EMU in 74
Euro-elections in 72–88; concentration on domestic issues in 72, 81; EU role in 76; volatility in 77; campaign and issues in 81–2: results of 82–4, **83**; threshold of votes required in 253; electoral systems in 260 party system, transformation of 77 procedural differences in 84 role in European construction 73 SPD–Green government, policies of 74–6 unification of, financial impact of 86 views of European integration in 73
Gilljam, Mikael 199, **208**
Gil-Robles, José Maria 235, 289
GLEI–GAP (*Déi Gréng* [The Greens]: Luxembourg) 46
globalisation 267
Goebbels, Robert 48, 51
Goerens, Charel 50, 51
González, Felipe 176, 190
GPV (Gere formeerd Politiek Verbond: Netherlands) 162, **168**
Greece
drachma entry into ERM 90

Greece (*cont.*)
  entry into EMU Stage 3 by 89–90, 93–4
  EU policy, stability of 89
  European elections in 89–99;
    coincidence with National elections 89; campaign and issues in 92–5, 97; funding of 92; advertising campaign in 92; national flavour of 93; overseas constituency in 95; results of 95–8, **96–7**; threshold of votes required in 253; compulsory voting in 253; electoral systems in 259, 260
  Europeanisation of domestic politics in 91
Green, Pauline 289
Green, Simon 72, 73, 77, 80
Green Left (Netherlands) 161, 162, 166, 167, 168, **168**
Green Party (Greece) 97
Green Party (Ireland) 141, 142, 144, **146**, 147
Green Party (UK) 215, 221, **224**
Greens (Austria) 34–43, **42**
Greens/European Free Alliance **157**
Greens (Finland) 106, 108
Greens (France) 117, 119, 121, 123, 124, 125, 127–8, **129**, 130, 134
Greens (Germany) 72, 73, 79, 80, 82, 83–4, **83**, **85**
Greens (Italy) **153**, **157**
Greens (Luxembourg) 49, **50**
Greens (Portugal) 187
growth hormones in beef 25
Guigou, Elisabeth 19
'guillotine clause' in budgetary allocations 16
Guterres, Antonio 172–3, **173**, 176, 180

Hague, William 227
Haider, Jörg 33, 38–9
Halonen, Tarja 103
Hänsch, Klaus, 235
Hautala, Heidi 109
HB (Basque Homeland: Spain) 187
health as electoral issue 221

Hollande, François 119, 123, 130
Holmberg, Sören 199, 208
human rights 25–7, 244
  clause in trade agreements 26–7

immigration as election issue 40, 66, 80, 130–1
individual freedom 244
Institute for Social Research and Analysis (Vienna) 38
Intergovernmental Conference
  (1996) 5
  (2000) 6–7, 8
  and EU system of governance 11
  forthcoming 209, 299
  preceding Amsterdam Treaty 18, 74
Ireland
  constituency arrangements in 141
  Euro-elections in 139–48; fought on domestic issues 139; actors in campaign for 141–4; simultaneous with local government elections 143–4, 146; results of 144–6, **146**; turnout for 147; aftermath of 146–7; disproportionality in 255; electoral systems in 255, 259
  roles in EP 148
Italy
  Euro-elections in 149–59; advertising for 155–5; campaign for 150–2; focus on domestic nature of 150, 158; as 'second-order' elections 155; turnout in 151; results of 152–8, **153**, **157**; turnover in Italian delegation to EP 152; compulsory voting in 253; electoral systems in 259, 260
  fragmented party system in 156
  membership of single currency 151
  pension and welfare reforms in 151
IU (United Left: Spain) 185, 186, 188, 191, 192, 193–4, 195, **196**, 197

Jakobsen, Mimi 65
Jospin, Lionel 123, 194
Juncker, Jean-Claude 47, 49, 50

June Movement (Denmark) 62, 66, **68**

kangaroo products, ban on 25
Karamanlis, Costas 91, 92, 93, 94–5, 98
Katzenstein, Peter 73, 78
Kazakhstan, democracy in 23
Kazamias, G. 91
Kd (Christian Democratic Party: Sweden) 201, 203, 206, **208**
Kellermann, P.A. 268
KESK (Centre Party of Finland) 101, **101**, 102, 105, **108**, 110, 111, 113
Kinnock, Neil 221
KIPU (Ecological Party: Finland) 101, 108
Kjaersgaard, Pia 63
KKE (Kommounistiko Komma-tis Elladas: Greece) 92–4, 96, 97, **97**
Klima, Viktor 37, 39
Kohl, Helmut 17, 72, 73, 74
Kok, Wim 160, 165
KOK (National Coalition: Finland) 100, **101**, 102, 105, 106–7, 108, **108**, 109, 110, 111, 113
Kollatos, Dimitris 97
Konstantopoulos, Nikos 98
Korean Peninsular Energy Development Organisation (KEDO) 24
Koskinen, Johannes 110
Kosovo crisis 39, 43, 75, 76, 78, 82, 89, 91, 93–4, 97, 105, 124, 142, 151, 153, 155, 176, 186, 188, 193–4, 204, 219, 252
Kossmann, E.H. 265
KPOe (Communist Party: Austria) 34–43
Krivine, Alain 119

Labour Party (Ireland) 141, 144, **146**
Labour Party (UK) 214, 215, 216, 217, 218, 223, **224**, 225, **225**, 226
Lafontaine, Oskar 74, 75–6, 78
Lagendijk, Joost 161
Laguiller, Arlette 119
Lalumière, Catherine 119, 235

law and order as election issue 55
Lega Nord (Northern League: Italy) **153**, **157**, 246
leg-hold traps, ban on animal fur from 25
Le Pen. Jean Marie 119
Leventis, Vassilis 96
Liberal Democrats (UK) 215, 217, 221, 223, 224, 226
Liberal Party (Denmark) 63, 67, **68**
Liberal Party (European) 125
Liberals (Greece) 92, 94
LIF (Liberal Forum: Austria) 33, 34–43, **42**
Linder, Staffan Burenstam 210
Lipponen, Paavo 102, 104
Lista Bonino *see* Emma Bonino List
LKP (Liberal People's Party: Finland) **101**
Lomé Convention 22
LRC (Revolutionary Communist League: France) 119
LSAP (Lëtzeberger Sozialistesch Arbechter Partei: Luxembourg) 46–51, 50
Lulling, Astrid 50
Luxembourg
  Chamber of deputies election results 50
  electoral system in 46–7
  Enregistrement affair 48
  EP delegation, strength of 51
  Euro-elections in 46–51; political situation in 45; obscured by national election 46; campaign issues in 49; results of 49–51, **50**; compulsory voting in 253; electoral systems in 259, 260
  government scandals in 47–8
  party system in 46
  political background of 47–9
  Santer–Poos government of 47
Luxemburger Wort 58
Luyckx, Theo 58

M (Moderate Party: Sweden) 201, 202, 205, 206, **208**, 211
Maastricht Treaty 6, 14, 17, 19, 21, 60, 214, 252

320 Index

Macmillan-Scott, Edward   219
Madelin, Alain   118, 131
Maij-Weggen, Hanja   161, 165, 167
Major, John   214
Malaga Declaration   227
Manos, Stefanos   92, 98
Marinho, Luis   182
market economy   244
Martens, Wilfred   53–4, 190, 206
Matikainen-Kallström, Marjo   109
Matutes, Abel   189
McKenna, Patricia   206
MDC (Citizens' Movement: France)   119, 120, 124, **129**, 132
Mediterranean countries, assistance to (MEDA)   24
Mégret, Bruno   119, 131
Meijer, Erik   161
MEPs
  democratic legitimacy of   285
  fraud by   166
  pensions of   166, 203
  salaries of   203
Michel, Louis   56
migration, trans-border work   42
Mitsotakis, Constantine   91
Mogens Camre (Denmark)   66
Móran, Fernando   189
Morillon, Philippe   123
Mp (Green Party: Sweden)   200–1, 205, 206, 208, **208**, 209
MPT (Movement of the Party of the Earth: Portugal)   174, **180**
MRPP/PCTP (Revolutionary Movement for the Reconstruction of the Party of the Proletariat/Communist Party of the Portuguese Workers)   174, 178, **180**, 181
MS–Tricolore (Social Movement-Tricolour: Italy)   153, **157**
Mundo, El   195
Myller, Riitta   109

National Alliance (Italy)   154
Natural Law Party   216
ND (New Democracy: Greece)   90–5, **97**, 98
neo-fascists (Italy)   157

Netherlands, The   4, 13, 16
  Euro-elections in   160–70;
    candidates and parties in domestic context   162; mobilising the vote   164–5; turnout   164; financing of   164; issues in   165–7; lack of policy choices in   169; results of   167–8, **168**; turnout in   167, 169; electoral systems in   259, 260
  minor parties in parliament   161–2
  political background   160–1
  role of EU, perceived   166
neutrality   139
New Zealand   26
Nieuwenkamp, Roel   162
Nogueira, Fernando   172
North Atlantic Treaty Organisation (NATO)   35, 37
  Kosovo intervention by   39, 176
Northern League (Italy)   154, 155
Nouvel Observateur, Le   124
nuclear energy as election issue   24, 42, 131
NUORS (Young Finns)   **101**

OeVP (People's Party: Austria)   33, 34–43, **42**
Oikonomou, Pantelis   90
openness in EP and Commission   7
Oreja, Commissioner   19
Organisation for Security and Co-operation in Europe (OSCE)   291–2
Overseas Development Assistance   27

PA (Andalucian Party: Spain)   187
Paasilinna, Reino   109
Pais, El   191
Palacio, Loyola de   188–9, 190, 191, 193
Pangalos, Theodoros   90, 91
Papandreou, Andreas   89, 90
Papandreou, Georgios   90
Papandreou, Vasso   90
Partito Socialista (Socialist Party: Italy)   **153**
Party of the Greek Hunters   96
PASOK (Panellinio Sosialistiko Kinima: Greece)   90–5, 97, 98

Pasqua, Charles   123, 132
Patijn, Schelto   272
PC (Plaid Cymru: UK)   216, 221, 223, **224**
PCF (Communist Party of France)   119, 124–5, 130
PDA (Democratic Party of the Atlantic: Portugal)   **180**
PdCI (Party of Italian Communists)   **153, 157**
PDS (Democratic Party of the Left: Italy)   156
PDS (Party of Democratic Socialists: Germany)   72, 73, 81, **83, 85**
Pensionati (Pensioners' Party: Italy)   153, 157
pensions as election issue   221
People's Movement Against the EC (Denmark)   62, 65, 66, 67, **68**
Pereira, José Pacheco   172, 173, 176, 178, 179
Pesälä, Mikko   109, 110
PES (Party of European Socialists)   93, **157**, 182, 205–6, 218, 231, 232, **233**, 235, 237, 238, 243, 244, 245, 246, 247, **247**
  election manifesto   244
Petersson, Olof   204, 211
PEV (Green Party: Portugal)   174
PfP (Partnership for Peace: Ireland)   14, 144, 147–8
Pimenta, Carlos   177, 182
Pinxten, Karel   56
Pires, L.M.   172
PNV (Basque Nationalist Party: Spain)   187, 192, 194, 195
Pohjamo, Samuli   109
Pol.An. (Politiki Anoixi: Greece)   96, **97**
Polfer, Lydie   50
Poos, Jacques   48, 50
Portas, Miguel   174, 176
Portas, Paulo   178, 179
Portugal   252
  contribution to Kosovo peace-keeping by   176
  economic benefit from EU to   171
  Euro-elections in   171–83;
    campaign and issues for   175–8;
    mobilising the vote   175; EP advertising in   176; opinion polls on   179; results of   179–82, **179**, **180, 181**; voter mobility in   **181**; electoral systems in   260
  lack of knowledge of EU in   175
  political background of   171–5
  secession from NATO   178
  unemployment in   172
POUS (Workers' Party for Socialist Unity: Portugal)   174, 178, **180**
PP (People's Party: Portugal)   171, **173**, 176, 179, **180**
PP (People's Party: Spain)   185, 188–90, 193, 194–5, **196**, 235
PPI (Italian Popular Party)   **153**, 155, **157**
PPM (People's Monarchic Party: Portugal)   174, **180**
PRI–Lib–ELDR (Republicans–Liberals–Elder: Italy)   **153, 157**
PRL/ULD (Liberal Party: Belgium)   52, **52**
Prodi, Romano   149–50, 221, 241, 288
pro-European Conservative Party (UK)   215, 220
Progress Party (Denmark)   63, 67, **68**
proportional representation   134, 158, 253, 272
  see also European
PSD (Social Democratic Party: Portugal)   171, 172, 173–4, **173**, 17, 179, 180, **180, 181**, 182, 232
PSE (European Socialist Party)   40, 175, 176
PSN (National Solidarity Party: Portugal)   174, **180**
PSOE (Socialist Workers' Party of Spain)   185, 186, 188–90, 191–2, 193, 194, 195, 196, **196**, 197–8
PSR (Socialist Revolutionary Party: Portugal)   174, 180
PS (Socialist Party: France)   117, 122–3, 124, **129**, 130
PS (Socialist Party: Portugal)   171, 172–3, 173, 177, 179, 180, **180, 181**
PS (True Finns)   101, **101**, 102, **108**
public ignorance of nature and purpose of voting   4

Puerta, Alonso 189
PvdA (Partij van de Arbeid: Netherlands) 160, 161, 163, 164, 166, 167, 168, **168**, 169

qualified majority voting (QMV) 74
Rainbow Group 234
Rauramo, Juhana 107
RC (Communist Refoundation: Italy) 149–50, **153**, 155, **157**
Reding, Viviane 50
Reif, Karlheinz 58, 199, 209
Research and Technology, 5th Framework Programme for 15–16
RI-Dini (Italian Renewal-Dini List) **153**, **157**
RKP (Swedish People's Party) 101, **101**, 102, 106–7, 108, **108**, 109, 110, 111, 112, 113
Rocard, Michel 123, 130, 132
Rodrigo, F. 185
Rome Treaty 22
RPFIE (Rally for France/Independence from Europe) 118, 120, 121, 125, 126, 127, **129**, 131, 133, 246
RPF (Rally for France) 131, 133
RPF (Reformatorische Politieke Federatie: Netherlands) 161, **168**
RPR–DL list (France) 120, 121, 126, **129**, 130, 131, 246
RPR (Rally for the Republic: France) 117, 118, 123, 124, 127, 130, 132, 133, 232

Saint-Josse, J. 119
Sala, Luc 162
Sänkiaho, Risto 113
Santer, Danièle 49
Santer, Jacques 17, 47–8, 49, 221, 241
Santer Commission, resignation of 18, 65, 75, 105, 125, 152, 167, 222, 252, 290
SAP (Social Democratic Party: Sweden) 200–1, 202, 203, 205, **208**, 209, 211
Sarkozy, Nicholas 118, 123, 124, 131
Scallon, Dana Rosemary 143, 145, 146, 147
Scharping, Rudolf 74, 82

Schäuble, Wolfgang 79
Schlüter, Poul 64
Schmidt, Heide 39
Schori, Pierre 203
Schröder, Gerhard 73, 74, 75, 78, 79, 82, 122, 132, 194
Schüssel, Wolfgang 37
SDI (Italian Democratic Socialists) **153**, **157**
SDLP (Social Democratic and Labour Party: UK) 216, **224**, **225**
SDP (Social Democratic Party: Finland) 100, **101**, 102, 103, 105, 106–7, **108**, 110, 111, 113, 114
seal products, ban on 25
seats in EP, ratio to population of 8
security policy in Europe 40
Segni Pact (Italy) 154
Séguin, Philippe 118, 123, 124, 125
Seppänen, Esko 106, 110
SGP (Staatkundig Gereformeerde Partij: Netherlands) 162, **163**, **168**
Silva, Anibel Cavaco 172
Simitis, Costas 89, 91, 97, 98
single currency 191, 201, 205, 219
Single European Act (SEA) 21, 60
Sinn Fein (Ireland) 141, 142, 144, 147, 225
SKL (Finnish Christian Union) 101, **101**, 102, 103, 106, 108, **108**, 110, 111
SKP (Communist Part: Finland) 101, **108**
SMP (Finland) **101**
SNP (Scottish Nationalist Party: UK) 216, 221, **224**
Soares, João 176
Soares, Mário 172–3, 173, 176, 180, 182
Social Democratic Party (Denmark) **68**
Socialist Group (EP) 160
Socialist Labour Party 215–16
Social Liberals (Denmark) 64–5, 66, **68**
Solana, Javier 76, 188, 194
Solbes, Pedro 193
Sommer, Franz 43
Sousa, Marcelo Rebelo de 172, 174

South Eastern Europe, Stability
   Pact 76
Southern Africa, aid to 25
SP (Socialistische Partij: Netherlands)
   161, 162, 163, **168**
Spaak, Antoniette 53
Spain 16, 252
   agricultural subsidies to 192, 193
   Basque separatism in 187, 189, 194
   Euro-election in 185–97; players in
      domestic context 187–8;
      coincidence with Andalusian
      elections 187–8; mobilising the
      vote in 188–91; electoral system
      used in 190; issues in 191–4;
      results of 194–6, **196**; electoral
      systems in 259, 260
   flax subsidies (*casa de lino*) scandal
      in 190, 193
   political background to 185–7
   views on further integration in 188
SPD (Social Democratic Party:
   Germany) 72, 73, 77, 80, 81,
   82–3, **83**, 85, **85**, 77–8, 235
SPOe (Social Democratic Party:
   Austria) 33, 34–43, **42**
SPP (Socialist People's Party:
   Denmark) 62, 66, 67, **68**
SP/PS (Socialist Party: Belgium) 52,
   52, 57–8, 57, **58**
Stenzel, Ursula 37, 40
Stoiber, Edmund 74, 79
Sundberg, Jan 113
Suominen, Ilkka 109, 110
SVP (South Tyrol People's Party:
   Italy) **153**, 157
Sweden 4, 17
   EU membership, attitudes to 200
   Euro-elections in 199–213; timing
      of 207; information campaign
      for 10, 202–3; electoral procedure
      for 207–8; players in domestic
      context 200–2; funding of
      202–3; media impact in 200;
      lack of media knowledge of
      EP 204; election campaign
      in 202–3; issues in 204–7;
      mobilising the vote for 202–4;
      results of 207–10, **208**;
   trans-national dimension of
      210–11; turnout for 199–211;
      threshold of votes required in
      253; electoral systems in 259,
      260
   neutrality of 200–1, 204
   participation in EMU 209
   political background of 200–2
   role of MEPs in 211
Swedish Trade Union
   Confederation 209
Sylla, Fodé 123
Synaspismos (Greece) 94, 96

tageblatt 58
Tapie, Bernard 117
Technical Group of Co-ordination
   and Defence of Independent
   MEPs 234
Theorin, Maj Britt 201
'Third Way', joint British–German
   paper on 78–9, 82, 194
Thors, Astrid 108, 109
timber, old-growth, ban on 25
Tindemans, Leo 53
trade disputes 25
traditional political processes, public
   disenchantment with 6
transparency in EU 166
Treaty on European Union (TEU) 23
Tsohatzopoulos, Akis 90–1, 93
Turkey
   custom union with EU and human
      rights 23–4
   reforms in 24
Turmes, Claude 51
turnout at elections 5–6, 11, 13, 38–9,
   60, **68**, 72, 84, 86, 95, 100, 107,
   108–10, 128, 139, 157, 179, 194
   compared with US Congressional
      elections 13
   and increase in EP powers 13
   and information campaigns 9–10
   reasons for poor 9

UDC (Unió Democrática de Catalunya:
   Spain) **196**
U.D. Eur. (Union of Democrats for
   Europe: Italy) **153**, **157**

UDF (Union for French Democracy) 117, 118, 120, 124, 127, **129**, 130, 131, 133
UDP (Democratic People's Union: Portugal) 120, 174, **180**
UEN (Union for the Europe of Nations) 146, **146**, **157**, 246, **247**
UFE (Union for Europe) 118, 127, 132, 232, **233**
UK 4, 13, 16
UK Independence Party 221, 226
unemployment as election issue 40, 78, 177, 205
Union Valdotaine (Valdotaine Union: Italy) **153**
United Kingdom
  and EMU 214
  Euro-election in 214–28; proportional representation in 215, 217; electoral boundaries for 215; issues in 216; campaign for 216–22; government's campaign for 217; Labour's campaign for 217–19; Conservative campaign for 219; Liberal Democrat campaign for 221; campaign in Northern Ireland 222; opinion polls in 226; funding for 216; timing of 220; results of 222–6, **223**, **224**, **225**; turnout in 216, 222–3, **223**, 226; electoral systems in 259, 260
  and further integration, attitude to 214–15
  and single currency 214
  Social Chapter, signing up to 214, 218
Unity List (Denmark) 62, **68**
'urgency' debates 25
UUP (Ulster Unionist Party: UK) **224, 225**

van den Berg, Max 160, 165, 167

van der Laan, Lousewies 161, 167
Vanguardia, La 190, 194
Vanhecke, Frank 55
van Raay, J. Janssen 162, 168
VAS (Left Alliance: Finland) 100, **101**, 102, 103, 106, 108, **108**, 110, 111, **111**, 112, 113
Vatanen, Ari 109
Väyrynen, Paavo 109, 110
Veen, Hans-Joachim 77
Veil, Simone 117
Verdes, los (Greens: Spain) **196**, 197
Verheugen, Günther 86
Verhofstadt, Guy 56
V (Green Group: Europe) **233**, 234, 237, 243, 244
VIHR (Green League of Finland) 100, **101**, 102, 106, 108, **108**, 109, 111, 112, 113
Villiers, Philippe de 117
Virrankoski, Kyösti 109
Vlams Bloc (Belgium) 52–5, 57–8, **57, 58**, 232, 246
V (Left Party: Sweden) 200, 205, 208, **208**, 209
voice telephony directive 14
voter apathy 4, 5, 81, 100, 103, 121, 122, 132, 145, 147–8, 199, 216
voters, characteristics of 5
VVD (Volkspartij voor Vrijjheid en Democratie SGP: Netherlands) 160, 161, 162, 163, 165, 166, 167, **168**, 169

Waechter, Antoine 119
Weber, Jup 49
Weins, Cornelia 77
Western European Union (WEU) 35
Wiebenga, Jan Kees 160
women
  in Euro-elections 110, 131–2
  position of 142
World Trade Organisation 25